AF780283

10th Anniversary of *Axioms*: Mathematical Physics

10th Anniversary of *Axioms*: Mathematical Physics

Editor

Hans J. Haubold

MDPI • Basel • Beijing • Wuhan • Barcelona • Belgrade • Manchester • Tokyo • Cluj • Tianjin

Editor
Hans J. Haubold
Office for Outer Space Affairs
United Nations
Vienna
Austria

Editorial Office
MDPI
St. Alban-Anlage 66
4052 Basel, Switzerland

This is a reprint of articles from the Special Issue published online in the open access journal *Axioms* (ISSN 2075-1680) (available at: www.mdpi.com/journal/axioms/special_issues/axioms_10th_anniversary_phy).

For citation purposes, cite each article independently as indicated on the article page online and as indicated below:

LastName, A.A.; LastName, B.B.; LastName, C.C. Article Title. *Journal Name* **Year**, *Volume Number*, Page Range.

ISBN 978-3-0365-6867-6 (Hbk)
ISBN 978-3-0365-6866-9 (PDF)

© 2023 by the authors. Articles in this book are Open Access and distributed under the Creative Commons Attribution (CC BY) license, which allows users to download, copy and build upon published articles, as long as the author and publisher are properly credited, which ensures maximum dissemination and a wider impact of our publications.
The book as a whole is distributed by MDPI under the terms and conditions of the Creative Commons license CC BY-NC-ND.

Contents

About the Editor

Hans J. Haubold

Hans J. Haubold studied physics at the Technical University, Chemnitz, Germany, and received a Ph.D. in 1980 and a D.Sc. in 1984, both on topics of stellar astrophysics. From 1974 to 1991, he worked at the Institute for Astrophysics, Potsdam, Germany, in the position of Professor of Theoretical Astrophysics. In 1988, he accepted a post at the United Nations Programme on Space Applications, Office for Outer Space Affairs, United Nations, New York, USA (relocated to the United Nations Office at Vienna, Austria, in 1993). Research interest in astrophysics, physics, mathematics, history of physics, and space education. Since 1983, he is also a Professor at the Centre for Mathematical and Statistical Sciences, Pala, Kerala, India. He edited more than 10 volumes of workshop proceedings, and is a co-author of more than 10 books and 350 papers.

axioms

MDPI

Editorial

Henri Poincaré's Comment on Calculus and Albert Einstein's Comment on Entropy: Mathematical Physics on the Tenth Anniversary of *Axioms*

Hans J. Haubold

United Nations Office for Outer Space Affairs, A-1220 Vienna, Austria; hans.haubold@gmail.com

This Special Issue of the journal *Axioms* collates submissions in which the authors report their perceptions and results in the field of mathematical physics and/or physical mathematics without any preconditions of the specific research topic. The papers are intended to provide the reader with a broad window into the status of the research field showing our understanding of how a known concept changes our thinking in that area of science.

The history of interactions between physics and mathematics is old and complex. Physics cannot flourish without mathematics and mathematics frequently takes its inspiration from physics. The inward-bound trajectory of 20th century physics towards the discovery of the most fundamental laws of physics resulted in the creation of quantum field theory (QFT) and string theory (ST). QFT/ST was a revelation of 20th century scientists well into the 21st century and is widely recognized as being far from fully understood. Research into QFT/ST has made use of ever more sophisticated mathematics, including cutting edge mathematics at the focus of present-day research. Conversely, many developments in QFT/ST have also led to profound new insights, constructions, and even entire subfields of mathematics (examples include vertex operator algebra theory and homological mirror symmetry). A community of scientists, involving both mathematicians and physicists, are vigorously engaged in the pursuit of investigating QFT/ST and its relationship with mathematics. There is dual and equal emphasis on both the discovery of the fundamental laws of nature as well as on mathematical discovery. This field of intellectual research has been termed physical mathematics. Physical mathematics is a subfield of the much broader field of mathematical physics [1].

check for updates

Citation: Haubold, H.J. Henri Poincaré's Comment on Calculus and Albert Einstein's Comment on Entropy: Mathematical Physics on the Tenth Anniversary of *Axioms*. *Axioms* **2023**, *12*, 83. https://doi.org/10.3390/axioms12010083

Received: 6 January 2023
Accepted: 10 January 2023
Published: 12 January 2023

Copyright: © 2023 by the author. Licensee MDPI, Basel, Switzerland. This article is an open access article distributed under the terms and conditions of the Creative Commons Attribution (CC BY) license (https://creativecommons.org/licenses/by/4.0/).

1. Solvay 1911: Poincaré (Leibniz Newton Calculus) and Einstein (Boltzmann Gibbs Entropy)

A similar situation occurred at the beginning of the 20th century in physics and mathematics. The situation was reflected in the proceedings of the first Solvay Council more than 100 years ago in 1911. The central topic at this time was not QFT but quantum mechanics and the Solvay Council proceeded in elaborating on questions concerning Planck's quantum of action in terms of mathematics and physics [2].

For this Special Issue of *Axioms*, the question of what the eminent physicists and mathematicians contributed to the deliberations of the first Solvay Council had in mind for their research to develop "the theory of [photon] radiation and quanta" into quantum physics on a similar standing as classical physics was considered. When planning this Special Issue of *Axioms* the question was asked if we were in a similar situation today asking for the discovery of "the theory of neutrino radiation and quanta" generalizing the mathematics and physics of the standard model of elementary particle interactions.

The Solvay Councils have been devoted to understanding preeminent open problems in physics by applying modern mathematics. Hendrik A. Lorentz was chairman of the first Solvay Conference on Physics, held in Brussels from 30 October to 3 November 1911. The subject was The Theory of Radiation and Quanta. This council looked at the problems of having two approaches, namely, classical physics and quantum physics.

On one side, at the first Solvay Council, a time between the discovery of Planck's quantum of action and the birth of Heisenberg's and Schrödinger's quantum mechanics, Poincaré asked if it was still possible to represent basic physical laws (mechanics) in terms of differential equations [3]?. Today a follow-up question to Poincare's question could be, in principle: What physics is behind the so-called fractional calculus and fractional differential equations [4]?

On the other side, although the equation on Boltzmann's grave at the Vienna Central Cemetery captures his insight into entropy, he never wrote it down himself. It was Planck who, in 1900, first wrote into the form that became Boltzmann's epitaph and supported the birth of quantum theory. In 1905, in one of his papers, Einstein termed it Boltzmann's principle. This equation reflects the fundamental insight that the second law of thermodynamics can only be understood in terms of a connection between entropy and probability and thus the second law is statistical in nature. Einstein's perspective on classical statistical mechanics and particularly on Boltzmann's principle is reflected by his written words:

> "What I find strange about the way Mr. Planck applies Boltzmann's equation is that he introduces a state probability W without giving this quantity a physical definition. If one proceeds in such a way, then, to begin with, Boltzmann's equation does not have a physical meaning. The circumstance that W is equated to the number of complexions belonging to a state does not change anything here; for there is no indication of what is supposed to be meant by the statement that two complexions are equally probable. Even if it were possible to define the complexions in such a manner that the S obtained from Boltzmann's equation agrees with experience, it seems to me that with this conception of Boltzmann's principle it is not possible to draw any conclusions about the admissibility of any fundamental theory whatsoever on the basis of the empirically known thermodynamic properties of a system." [5].

Even earlier, Einstein emphasized with respect to the equation $S = (R/N)\,\lg W + \text{const.}$, that:

> "Neither Herr Boltzmann nor Herr Planck gave a definition of W. They put formally W = number of complexions of the state under consideration".

2. Mathematical Physics

What is the meaning and purpose of the field of mathematical physics or physical mathematics? The common understanding is that mathematical physics applies rigorous mathematical ideas to problems inspired by physics and to investigate the mathematical structure of physical theories, and vice versa. As such, it is a remarkably broad subject. Mathematics and physics are traditionally tightly linked subjects and many historical figures, such as Isaac Newton and Carl Friedrich Gauss, were both physicists and mathematicians. Traditionally mathematical physics has been closely associated with ideas in calculus, particularly those of differential equations.

In recent years, in part due to the rise of QFT, quantum gravity, and cosmology, many more branches of mathematics have become major contributors to physics. The section Mathematical Physics of the Journal *Axioms* covers a wide field for research in the mathematical and physical sciences and their applications, including applications in chemistry, biology and the social sciences. Depending on the inclination of the authors of research papers in *Axioms*, one may prefer mathematics from the point of view of physics or vice versa.

3. From Solvay 1911 to *Axioms* 2022: Fractional Calculus and Non-Additive Entropy

Mathematical structures entered the development of physics, and problems emanating from physics influenced developments in mathematics. Examples include the role of Riemann's differential geometry in Einstein's general relativity, the dynamical theory of space and time, and the influence of Heisenberg's quantum mechanics in the development of functional analysis built on the understanding of Hilbert spaces. A prospective similar development occurred a couple of decades ago when non-Abelian gauge theories emerged as QFTs for describing fundamental particle interactions. Recently, attention has turned to the application of Riemann–Liouville fractional calculus [6–8] to physics, including Tsallis non-additive entropy [9,10]. Statistical mechanics concerns mechanics (classical, quantum, special or general relativistic) and the theory of probabilities through the adoption of a specific entropic functional [11]. Connection with thermodynamics and its macroscopic laws is established through this function.

Institutional Review Board Statement: Not applicable.

Data Availability Statement: Not applicable.

Acknowledgments: The Guest Editor would like to thank all authors for their generous contributions to this Special Issue focusing on mathematical physics or physical mathematics. Gratitude is paid to all the reviewers.

Conflicts of Interest: The author declares no conflict of interest.

List of Papers in Special Issue:

1. Sternheimer, D. Some Multifaceted Aspects of Mathematical Physics, Our Common Denominator with Elliott Lieb. [12]
The field of mathematical physics is driven by ingenious individuals and deep international cooperation.
2. Fedorov, V.E.; Turov, M.M.; Kien, B.T. A Class of Quasilinear Equations with Riemann–Liouville Derivatives and Bounded Operators. [6]
Fractional integro-differential calculus is an important tool in modelling various phenomena that arise in natural sciences and engineering. The studies the local unique solvability of initial value problems for multi-term equations in Banach spaces with fractional Riemann–Liouville derivatives, fractional Riemann–Liouville integrals, and with nonlinearity, which depends on fractional derivatives of lower orders.
3. Tudorache, A.; Luca, R. Positive Solutions for a System of Fractional Boundary Value Problems with r-Laplacian Operators, Uncoupled Nonlocal Conditions and Positive Parameters. [7]
An investigation is pursued for existence and nonexistence of positive solutions for a system of Riemann–Liouville fractional differential equations with r-Laplacian operators, subject to nonlocal uncoupled boundary conditions that contain Riemann–Stieltjes integrals, various fractional derivatives, and positive parameters.
4. Chinchane, V.L.; Nale, A.B.; Panchal, S.K.; Chesneau, C.; Khandagale, A.D. On Fractional Inequalities Using Generalized Proportional Hadamard Fractional Integral Operator. [8]
A contribution is made to fractional calculus by using the generalized proportional Hadamard fractional integral operator to establish some new fractional integral inequalities for extended Chebyshev functionals.
5. Borges, E.P.; da Costa, B.G. Deformed Mathematical Objects Stemming from the q-Logarithm Function. [11]
The long-standing question on the emergence of nonadditive entropies based on mathematical formalism has been solved.
6. Kichenassamy, S. Hot Spots in the Weak Detonation Problem and Special Relativity. [13]
Special relativistic physics and mathematics is developed for the initiation of a detonation in an explosive gaseous mixture in the high activation energy regime, in three space dimensions, that typically leads to the formation of a singularity at one point, the "hot spot".
7. Redkina, T.V.; Zakinyan, R.G.; Zakinyan, A.R.; Novikova, O.V. Bäcklund Transformations for Liouville Equations with Exponential Nonlinearity. [14]
New results for Bäcklund transformations, that are an example of differential geometric structures generated by differential equations, to study solutions for nonlinear partial differential equations, are reported.
8. Ernst, T. A New q-Hypergeometric Symbolic Calculus in the Spirit of Horn, Borngässer, Debiard and Gaveau. [15]
The paper introduces a new complete multiple q-hypergeometric symbolic calculus, which leads to q-Euler integrals and a very similar canonical system of q-difference equations for multiple q-hypergeometric functions.
9. Barseghyan, V.; Solodusha, S. Control of String Vibrations by Displacement of One End with the Other End Fixed, Given the Deflection Form at an Intermediate Moment of Time. [16]
Mathematical modelling is studied of boundary-controlled processes with the equation of string vibration with given initial and final conditions and given the deflection form at an intermediate moment of time leads to wave equations.
10. Morgulis, A. Waves in a Hyperbolic Predator–Prey System. [17]
Mathematical results are derived for a hyperbolic predator–prey model, which is formulated with the use of the Cattaneo model for chemo sensitive movement.

11. Fu, J.; Zhang, L.; Cao, S.; Xiang, C.; Zao, W. A Symplectic Algorithm for Constrained Hamiltonian Systems. [18]
Using the symplectic method of constrained Hamiltonian systems, singular dynamic problems of nonconservative constrained mechanical systems, nonholonomic constrained mechanical systems as well as physical problems in quantum dynamics can be solved.
12. Mercorelli, P. A Theoretical Dynamical Noninteracting Model for General Manipulation Systems Using Axiomatic Geometric Structures. [19]
A new theoretical approach is developed to the study of robotics manipulators dynamics, based on the geometric approach to system dynamics, according to which some axiomatic definitions of geometric structures concerning invariant subspaces are used.
13. Steinle, R.; Kleiner, T.; Kumar, P.; Hilfer, R. Existence and Uniqueness of Nonmonotone Solutions in Porous Media Flow. [20]
Mathematical models exhibiting nonlinearity and hysteresis are longstanding "hot topics" that continue to generate fundamental insights and progress in mathematics, physics, and engineering.
14. Apostol, B.F. Near-Field Seismic Motion: Waves, Deformations and Seismic Moment. [21]
A tensorial force acting in a localized seismic focus is introduced and the corresponding seismic waves as solutions of the elastic wave equation in a homogeneous and isotropic body are derived.

References

1. Bah, I.; Freed, D.; Moore, G.W.; Nekrasov, N.; Razamat, S.S.; Schaefer-Nameki, S. Snowmass Whitepaper: Physical Mathematics. *arXiv* **2021**, arXiv:2203.05078. [CrossRef]
2. Marage, P.; Wallenborn, R. *The Solvay Councils and the Birth of Modern Physics, Birkhäuser*; Basel-Boston-Berlin: Berlin, Germany, 1999.
3. Poincaré, H. Die Theorie der Strahlung und der Quanten, Verhandlungen auf einer von E. Solvay einberufenen Zusammenkunft (30. Oktober bis 3. November 1911). In *Mit einem Anhang über die Entwicklung der Quantentheorie vom Herbst 1911 bis zum Sommer 1913*; University of Michigan Library: Ann Arbor, MI, USA, 2014; p. 365.
4. Hilfer, R. Mathematical and physical interpretations of fractional derivatives and integrals. In *Handbook of Fractional Calculus with Applications, Volume 1: Basic Theory*; Kuchubei, A., Luchko, Y., Eds.; de Gruyter: Berlin, Germany, 2019; pp. 47–85.
5. Einstein, A. Die Theorie der Strahlung und der Quanten, Verhandlungen auf einer von E. Solvay einberufenen Zusammenkunft (30. Oktober bis 3. November 1911). In *Mit einem Anhang über die Entwicklung der Quantentheorie vom Herbst 1911 bis zum Sommer 1913*; University of Michigan Library: Ann Arbor, MI, USA, 2014; p. 95.
6. Fedorov, V.E.; Turov, M.M.; Kien, B.T. A Class of Quasilinear Equations with Riemann–Liouville Derivatives and Bounded Operators. *Axioms* **2022**, *11*, 96. [CrossRef]
7. Tudorache, A.; Luca, R. Positive Solutions for a System of Fractional Boundary Value Problems with r-Laplacian Operators, Uncoupled Nonlocal Conditions and Positive Parameters. *Axioms* **2022**, *11*, 164. [CrossRef]
8. Chinchane, V.L.; Nale, A.B.; Panchal, S.K.; Chesneau, C.; Khandagale, A.D. On Fractional Inequalities Using Generalized Proportional Hadamard Fractional Integral Operator. *Axioms* **2022**, *11*, 266. [CrossRef]
9. Tsallis, C. Entropy. *Encyclopaedia* **2021**, *1*, 1–44.
10. Mathai, A. A pathway to matrix-variate gamma and normal densities, Linear Algebra and its Applications. *Linear Algebra Its Appl.* **2005**, *396*, 317–328. [CrossRef]
11. Borges, E.P.; da Costa, B.G. Deformed Mathematical Objects Stemming from the q-Logarithm Function. *Axioms* **2022**, *11*, 138. [CrossRef]
12. Sternheimer, D. Some Multifaceted Aspects of Mathematical Physics, Our Common Denominator with Elliott Lieb. *Axioms* **2022**, *11*, 522. [CrossRef]
13. Kichenassamy, S. Hot Spots in the Weak Detonation Problem and Special Relativity. *Axioms* **2021**, *10*, 311. [CrossRef]
14. Redkina, T.V.; Zakinyan, R.G.; Zakinyan, A.R.; Novikova, O.V. Bäcklund Transformations for Liouville Equations with Exponential Nonlinearity. *Axioms* **2021**, *10*, 337. [CrossRef]
15. Ernst, T. A New q-Hypergeometric Symbolic Calculus in the Spirit of Horn, Borngässer. *Axioms* **2022**, *11*, 64. [CrossRef]
16. Barseghyan, V.; Solodusha, S. Control of String Vibrations by Displacement of One End with the Other End Fixed, Given the Deflection Form at an Intermediate Moment of Time. *Axioms* **2022**, *11*, 157. [CrossRef]
17. Morgulis, A. Waves in a Hyperbolic Predator–Prey System. *Axioms* **2022**, *11*, 187. [CrossRef]
18. Fu, J.; Zhang, L.; Cao, S.; Xiang, C.; Zao, W. A Symplectic Algorithm for Constrained Hamiltonian Systems. *Axioms* **2022**, *11*, 217. [CrossRef]
19. Mercorelli, P. A Theoretical Dynamical Noninteracting Model for General Manipulation Systems Using Axiomatic Geometric Structures. *Axioms* **2022**, *11*, 309. [CrossRef]

20. Steinle, R.; Kleiner, T.; Kumar, P.; Hilfer, R. Existence and Uniqueness of Nonmonotone Solutions in Porous Media Flow. *Axioms* **2022**, *11*, 327. [CrossRef]
21. Apostol, B.F. Near-Field Seismic Motion: Waves, Deformations and Seismic Moment. *Axioms* **2022**, *11*, 409. [CrossRef]

Disclaimer/Publisher's Note: The statements, opinions and data contained in all publications are solely those of the individual author(s) and contributor(s) and not of MDPI and/or the editor(s). MDPI and/or the editor(s) disclaim responsibility for any injury to people or property resulting from any ideas, methods, instructions or products referred to in the content.

axioms MDPI

Perspective

Some Multifaceted Aspects of Mathematical Physics, Our Common Denominator with Elliott Lieb †

Daniel Sternheimer [1,2,‡]

1 Department of Mathematics, Rikkyo University, Tokyo 171-8501, Japan; dasternh@gmail.com
2 Institut de Mathématiques de Bourgogne, 21078 Dijon, France
† Dedicated to our friend Elliott Lieb on the occasion of the ninetieth anniversary of his birth.
‡ Honorary Professor, St Petersburg University (Russia) and Member of the Board of Governors, Ben Gurion University (Israel).

Abstract: Mathematical physics has many facets, of which we shall briefly give a (very partial) description, centered around those of main interest for Elliott and us (Moshe Flato and I), and around the seminal scientific and personal interactions that developed between us since the sixties until Moshe's untimely death in 1998. These aspects still influence my scientific activity and my life. They also had as a corollary a variety of "parascientific activities", in particular, the foundation of IAMP (the International Association of Mathematical Physics) and of the journal LMP (Letters in Mathematical Physics), both of which were strongly impacted by Elliott, and Elliott's long insistence that publishers do not demand "copyright transfer" as a precondition for publication but are satisfied with a "consent to publish", which is increasingly becoming standard. This article being mainly a testimony to the huge scientific impacts of Elliott and also of Moshe, their intertwined aspects constitute the core of the present contribution. The last part deals briefly with metaphysical and metamathematical considerations related to axioms.

Keywords: mathematical physics; Elliott Lieb; the International Association of Mathematical Physics (IAMP, history and development); Letters in Mathematical Physics (LMP); history; God as an (optional) axiom; metamathematics; symmetries; particle physics

MSC: 01A99; 01A85; 01A80; 00A79; 00A30; 03F40

Citation: Sternheimer, D. Some Multifaceted Aspects of Mathematical Physics, Our Common Denominator with Elliott Lieb. *Axioms* **2022**, *11*, 522. https://doi.org/10.3390/axioms11100522

Academic Editor: Hans J. Haubold

Received: 29 July 2022
Accepted: 9 September 2022
Published: 1 October 2022

Publisher's Note: MDPI stays neutral with regard to jurisdictional claims in published maps and institutional affiliations.

Copyright: © 2022 by the authors. Licensee MDPI, Basel, Switzerland. This article is an open access article distributed under the terms and conditions of the Creative Commons Attribution (CC BY) license (https://creativecommons.org/licenses/by/4.0/).

1. Some History and Related Material

1.1. The Context of Our First Interactions with Elliott Lieb

a. *Moshe Flato and I* first met at HUJI (the Hebrew University of Jerusalem) in 1958–1961, when we were M.Sc. students, he of Giulio (Joel) Racah and I of Shmuel Agmon, whose original lecture on the theory of distributions we both attended in 1958–1959. Agmon turned 100 in February 2022 and still lives in Jerusalem, where I visited him in early June; on that occasion we discovered that his French D.Sc. thesis had been edited for French language by a common friend (the wife of a leading French astrophysicist), whose father was a friend of my mother and had found for me a room in the apartment of a neighbor when I was student at HUJI. That year, I also visited Moshe's home in Tel Aviv to borrow the excellent notes he had taken at the lecture on hydrodynamics, given the year before by Abraham Robinson, who was then back at HUJI and is better known at present as the founder of "nonstandard analysis". (After getting a B.Sc. at the Hebrew University and, at the beginning of WWII, being stuck in France with a German passport, Robinson managed to reach London and became an expert on the airfoils used in the wings of fighter planes (essentially, conformal mapping) which he taught to himself while serving with the Free French Air Force). The lecture on hydrodynamics was a prerequisite for the "graduate seminar" on aerodynamics where Robinson had given me a talk and which I was attending at HUJI after immigrating to a kibbutz in Israel in October 1958 with a "Licence" (B.Sc.) from Lyon in France.

Quite naturally I was asked (by a relative of Moshe in Lyon who was my mother's best friend and her deputy in the local branch of the Women's International Zionist Organization, it is a small world) to help Moshe Flato after he arrived in Paris in October 1963, before defending the Ph.D. on group theory in nuclear physics, which he had prepared under Racah. Our interaction became much stronger, and its impact is felt after his untimely death in 1998. In Paris, since October 1961, I was "attaché de recherche" at CNRS, with Szolem Mandelbrojt (who had been Agmon's D.Sc. adviser) as adviser and his former student Jean-Pierre Kahane as deputy.

Incidentally, I had met Kahane at the first scientific meeting I attended, an "International Symposium on linear spaces", 5–12 July 1960 in Jerusalem. Much later, Lech Maligranda (a former student of Orlicz, who appears in the photo) published on arXiv [1] a very interesting historical note consisting of the photo below, accompanied by a brief description of the meeting. It shows the participants, most of them famous mathematicians (including Agmon); I still remember many of them. For the benefit of the reader, I insert in this paper, a semi-final version of the photo that was sent to me by Maligranda, with a request to help identify some of the few names that were not handwritten by him on the Figure 1.

Figure 1. Jerusalem 1960, Linear Spaces Symposium.

In 1964, Moshe and I started to work on applications of group theory to physics (a major topic for Wigner and Racah), in particular, to particle physics. That evolved into "team work" (see, e.g., [2]), which lasted 34 years, until his death, and which I have been developing ever since.

b. The first *interactions* we (Moshe Flato and I) had with Elliott, beyond those which happen normally between scientists, date (if I remember well) to the mid seventies, after Elliott arrived at Princeton. It seems to me important to describe briefly how (and in which context) this happened. However, before this, I want to extend my most heartfelt congratulations to Elliott for being awarded (in January 2022) the highest honor bestowed by the American Physical Society, the APS Medal for Exceptional Achievement in Research for *"major contributions to theoretical physics through obtaining exact solutions to important physical problems, which have impacted condensed matter physics, quantum information, statistical mechanics and atomic physics"*. (The Medal was awarded for the first time in 2016 to Edward Witten.) And for being awarded the 2022 Gauss Prize at the International Congress of Mathematicians *"for deep mathematical contributions of exceptional breadth which have shaped* *the fields of quantum mechanics, statistical mechanics, computational chemistry, and quantum information theory."* (Since 2006 the Gauss Prize is awarded every four years at the IMU Congress to honor scientists whose mathematical research has had an impact outside mathematics.)

Flato and I have been visiting Princeton, frequently for the time, since our first visit from France to the US in 1966, at the invitation of Eugene Wigner. Establishing a connection with Wigner was a natural step for Moshe. At Princeton, Wigner held the Higgins Professorship in physics, a named chair, which (after Wigner's retirement) Elliott inherited. The chair, in Wigner's time, was not subjected to regular teaching duties. However, when Elliott received it, and was asked by his colleagues about retirement, he declared very honestly that he does not intend to retire. Since he had only turned 43 in July 1975, the "perk" in teaching duties (not the named chair) became limited to 5 years! (Elliott transferred to Emeritus status only 42 years later.)

Already in the seventies, Moshe and I had experience in a wide variety of areas of mathematics (both applied and pure) and physics. This certainly helped us to then become friends with Elliott after his arrival at Princeton. A main factor for that "chemistry" was also that, though specializing in different aspects of mathematical physics, we shared a love for Science in all its aspects, especially in mathematics, physics and their interrelation.

Moshe's uncompromising attitude made a strong impact in the French scientific community immediately after his arrival in October 1963 (as a theoretical physicist and for some reasons in the group of Louis de Broglie). Eventually (and quite naturally), he found "scientific asylum" in the (more open) French mathematical school.

As mentioned above, I was, since 1961, a member of CNRS and a D.Sc. student in complex analysis, participating in the vivid mathematical life at Institut Henri Poincaré. That included, in 1963–1964, participating (together with many who eventually became leading French mathematicians in our generation) in the Cartan–Schwartz seminar [3] on the (then new) important Atiyah–Singer index theorem, which was held in parallel with another seminar on the same topic at Princeton organized by Richard Palais [4], exchanging by airmail roneotyped documents. My share (Exp. No. 22, 23p.) in the "Schwartz" side of it consisted of two talks on the multiplicative property of the analytic index, a crucial and nontrivial step needed for "dimensional reduction" (to varieties of dimension 2 and 1, and to the Laplacian and Dirac operators, respectively). My interests shifted to group theory in (particle) physics in 1965 because of my increasingly close collaboration with Flato.

1.2. Our Mathematical Physics around the Seventies

The main topics treated by Moshe and I in the 1970s dealt with two main intertwined aspects of physics, in general, and its mathematical formulation, in particular. They are, on the one hand, the importance (and an original use) of symmetries, especially in particle physics and in connection with relativity (a natural development of our work in the 1960s). That included the use of the AdS (Anti de Sitter) deformation of the Poincaré (inhomogeneous Lorentz) group and of its two special (most degenerate) representations, discovered in 1963 by Wigner's brother-in-law Dirac [5], who called them "singletons" and which we called Di and Rac (on the basis of Dirac's Bra and Ket); we used them, in particular, to interpret (also dynamically [6]) the photon as a composite of two singletons.

On the other hand, an original mathematical interpretation of quantization as a deformation of classical theories, now widely called "deformation quantization" [7,8]. The latter aspect relates to (and if I may say, constitutes) a conceptual basis for quantum physics, which has been, in a variety of ways, a leitmotif in Elliott's huge scientific production.

The "chemistry" that immediately developed with Elliott was increased by the fact that some of Elliott's first papers have a common background with Moshe's first interests in physics, in particular with Moshe's 1960 M.Sc. Thesis with Racah on ionic energy levels in crystals [9,10].

1.2.1. A Couple of Short Explanations

a. As was observed already in 1964–1965, the Poincaré symmetry group of special relativity $SO(3,1) \cdot \mathbb{R}^4$ can be viewed as a deformation (in the sense that had been defined then by Gerstenhaber [11]) of the Euclidean symmetry group $SO(4) \cdot \mathbb{R}^4$ of Newtonian mechanics. The mathematically precise notion of deformation of groups and algebras is, in a way, an inverse operation to the "physical" notion of "contractions", which had been introduced in a more limited context in the 1950s, "in physics" by E.P. Wigner and E. Inonu [12], and by I.E. Segal in a side remark at the end of an article [13]. The latter notion has been studied and generalized by a number of people (for an informative more recent paper, see, e.g., [14]).

In those days, a natural question was asked, whether there is any connection between the experimentally guessed (by analogy with spectroscopic symmetries, of which Racah and Wigner had made ample use) unitary symmetries of elementary particles, especially the $SU(3)$ "internal" symmetry of the "eightfold way" (of Gell'mann and Neeman), and the Poincaré "external" symmetry. It would have made life easier for many at the time that there be no connection, but we objected [15] on mathematical grounds (dear to Elliott in other areas of physics) to "proofs" that the only connection possible was a direct product [16]. The "theorem" of Lochlainn O'Raifeartaigh was formulated at the Lie algebra level, where the proof is not correct because it implicitly assumes that there is a common dense domain of analytic vectors for all the generators of an algebra containing both symmetries. (Some later attempts by physicists to obtain similar results in a variety of contexts also contained implicit assumptions.) In fact, as it was formulated, the result is wrong, which we exemplified later with counterexamples. That did not prevent O'Raifeartaigh from becoming a friend. A direct product result was proved shortly afterward by Res Jost [17] and, independently, by Irving Segal [18] but only in the more limited context of unitary representations of Lie groups.

b. In recent years, it dawned on me, based on the fact that *deformations of algebraic structures play a major role in physics*, that the above could be a false question. In a nutshell, we know that the Euclidean symmetry can be mathematically deformed to the Poincaré group of relativity by introducing a nonzero parameter $1/c$ (where c is the velocity of light in vacuum), and that, in turn, the latter can be deformed into AdS (the anti De Sitter symmetry $SO(3,2)$) by introducing a tiny negative curvature in space-time. This permitted us to show that the photon may be considered [6] as dynamically composite (of two Dirac singletons) and that so can the leptons [19].

Moreover, we know that Lie groups and algebras may be deformed into the so-called "quantum groups", but as Hopf algebras. The axioms for the latter were written in the 1940s, well before truly representative examples emerged from physics in the 1980s (in particular in Faddeev's Leningrad group in relation with quantum integrable systems). These "quantum groups" have an additional Hopf algebra structure, which makes them, in a way, analogous to Lie groups. They have been extensively studied in the past 40 years or so, and applied in various areas of physics. We also know since the 1990s that the latter "at root of unity" (i.e., when the deformation parameter is a root of unity) have finite-dimensional unitary representations, a property that was at the base of the introduction of compact "internal symmetries" to organize the experimentally discovered multiplets of elementary particles (not to mention the so-called "quarks", proposed already in 1964, but that is another story).

It is, thus, tempting to try and consider some form of quantum AdS as a candidate for these mysterious internal symmetries, even more so since it arises from relativity by deformations. That is what I have suggested in recent papers and talks (see, e.g., [20]).

1.2.2. The Context around This Contribution

As mentioned above, some of Elliott's works deal with topics that have a nonzero intersection with Moshe's early works (with Racah). More importantly, in most of his works and in ours, paying attention as much as possible to mathematical rigor is essential (which for Elliott may include finding the best constant in inequalities). Admittedly, a number of physicists do not understand the point in working so hard to prove mathematically results that have been "known" for a very long time using handwaving arguments, and have often been "confirmed experimentally". However, Elliott and us are convinced that achieving as much mathematical rigor as possible is of utmost importance, can prevent drawing erroneous conclusions (as shown also by our above-mentioned counterexamples) and may even lead to the discovery of new phenomena.

Furthermore, Elliott was never narrow minded and is (like us) interested in a large variety of scientific domains. His position at Princeton gave him superb occasions to satisfy his scientific curiosity. (For us, extensive traveling and inviting a wide spectrum of mathematicians and physicists, which Elliott also practiced a lot, achieved similar results). I remember that, during one of my visits to Princeton this century, he was amused by the fact that "stringies", as he called the many people working in and around the so-called string theory (which is more a framework as David Gross remarked), had become very excited by our works on singletons (see, e.g., [6]). In October 2015, he (together with Michael Aizenman) invited me to deliver a talk at the Princeton Mathematical Physics seminar (which meets at irregular intervals on Tuesdays). That talk was very well attended by leading people from the University and the Institute. My title (based in part on [20]) was (in obvious allusion to a popular paper by Wigner): "The reasonable effectiveness of mathematical deformation theory in physics, especially quantum mechanics and maybe elementary particle symmetries".

Our interactions with Elliott were not limited to the professional side, but here is not the place to expand on that. There were many occasions for interactions since he has always been an avid traveler, also for non scientific reasons (which did happen rarely to us and, the pandemic and age getting in the way, regretfully happens less to me in recent years).

2. Elliott and Us, the Science and Society Aspect

In the following, I shall therefore, concentrate on three main aspects of our interactions, which relate to physics and (the scientific) society: the birth of IAMP (the International Association of Mathematical Physics) in the 1970s; Elliott's impact on LMP (Letters in Mathematical Physics), the scientific journal initiated by Moshe, both in the mid 1970s; and his largely victorious battle with publishers on the Copyright issue.

LMP started with the relatively small D. Reidel publishing company, later included in the mathematical section of Kluwer. Eventually, LMP became affiliated at Kluwer with physics (for a variety of "corporate" reasons), after Elliott's time as one of the Editors. It remained there when the scientific part of Kluwer was merged by new owners with Springer into a new company, still named Springer, which is now part of the huge Springer–Nature. That is a typical example of the acquisition-merger trend, which pervaded also the scientific publications world. It has some benefits (of scale in particular) but it is potentially dangerous in our increasingly digitalized era. For example, what will happen to the few platforms that host most scientific publications, paid for by the work of dedicated scientists and the (large) contributions of their institutions, if the corporations who own and manage these platforms become bankrupt and the platforms become suddenly dark? It would take time for the scientific societies to get around the technical and legal problems involved, and in the meantime, most of the past scientific work will be available only in the libraries of a few institutions, bringing (especially in theoretical domains) research tools back at least half a century.

2.1. The Birth of IAMP

(a). The prehistory of $M \cap \Phi$. In April 1966, the CNRS organized in Gif-sur-Yvette an international conference titled "Extension du groupe de Poincaré aux symétries internes des particules élémentaires". Flato was an initiator and a co-organizer, together with Louis Michel and Jean-Pierre Vigier. At the conference dinner, he sat next to Gunnar Källén, an auxiliary member of the Nobel Committee for physics, who, like Flato, had (mildly

speaking) a sharp tongue. (I was seated on the other side of Källén and remember well some of their exchanges, which are not for publication, even now when those involved are no longer with us). Between the two developed an immediate and strong empathy. A year later, our friend Gilbert Karpman was appointed scientific attaché of France in Stockholm and a Franco-Swedish conference on mathematical physics was planned with Flato and Källén. After Källén's death in the crash of the plane, he was piloting on his way to CERN in October 1968, that series of meetings (the first was held in Stockholm in December 1968) was given Källén's name. A second one was held in Paris in June 1970 and a third in Göteborg in June 1971. In December 1972, that became a franco-polish-swedish conference on fundamental problems in elementary particles physics, held in Warsaw just before the "International Conference on Mathematical Problems of Quantum Field Theory and Quantum Statistics" magnificently organized in Moscow by our friend Nikolay Nikolayevich (N.N.) Bogoliubov (with a ceremony at the Kremlin). The latter was eventually considered as the First IAMP Congress. That is where the symbol $M \cap \Phi$ was introduced.

It was at the International Congress of Mathematicians held in Moscow in August 1966, which Moshe and I attended as part of the (relatively large) French delegation, that we first met N.N. Bogolyubov and many other Soviet mathematicians with whom Moshe established friendly relations from the beginning. Among them were the young Ludwig Dmitriyevich Faddeev and the older Israïl Moyseyovich Gelfand, who invited Moshe to deliver a talk at his celebrated seminar. There was immediate empathy with them, facilitated by the fact that Moshe could speak quite fluently Russian with an almost native accent which his perfect ear had caught from his family (he had, however, to rely on me to read Cyrillic). At ICM66 N.N. (Nikolay Nikolayevich) invited Moshe and me to come after ICM66 to the Laboratory of Theoretical Physics (which he created and now bears his name) in the relatively new J.I.N.R. (Joint Institute for Nuclear Research) established in 1956 in the new "town of science" Dubna. Which we did, and we visited Dubna several times after that. I continued the tradition this century until the pandemic (and more) got in the way.

In December 1972, after the extension to Poland of our Gunnar Källén meetings, almost all participants of it continued to Moscow to $M \cap \Phi$. A cocasse anecdote occurred in Warsaw while we were there for Christmas 1972 at the home of our friend, the late Ryszard Rączka. He had direct phone connection to Moscow, so Moshe called I.M. Gelfand's home and the call started as follows: "Merry Christmas Israïl Moyseyovich" said Moshe, to which the latter replied "Merry Christmas Moysey Salomonovich".

(b). In March 1974, a continuation and extension of the 1972 franco-polish-swedish meeting was held in Warsaw, an International Symposium on Mathematical Physics (eventually considered as the second IAMP Congress). That is where we launched the ideas of both a mathematical physics society and a new scientific journal, of shorter publications, a somewhat mathematical physics analog of the Physical Reviews Letters.

2.2. The Development of the Concept of IAMP

Though a European, I had become an individual member of the APS (American Physical Society) in 1967, after our first visit to US in 1966 during which we visited at BNL (Brookhaven National Laboratory) my direct cousin Rudolph Sternheimer, who sponsored me. APS is, by nature, much larger than the AMS (American Mathematical Society) and organized differently. A European Physical Society (EPS) was created only in 1968, mainly as a federation of national societies (individual membership is possible, but relatively rare and powerless). The European Mathematical Society (EMS) was founded much later, in 1990. It is also a federation of about 60 national societies, but has many individual members electing representatives who participate in the general assemblies alongside with delegates of national societies.

After the creation of the EPS, it seemed natural to me, since mathematical physics was more developed in Europe, that a European Mathematical Physics society be created, but (like the APS) with mainly individual members, who could be coming from all over the world. That is why, together with Flato, we suggested just that in 1974 in Warsaw. There was an immediate reaction, in particular from our friends Elliott and Huzihiro Araki (from Japan, who also turned 90 in July 2022) who said: the idea is good but should by no means be restricted to Europe in its denomination. We obviously agreed. The development took some time, in particular because a number of leading mathematical physicists did not see the necessity and were afraid that Moshe would use it for personal enlargement, which had never been his intention.

Rudolf Haag, who, at first, had been reluctant to the idea, eventually became convinced that Moshe was not looking to use IAMP. That happened in 1975, at a meeting in Lausanne (managed by Marcel Guenin, who had played a role in the early EPS). He described the events in a historical paper [21] from which I extract the following part:

> *"An important development for mathematical physics taking shape in 1975 was the foundation of the International Association of Mathematical Physics. The creation of such an organization had been proposed for several years by Moshe Flato and pushed very vigorously by him against some opposing faction of scientists which included me. The controversy was in part due to lack of clarity about the objectives of the proposed organization but in part also due to personal animosities. Some of us had begun our scientific life before the great inflation in numbers at a time when the theoretical physics community was a rather tightly knit group, inspired by great masters like Lorentz, Planck, Einstein, Bohr, Sommerfeld, ... We did not see any need for a new organization outside the existing mathematical and physical societies and feared the spectre of a public relations oriented lobby engaged in fund raising for some pet projects. The somewhat flamboyant and aggressive manners of Moshe Flato had earned him quite a number of enemies. Res Jost had published* [17] *an unusually sharp reply to a criticism by Flato; Louis Michel had had some clashes with him; Daniel Kastler and myself were embarrassed and annoyed when at a party in Moscow, while we were talking with Bogolubov, Flato came up raising his glass, slapping Daniel on the shoulder exclaiming: "Don't you think Daniel, that we should see to it that Bogolubov gets the next Nobel Prize? But Flato had also friends who appreciated his unconventional ways and his generosity. A 1974 attempt to create the organization by an overwhelming vote of the participants at an international congress on mathematical physics in Warsaw failed, mainly because the Russian delegation was uncertain whether this was politically correct. So in Fall 1975 it was decided that a few representatives of the opposing groups should get together and settle the issue. We met in Lausanne. On one side there was Flato and Piron, on the other side Hepp and myself and, if I remember correctly, Borchers as a neutral witness. In the course of the discussion Flato succeeded in convincing me that he was not a bad guy and we ultimately agreed that the organization should be created, that the first president should be Walter Thirring and that in the executive board there should be no person who had played any role in the previous controversy. Thirring accepted the task and appointed a committee of four persons, consisting of Araki, Piron, Ruelle and Streater, to work out the statutes of the organization. Araki in his usual careful, conscientious way wrote the final version of the statutes, which were approved by the vote of the inscribed members in July 1977. Thus the organization could start its life".*

The above-mentioned paper by Haag was initiated by (and with) Araki and me, for a different purpose that was eventually discontinued by the publisher and the paper was published in the H (history) section of the EPS Journal. It basically reflects the events of that time, but, in my opinion, a number of statements could require "footnotes", especially concerning the description of events involving Flato and me. Most of that is another matter (see, e.g., [2]), but some precisions can be made, in addition to the fact that I came to Moshe with the idea of an Association, which he liked and adopted as his: we often operated like that, as a team. (The issue of the Nobel prize for Bogolubov, which did not happen for reasons related to the way the Soviet Union operated, is still restricted.)

As a "sabra", Moshe seemed quite extroverted with a flamboyant and often aggressive style. However, that was his way to hide shyness and humility, as people who got to know him well eventually realized, including Yvette Chassagne, the first woman to become "préfet" in France, after which she was chairperson of Union des Assurances de Paris (UAP) between 1983 and 1987, then the largest French insurance company. There, she asked Flato to establish a Scientific Council which included many VIPs, and a scientific prize, that was awarded by that Council, one of the first being given to Lieb in 1985.

Moshe did not leave anybody indifferent. Most of those who got to know him became friends, which includes Elliott of course. Moshe's extraordinary personality and ability to develop contacts in many areas and countries gave him a special status. Mutatis mutandi, similar things can be said of Elliott.

2.3. The Prolonged Impact of Elliott on IAMP

2.3.1. The Evolution of IAMP

As attested by Haag, Elliott's friend Walter Thirring played a crucial role in the effective start of IAMP and was the president during its first 3 years of existence (1976–1978), followed by Araki and Elliott, who was always in the background and is the first scientist to have been president twice (1982–1984 and 1997–1999). As a matter of fact, the indirect impact of Elliott on IAMP never ceased to be felt and remains a kind of watermark, though he himself would have preferred, and been more comfortable with, a greater diversity in the definition of "IAMP mathematical physics", which represents only a part of his wide scientfic interests. Elliott (private communications) is fully aware that this is not a healthy situation, and so are many members of the Executive Committees over the years. Serious efforts towards increasing diversity are made, but so far inner dynamics make these efforts insufficient. Both Elliott and I feel it would be good for the spectrum represented in the IAMP institutions and activities to be more inclusive.

2.3.2. A Perverse Effect of Democracy

As one can read in its statutes (which appear as a page in the IAMP site), IAMP is governed by an "Executive Committee" elected by a ballot of the General Assembly (in practice, electronically before an ICMP) for a term of three years (renewable only once) by its ordinary members who have paid their (modest) dues. The problem, which happens to various degrees in many democratic institutions, is that many mathematical physicists (an admittedly imprecise notion) do not pay dues or do not bother to vote for a variety of reasons, and that those who do vote represent only a fraction of the spectrum of mathematical physics, concentrated in parts of the spectrum. The net result is that the first twelve scientists appearing on the list after the vote increasingly reflect these parts, which contributes to discouraging people from other areas.

It is natural that scientists tend to vote for scientists they know, which often means people close to their fields. Not many have as wide a vision of Science as Elliott. Being at the origin of the creation of IAMP, many of its first members knew me well. That is how I was elected to its first two Executive Committees, and in that capacity, became also a member of the Commission on Mathematical Physics (C18) established by the IUPAP (the International Union of Pure and Applied Physics) in 1981. However, a perverse effect manifested itself with the passing of time. The Executive Committees tried to suggest that voters take into account the diversity of mathematical physics, which, in fact, might be better represented by the symbol $M \cup \Phi$, in order to make more clear that it includes also what we call "physical mathematics", i.e., mathematics motivated by physics. Incidentally, in the UK, for a long time, mathematical physics referred mainly to the study of partial differential equations.

The problem recently became more acute with the pandemic and the overwhelming use of talks via Zoom or similar, which made an increasing number of talks accessible to a large audience (if people have time for that). As stated on the IAMP site, a long list of talks can be found on the site researchseminars.org, where the official "One World" IAMP seminars are listed. That long list is not easy to use, and (at present) misses a number of important seminar series that give perspectives on wide areas. For instance, the Rutgers

Mathematical Physics seminars of Elliott's friend Joel Lebowitz who (at 91) manages to bring, week after week, review talks by leading speakers covering important sectors of mathematical physics and applied mathematics in a very broad sense.

2.4. Elliott and LMP

Concommitant to, but independant of, the creation of IAMP is the start of LMP (Letters in Mathematical Physics), which initially was meant to be a journal of important short contributions, being to CMP (Communications in Mathematical Physics) a kind of analog of what is PRL (the Physical Review Letters) is to the remainder of the Physical Review. It was also initiated by Moshe in Warsaw in 1974. The first contacts were made (with the help of Ryszard Rączka) with PWN (Polish Scientific Publishers), but it soon became clear that, given the delays that regular mail would bring (in particular due to political censorship in Poland), that was not a realistic option. However, through PWN, contacts were made with the (then small) D. Reidel Publishing Company, based in Dordrecht, whose owner Anton Reidel liked to come to Poland in order to buy icons.

That is how the first issue of LMP appeared at the end of 1975, with at first four Editors (Flato, Ryszard Rączka, Stan Ulam and Marcel Guenin) and a diversified Editorial Board. For a short while thereafter, in order to speed up publication, it was decided that proofs would not be sent to the authors unless they insisted on it. After a year or so, during which errors were introduced at the production level by an automated text editor (e.g., "conformal" becoming "conformational" in a title), a problem we all encounter until now with automatic corrections on many devices, that practice was abandoned. Eventually the publisher grew in size, becoming one of the two major Dutch publishing houses in Science, and hired for LMP a superb dedicated editor for the papers (a former British physics student named Richard Freeman). LMP also grew in size and so did its backlog, but the quality remained high. Contrarily to what some had feared, there was no competition with the "classic" journal Communications in Mathematical Physics (CMP, published by Springer in Heidelberg), rather complementarity. The structure of the journal evolved somewhat with time, but its way of operating remained.

Following our suggestion, Elliott joined LMP as one on the main Editors starting with volume 8 (issue 1), January 1984. That was announced in an Editorial signed by Moshe in the preceding issue. His first duty as Editor was to cosign with Moshe an obituary for Mark Kac, who had been a member of the Editorial Board of LMP. Elliott's impact was felt in many areas, both scientific and concerning the journal governance. He formally ceased being one of the Editors after volume 39 (issue 4), March 1997. When Flato died suddenly (at the age of 61) in November 1998, it fell on me to manage a smooth transition, which I did, with the help of the team he had built. Elliott's input has been very important in those circumstances, and he continues informally to be of help in many matters concerning the Journal.

2.5. Copyright Transfer vs. Consent to Publish

Before, during and after, our interactions around LMP, we became involved in the uphill battle Elliott has been waging, for years, with many publishers concerning the latters' demand of transfer of copyright for scientific texts that were published, mainly in journals. Without such a transfer, papers could not be published. (The case of books by one or few authors may be different). We supported Elliott's point and made it ours, including for our publications. With good reason, he considered that such a demand is outrageous, dealing with work performed by scientists (usually paid by academic institutions, mostly with public support and often outside of their obligations) and was the fruit of their minds, formed by years of studies. One of the arguments of the publishers, which we encountered both as authors and as editors, was that they are better equipped than scientists to defend the copyright (from possible plagiarism) than individual authors. In mathematical physics (and other areas), such an argument is largely hypothetical. We were sometimes told (orally) that there is a notion of "fair use" permitting authors to use (up to about 20% of) their own work in a later work. That would not be needed if only Consent to Publish is granted. However, even if Copyright is transfered, no sensible publisher would count words and sue an author if it determines that the use of previous material exceeds the interpretation of "fair use".

My conjecture is that a hidden reason for the demand of copyright transfer, across the board of the large spectrum of a publisher (without consideration of the fact that domains like mathematical physics can be special cases), is that some "smart" managers assumed that the amount of copyrights they have adds to the market value of the publisher, if and when (as has happened an increasing number of times) some financial institution wants to buy it. This argument was probably found fallacious, the consent to publish, which Elliott has been insisting on since many years, being sufficient. At first, a compromise was found, but not publicized for a long time, that authors may keep Copyright to themselves and give the publisher only "Consent to Publish" of the text in a given journal (a kind of "Lieb exception"). That is now increasingly the case of most of our publishers. Various juridical formulations are given to that, depending on the lawyers who play an increasing role with publishers. Here also, Elliott's vision and persistence have been rewarded.

3. Metaphysical—"Metamathematical" Remarks

3.1. Preparatory Introduction

On the web site of the journal "Axioms", its aims start with the sentence *"Axioms (ISSN 2075-1680) is an international, open access journal which provides an advanced forum for studies related to axioms"*. The following remarks are motivated by that sentence, though they are largely disconnected from the remainder of that contribution, and only vaguely related to the very general definition of the journal as "a journal of mathematics, mathematical logic and mathematical physics".

These remarks were, in part, triggered by a question that is often asked in various forms. That kind of question was recently (on 22 January 2020 to be precise) asked to me by the most tolerant rabbi of one of the two branches of Chabad in Tokyo: whether I believe in a God or at least in a kind of supreme being.

My answer started with the 1931 incompletemess theorems of Gödel in axiomatic set theory (more precisely, about natural numbers). The interested reader can find online precise formulations of these theorems. For the purpose of the present remarks, we can be satisfied with the somewhat vague statement that the (second) incompleteness theorem, an extension of the first, shows that such a system (of axioms) cannot demonstrate its own consistency.

My answer to the Chabad rabbi continued with a far reaching extension of the above, which, in a way, is the basis of my "Weltanschauung" (comprehensive conception of the world), according to which the notion of a God, or supreme being, is an axiom that may, or not, be added to a Weltanschauung. That attitude may be called "atheism" if one takes at its face value the prefix "a", in contradistinction with what Proudhon and others called "anti-theism" (a notion which appeared already in 1788 in the Oxford English dictionary).

For me the notion of a God (or supreme being) is an axiom which may, but need not, be added to whatever "axioms" you choose to govern your life. Both alternatives are self-consistent. It is then a matter of choice whether one chooses to add such an axiom to one's own system of axioms. That is, in a way, similar to the so-called "axiom of choice" in the most commonly accepted (Zermelo–Fraenkel) set theory, which is at the basis of most parts of modern mathematics. The analogy (imperfect as all analogies are) goes even further, because most humans use, in their everyday language, an implicit reference to some God. For example, I have often heard Japanese (most of whom have a very relaxed attitude toward religions) say "I *play* to God" (Asians often do not distinguish between r and l) merely to express some wish.

Incidentally, when in Fall 1958, I arrived from France at the Hebrew University (HUJI) as a (third year mathematics) student, Avraham (Halevi, born Adolph) Fraenkel (1891–1965) had just retired. My mother met him there in 1957, which was instrumental in her decision to let me immigrate to Israel in 1958 (at that time I was only 20 and still a minor by French law). Fraenkel was succeeded in his chair by Abraham Robinson (mentioned in the historical introduction above). There are many anecdotes related to Fraenkel from the time when he was teaching at HUJI (one of which I used in a small humoristic column that got published in "Le Monde" in the late 1950s), but that would take us too far off.

3.2. God as an Axiom—And Some Corollaries

My purpose here is not to add a tiny fraction, which some may consider as blasphematory, to the many (often millenary) discussions around the notions of God (sometimes written G*d, especially by "Haredi" Jews, a term derived from the Biblical verb *hared*, which appears in the Book of Isaiah and is translated as "[one who] trembles" at the word of God) and around the various religions derived therefrom.

My late mother, whom I respect immensely (fifth commandment of Judaism), used to tell me that a main difference between Judaism and all other religions is that every Jew has a direct line to God (if I may add, whatever that means), while in all other religions, there must be some intermediary (whatever name is given to that notion). The problem is that in most "sects" (including of Jews) the members feel the need to follow some guide. If that is their free choice, let it be.

My very personal (rather iconoclastic and deliberately carricatural) definition of religion is that it is a sociological phenomenon of sado-masochistic nature, in which those I will call here (for the purpose of cartoon)"sadists", avatars of those who claim to speak in the name of their god (and would like to impose their view on others) fix rules, which "masochists" enjoy imposing upon themselves. In "Le médecin malgré lui" by Molière (act 1, scene 2) a bigot (Robert) attempts to interfere in the life of a couple of servants (Martine and Sganarelle) because the latter beats Martine, only to be answered by her: "I want him to beat me".

Coming back to axioms, as stated above, my personal claim is that, by an audacious generalization of Gödel's incompleteness theorem, the notion of God is an axiom which may, but need not, be added to whatever set of axioms one uses as a guide. Many prefer to add such an axiom, and to follow one of the manifold variations on that theme that have been added over time. In many modern societies one is free not to do so, but that does not mean that one is "off the hook", on the contrary. When one lives in a society (which is the case of almost all humankind), there are limitations to the freedom of each individual person, when (and hopefully only when) it infringes upon the freedom of others. These limitations, or at least many of them, may be easier to accept when it is claimed that they come from some supreme being. However, claiming that the existence of such an axiom is not your problem does not free you from its corollaries.

On the contrary, it renders each and every one responsible to make life in society as harmonious as possible, without having to fear a "Père Fouettard", the character who supposedly accompanies Saint Nicholas on his rounds during Saint Nicholas Day (6 December). "Vaste programme" as is supposed to have said, in a very different context, Général De Gaulle after seeing on the first Jeep who entered Paris on 24 August 1944 an inscription "death to idiots" ("mort aux cons"). Indeed, neither alternative is easy to achieve, but these are not "mission impossible". Both Theodor Herzl and Walt Disney, and many others, have quotes going in similar directions, e.g., "If you will it, it is no dream" and "If you can dream it, you can do it. Remember it all started with a mouse".

Funding: This research received no external funding.

Institutional Review Board Statement: Not applicable.

Data Availability Statement: Not applicable.

Conflicts of Interest: The author declares no conflict of interest.

References

1. Maligranda, L. Fifty two years ago in Jerusalem. *arXiv* **2013**, arXiv:1303.0570.
2. Sternheimer, D. Moshe's Memoirs, as told, from time to time, to Daniel Sternheimer. In *Conférence Moshé Flato 1999*; Dito, G., Sternheimer, D., Eds.; Kluwer Academic Publishers (KAP): Dordrecht, The Netherlands, 1999; Volume 1, pp. 23–30.
3. Cartan, H. *Théorème d'Atiyah-Singer sur l'indice d'un opérateur différentiel elliptique*; 16e année 1963/64 du Séminaire, dirigée par Henri Cartan et Laurent Schwartz. Fasc. 1 (Exp. Nos. 1 à 15), Fasc. 2 (Exp. Nos. 16 à 25); École Normale Supérieure, Secrétariat mathématique: Paris, France, 1965; Zbl 0149.41102. (In French)
4. Palais, R.S. *Seminar on the Atiyah-Singer Index Theorem*; Princeton University Press: Princeton, NJ, USA, 1965.
5. Dirac, P.A.M. A remarkable representation of the 3 + 2 de Sitter group. *J. Math. Phys.* **1963**, *4*, 901–909. [CrossRef]
6. Flato, M.; Fronsdal, C. Composite Electrodynamics. *J. Geom. Phys.* **1988**, *5*, 37–61. [CrossRef]
7. Bayen, F.; Flato, M.; Fronsdal, C.; Lichnerowicz, A.; Sternheimer, D. Deformation theory and quantization—I: Deformations of symplectic structures. *Ann. Phys.* **1978**, *111*, 61–110. [CrossRef]
8. Bayen, F.; Flato, M.; Fronsdal, C.; Lichnerowicz, A.; Sternheimer, D. Deformation theory and quantization—II: Physical applications. *Ann. Phys.* **1978**, *111*, 111–151. [CrossRef]
9. Flato, M. Ionic energy levels in trigonal and tetragonal fields. *J. Mol. Spectrosc.* **1965**, *17*, 300–324. [CrossRef]
10. Kibler, M. The Master thesis of Moshé Flato. In *Conférence Moshé Flato 1999*; Dito, G., Sternheimer, D., Eds.; Kluwer Academic Publishers (KAP): Dordrecht, The Netherlands, 1999; Volume 2, pp. 177–184.
11. Gerstenhaber, M. On the deformation of rings and algebras. *Ann. Math.* **1964**, *79*, 59–103. [CrossRef]
12. Inonu, E.; Wigner, E.P. On the contraction of groups and their representations. *Proc. Natl. Acad. Sci. USA* **1953**, *39*, 510–524. [CrossRef] [PubMed]
13. Segal, I. A class of operator algebras which are determined by groups. *Duke Math. J.* **1951**, *18*, 221–265. [CrossRef]
14. Weimar-Woods, E. Contractions, generalized Inönü–Wigner contractions and deformations of finite-dimensional Lie algebras. *Rev. Math. Phys.* **2000**, *12*, 1505–1529. [CrossRef]
15. Flato, M.; Sternheimer, D. Remarks on the connection between external and internal symmetries. *Phys. Rev. Lett.* **1965**, *15*, 934–936. [CrossRef]
16. O'Raifeartaigh, L. Mass differences and Lie algebras of finite order. *Phys. Rev. Lett.* **1965**, *14*, 575–577. [CrossRef]
17. Jost, R. Eine Bemerkung zu einem "Letter" von L. O'Raifeartaigh und einer Entgegnung von M. Flato und D. Sternheimer. *Helv. Phys. Acta* **1966**, *39*, 369–375.
18. Segal, I. An extension of a theorem of L.O'Raifeartaigh. *J. Funct. Anal.* **1967**, *1*, 1–21. [CrossRef]
19. Frønsdal, C. Singletons and neutrinos. *Lett. Math. Phys.* **2000**, *52*, 51–59. [CrossRef]
20. Sternheimer, D. "The important thing is not to stop questioning", including the symmetries on which is based the standard model. In *Geometric Methods in Physics*; Birkhäuser: Basel, Switzerland, 2014; pp. 7–37.
21. Haag, R. Some people and some problems met in half a century of commitment to mathematical physics. *Eur. Phys. J. H* **2010**, *35*, 263–307. [CrossRef]

axioms MDPI

Article

A Class of Quasilinear Equations with Riemann–Liouville Derivatives and Bounded Operators

Vladimir E. Fedorov [1,2,*], **Mikhail M. Turov** [1] and **Bui Trong Kien** [3]

[1] Department of Mathematical Analysis, Mathematics Faculty, Chelyabinsk State University, Kashirin Brothers St., 129, 454001 Chelyabinsk, Russia; turov_m_m@mail.ru
[2] Laboratory of Functional Materials, South Ural State University (National Research University), Lenin Av., 76, 454080 Chelyabinsk, Russia
[3] Department of Optimization and Control Theory, Institute of Mathematics of the Vietnam Academy of Sciences and Technologies, 8 Hoang Quoc Viet Road, Caugiay District, Hanoi 10307, Vietnam; btkien@math.ac.vn
* Correspondence: kar@csu.ru

Abstract: The existence and uniqueness of a local solution is proved for the incomplete Cauchy type problem to multi-term quasilinear fractional differential equations in Banach spaces with Riemann–Liouville derivatives and bounded operators at them. Nonlinearity in the equation is assumed to be Lipschitz continuous and dependent on lower order fractional derivatives, which orders have the same fractional part as the order of the highest fractional derivative. The obtained abstract result is applied to study a class of initial-boundary value problems to time-fractional order equations with polynomials of an elliptic self-adjoint differential operator with respect to spatial variables as linear operators at the time-fractional derivatives. The nonlinear operator in the considered partial differential equations is assumed to be smooth with respect to phase variables.

Keywords: multi-term fractional differential equation; quasilinear equation; Riemann–Liouville fractional derivative; defect of Cauchy type problem; fixed point theorem; initial-boundary value problem

Citation: Fedorov, V.E.; Turov, M.M.; Kien, B.T. A Class of Quasilinear Equations with Riemann–Liouville Derivatives and Bounded Operators. *Axioms* **2022**, *11*, 96. https://doi.org/10.3390/axioms11030096

Academic Editor: Chris Goodrich

Received: 31 January 2022
Accepted: 23 February 2022
Published: 24 February 2022

Publisher's Note: MDPI stays neutral with regard to jurisdictional claims in published maps and institutional affiliations.

Copyright: © 2022 by the authors. Licensee MDPI, Basel, Switzerland. This article is an open access article distributed under the terms and conditions of the Creative Commons Attribution (CC BY) license (https://creativecommons.org/licenses/by/4.0/).

1. Introduction

In recent decades, problems with fractional derivatives have been studied by many authors [1–5]. Now fractional integro-differential calculus is an important tool in modeling various phenomena that arise in physics, chemistry, mathematical biology, engineering, etc. (see e.g., [6,7]).

The purpose of this paper is to study the local unique solvability of initial value problems for multi-term equations in Banach spaces with fractional Riemann–Liouville derivatives $D_t^\beta z$, $\beta > 0$, fractional Riemann–Liouville integrals $J_t^\beta z$, $\beta \geq 0$, and with nonlinearity, which depends on fractional derivatives of lower orders

$$D_t^\alpha z(t) = \sum_{j=1}^{m-1} A_j D_t^{\alpha-m+j} z(t) + \sum_{l=1}^{n} B_l D_t^{\alpha_l} z(t) + \sum_{s=1}^{r} C_s J_t^{\beta_s} z(t)$$
$$+ F(t, D_t^{\alpha-m} z(t), D_t^{\alpha-m+1} z(t), \ldots, D_t^{\alpha-1} z(t)). \quad (1)$$

Operators A_j, $j = 1,2,\ldots,m-1$, B_l, $l = 1,2,\ldots,n$, C_s, $s = 1,2,\ldots,r$ are supposed to be bounded on a Banach space $\mathcal{Z}$, a nonlinear map $F \in C(Z;\mathcal{Z})$, where Z is an open set in $\mathbb{R} \times \mathcal{Z}^m$.

Note that unique solvability issues for the Cauchy problem to multi-term linear equation of form (1) with Gerasimov—Caputo derivatives and bounded operators at them were studied in [8], various classes of nonlinear equations with Gerasimov—Caputo derivatives [9–11], or with a unique Riemann–Liouville derivative in a linear part of an equation [12,13] have been studied before.

Linear equations of form (1) with Riemann–Liouville derivatives were studied in the work [14] in the case of bounded operators in the equation, and in [15] in the case of closed operators. In [14] it was shown, that the Cauchy type problem for an equation with several Riemann–Liouville derivatives has the so-called defect m^*, when several initial data must be zero in the lower order initial conditions for the solvability of the problem. So, a natural initial value problem for a multi-term equation of such type is, generally speaking, the incomplete Cauchy problem

$$D_t^{\alpha-m+k}z(t_0) = z_k,\ k = m^*, m^*+1, \dots, m-1. \tag{2}$$

Section 2 of this work contains the unique solvability theorem for linear ($F \equiv f(t)$) problem (1), (2) from the work [14].

In Section 3, firstly problem (1), (2) is reduced to the integro-differential equation

$$z(t) = \sum_{p=m^*}^{m-1} Z_p(t-t_0)z_p + \int_{t_0}^{t} Z_{m-1}(t-s)F(s, D_s^{\alpha-m}z(s), \dots, D_s^{\alpha-1}z(s))ds, \tag{3}$$

where $\{Z_p(t) \in \mathcal{L}(\mathcal{Z}) : t > 0\}$, $p = m^*, m^*+1, \dots, m-1$ are the p-resolving families of operators for linear Equation (1). Next, under the condition of Lipschitzian continuity of the nonlinear operator F, using the theorem of contraction mapping for Equation (3), we prove the unique solvability of problem (1), (2) on a small enough interval.

Finally, in the last section a theorem of a local in time unique solution existence is obtained for initial-boundary value problems to a class of quasilinear equations with time-fractional derivatives, where linear operators are polynomials of an elliptic self-adjoint operator, which is differential with respect to spatial variables.

2. Preliminary Results

Let us consider the fractional integral and fractional derivative of Riemann–Liouville with the initial point at $t_0 \in \mathbb{R}$:

$$J_t^{\alpha}h(t) := \int_{t_0}^{t} \frac{(t-s)^{\alpha-1}}{\Gamma(\alpha)}h(s)ds, \quad D_t^{\alpha}h(t) = D_t^m J_t^{m-\alpha}h(t), \quad t > t_0,$$

where $m-1 < \alpha \le m \in \mathbb{N}$, i.e., $m := \lceil \alpha \rceil$.

By $\mathfrak{L}[h]$ denote the Laplace transform of a function $h : \mathbb{R}_+ \to \mathcal{Z}$. For the fractional integral and the fractional derivative of Riemann–Liouville we have the equalities [2]

$$\mathfrak{L}[J_t^{\alpha}h](\lambda) = \lambda^{-\alpha}\mathfrak{L}[h](\lambda), \quad \mathfrak{L}[D_t^{\alpha}h](\lambda) = \lambda^{\alpha}\mathfrak{L}[h](\lambda) - \sum_{k=0}^{m-1}\lambda^{m-1-k}D_t^{\alpha-m+k}h(0),$$

Hereafter $D_t^{\alpha-m+k}h(0) := \lim\limits_{t\to 0+} D_t^{\alpha-m+k}h(t)$.

Let $\mathcal{Z}$ be a Banach space, $\mathcal{L}(\mathcal{Z})$ be the Banach space of bounded linear operators on $\mathcal{Z}$, $T > t_0$. Consider the inhomogeneous equation

$$D_t^{\alpha}z(t) = \sum_{j=1}^{m-1} A_j D_t^{\alpha-m+j}z(t) + \sum_{l=1}^{n} B_l D_t^{\alpha_l}z(t) + \sum_{s=1}^{r} C_s J_t^{\beta_s}z(t) + f(t),\ t \in (t_0, T). \tag{4}$$

Here $0 < \alpha_1 < \alpha_2 < \dots < \alpha_n < \alpha$, $m_l := \lceil \alpha_l \rceil$, $m := \lceil \alpha \rceil$, $\alpha_l - m_l \ne \alpha - m$, $l = 1, 2, \dots, n$, $\beta_1 > \beta_2 > \dots > \beta_r \ge 0$, operators A_j, $j = 1, 2, \dots, m-1$, B_l, $l = 1, 2, \dots, n$, C_s, $s = 1, 2, \dots, r$, are linear and bounded in $\mathcal{Z}$. Let

$$\underline{\alpha} := \max\{\alpha_l : l \in \{1,2,\ldots,n\},\ \alpha_l - m_l < \alpha - m\}, \quad \underline{m} = \lceil \underline{\alpha} \rceil,$$

$$\overline{\alpha} := \max\{\alpha_l : l \in \{1,2,\ldots,n\},\ \alpha_l - m_l > \alpha - m\}, \quad \overline{m} = \lceil \overline{\alpha} \rceil.$$

Denote by $m^* := \max\{\underline{m} - 1, \overline{m}\}$ the defect of the Cauchy type problem for Equation (4) [14].

A solution of the incomplete Cauchy type problem

$$D_t^{\alpha-m+k} z(t_0) = z_k,\ k = m^*, m^*+1, \ldots, m-1, \tag{5}$$

for (4) is a function $z : (t_0, T] \to \mathcal{Z}$ such that $J_t^{m-\alpha} z \in C^m((t_0,T];\mathcal{Z}) \cap C^{m-1}([t_0,T];\mathcal{Z})$, $J_t^{m_l-\alpha_l} z \in C^{m_l}((t_0,T];\mathcal{Z})$, $l = 1,2,\ldots,n$, $J_t^{\beta_s} z \in C((t_0,T];\mathcal{Z})$, $s = 1,2,\ldots,r$, while equality (4) for $t \in (t_0,T]$ and (5) hold.

Put $\Gamma = \Gamma_+ \cup \Gamma_- \cup \Gamma_0$, $\Gamma_0 = \{\lambda \in \mathbb{C} : |\lambda| = r_0, \arg\lambda \in (-\pi,\pi)\}$, $\Gamma_\pm = \{\lambda \in \mathbb{C} : \arg\lambda = \pm\pi, |\lambda| \in [r_0,\infty)\}$,

$$R_\lambda := \left(I - \sum_{j=1}^{m-1} \lambda^{j-m} A_j - \sum_{l=1}^{n} \lambda^{\alpha_l-\alpha} B_l - \sum_{s=1}^{r} \lambda^{-\beta_s-\alpha} C_s \right)^{-1},$$

$$Z_p(t) = \frac{1}{2\pi i} \int_\Gamma \lambda^{-\alpha} R_\lambda \cdot \left(\lambda^{m-1-p} I - \sum_{j=p+1}^{m-1} \lambda^{j-1-p} A_j \right) e^{\lambda t} d\lambda,\ p = 0,1,\ldots,m-1,\ t > 0.$$

Substitute in ([14], Theorem 2) $t - t_0$ instead of t and obtain the next result.

Theorem 1 ([14]). *Let $m-1 < \alpha \le m \in \mathbb{N}$, $0 < \alpha_1 < \alpha_2 < \cdots < \alpha_n < \alpha$, $m_l - 1 < \alpha_l \le m_l \in \mathbb{N}$, $\alpha_l - m_l \ne \alpha - m$, $l = 1,2,\ldots,n$, $\beta_1 > \beta_2 > \cdots > \beta_r \ge 0$, $A_j \in \mathcal{L}(\mathcal{Z})$, $j = 1,2,\ldots,m-1$, $B_l \in \mathcal{L}(\mathcal{Z})$, $l = 1,2,\ldots,n$, $C_s \in \mathcal{L}(\mathcal{Z})$, $s = 1,2,\ldots,r$, $z_k \in \mathcal{Z}$, $k = m^*, m^*+1, \ldots, m-1$, $f \in C((t_0,T);\mathcal{Z}) \cap L_1(t_0,T;\mathcal{Z})$. Then there exists an unique solution to (4), (5). It has the form*

$$z(t) = \sum_{p=m^*}^{m-1} Z_p(t-t_0) z_p + \int_{t_0}^{t} Z_{m-1}(t-s) f(s) ds.$$

3. Quasilinear Equation

Let Z be an open set in $\mathbb{R} \times \mathcal{Z}^m$, $F : Z \to \mathcal{Z}$, consider the quasilinear equation

$$D_t^\alpha z(t) = \sum_{j=1}^{m-1} A_j D_t^{\alpha-m+j} z(t) + \sum_{l=1}^{n} B_l D_t^{\alpha_l} z(t) + \sum_{s=1}^{r} C_s J_t^{\beta_s} z(t)$$
$$+ F(t, D_t^{\alpha-m} z(t), D_t^{\alpha-m+1} z(t), \ldots, D_t^{\alpha-1} z(t)). \tag{6}$$

A solution of the incomplete Cauchy type problem

$$D_t^{\alpha-m+k} z(t_0) = z_k,\ k = m^*, m^*+1, \ldots, m-1, \tag{7}$$

for Equation (6) on $(t_0, t_1]$ will be called such function $z \in C((t_0,t_1];\mathcal{Z})$, that $J_t^{m-\alpha} z \in C^m((t_0,t_1];\mathcal{Z}) \cap C^{m-1}([t_0,t_1];\mathcal{Z})$, $J_t^{m_l-\alpha_l} z \in C^{m_l}((t_0,t_1];\mathcal{Z})$, $l = 1,2,\ldots,n$, and $J_t^{\beta_s} z \in C((t_0,t_1];\mathcal{Z})$, $s = 1,2,\ldots,r$, the inclusion $(t, D_t^{\alpha-m} z(t), D_t^{\alpha-m+1} z(t), \ldots, D_t^{\alpha-1} z(t)) \in Z$ and equality (6) are valid for all $t \in (t_0,t_1]$, conditions (7) are fulfilled.

Let us introduce the notations $\overline{x} := (x_0, x_1, \ldots, x_{m-1}) \in \mathcal{Z}^m$, $S_\delta(\overline{x}) = \{\overline{y} \in \mathcal{Z}^{m-1} : \|y_k - x_k\|_{\mathcal{Z}} \le \delta, k = 0,1,\ldots,m-1\}$.

A mapping $F : Z \to \mathcal{Z}$ is called locally Lipschitzian in $\overline{x}$, if for every $(t, \overline{x}) \in Z$ there exist such $\delta > 0$, $l > 0$, that $[t - \delta, t + \delta] \times S_\delta(\overline{x}) \subset Z$, and for all $(s, \overline{y}), (s, \overline{v}) \in [t - \delta, t + \delta] \times S_\delta(\overline{x})$ the inequality

$$\|F(s, \overline{y}) - F(s, \overline{v})\|_{\mathcal{Z}} \leq l \sum_{k=0}^{m-1} \|y_k - v_k\|_{\mathcal{Z}}$$

is satisfied.

Lemma 1. *Let* $0 < \alpha_1 < \alpha_2 < \cdots < \alpha_n < \alpha$, $m = \lceil \alpha \rceil$, $m_l = \lceil \alpha_l \rceil$, $\alpha_l - m_l \neq \alpha - m$, $l = 1, 2, \ldots, n$, $\beta_1 > \beta_2 > \cdots > \beta_r \geq 0$, $A_j \in \mathcal{L}(\mathcal{Z})$, $j = 1, 2, \ldots, m-1$, $B_l \in \mathcal{L}(\mathcal{Z})$, $l = 1, 2, \ldots, n$, $C_s \in \mathcal{L}(\mathcal{Z})$, $s = 1, 2, \ldots, r$, $z_k \in \mathcal{Z}$, $k = m^*, m^* + 1, \ldots, m-1$, *Z be an open set in* $\mathbb{R} \times \mathcal{Z}^m$, $(t_0, 0, 0, \ldots, 0, z_{m^*}, z_{m^*+1}, \ldots, z_{m-1}) \in Z$, $F \in C(Z; \mathcal{Z})$. *Then a function* $z : (t_0, t_1] \to \mathcal{Z}$ *is a solution of problem* (6), (7) *on* $(t_0, t_1]$, *if and only if* $J_t^{m-\alpha} z \in C^{m-1}([t_0, t_1]; \mathcal{Z})$ *and for all* $t \in (t_0, t_1]$

$$z(t) = \sum_{p=m^*}^{m-1} Z_p(t - t_0) z_p + \int_{t_0}^{t} Z_{m-1}(t - s) F(s, D_s^{\alpha-m} z(s), \ldots, D_s^{\alpha-1} z(s)) ds. \tag{8}$$

Proof. If z is a solution of problem (6), (7), then the mapping

$$t \to F(t, D_t^{\alpha-m} z(t), D_t^{\alpha-m+1} z(t), \ldots, D_t^{\alpha-1} z(t))$$

acts continuously from $[t_0, t_1]$ into $\mathcal{Z}$ due to the definition of the solution at small enough $t_1 - t_0$. By Theorem 2 (see [14]) a solution satisfies Equation (8).

Let z satisfy Equation (8), then one can verify that z is a solution to problem (6), (7) due to Theorem 1 [14] and by repeating word to word the proof of Lemma 3 in [14]. □

Theorem 2. *Let* $m - 1 < \alpha \leq m \in \mathbb{N}$, $0 < \alpha_1 < \alpha_2 < \cdots < \alpha_n < \alpha$, $m_l - 1 < \alpha_l \leq m_l \in \mathbb{N}$, $\alpha_l - m_l \neq \alpha - m$, $l = 1, 2, \ldots, n$, $\beta_1 > \beta_2 > \cdots > \beta_r \geq 0$, $A_j \in \mathcal{L}(\mathcal{Z})$, $j = 1, 2, \ldots, m-1$, $B_l \in \mathcal{L}(\mathcal{Z})$, $l = 1, 2, \ldots, n$, $C_s \in \mathcal{L}(\mathcal{Z})$, $s = 1, 2, \ldots, r$, $z_k \in \mathcal{Z}$, $k = m^*, m^* + 1, \ldots, m-1$, *Z be an open set in* $\mathbb{R} \times \mathcal{Z}^m$, $(t_0, 0, 0, \ldots, 0, z_{m^*}, z_{m^*+1}, \ldots, z_{m-1}) \in Z$, *a mapping* $F \in C(Z; \mathcal{Z})$ *is locally Lipschitzian in* $\overline{x}$. *Then there exists such* $t_1 > t_0$, *that problem* (6), (7) *has an unique solution on* $(t_0, t_1]$.

Proof. Take $y := J_t^{m-\alpha} z \in C^{m-1}([t_0, t_1], \mathcal{Z})$, then $y^{(k)} = D_t^{\alpha-m+k} z$, $k = 1, 2, \ldots, m-1$. Then the mapping $t \to F(t, y(t), y^{(1)}(t), \ldots, y^{(m-1)}(t))$ acts continuously from $[t_0, t_1]$ into $\mathcal{Z}$. By Lemma 1 it suffices to show that the equation

$$y(t) = \sum_{p=m^*}^{m-1} J_t^{m-\alpha} Z_p(t - t_0) z_p + J_t^{m-\alpha} \int_{t_0}^{t} Z_{m-1}(t - s) F(s, y(s), y^{(1)}(s), \ldots, y^{(m-1)}(s)) ds \tag{9}$$

has an unique solution $y \in C^{m-1}([t_0, t_1], \mathcal{Z})$ for some $t_1 > t_0$.

It was proved in Theorem 1 [14] that $D_t^{\alpha-m+n} Z_{m-1}(0) = 0$, $n = 0, 1, \ldots, m-2$. Since for all $p = 0, 1, \ldots, m-1$

$$\left\| \frac{\lambda^{-\alpha} R_\lambda}{\mu - \lambda} \left(\lambda^{m-1-p} I - \sum_{j=p+1}^{m-1} \lambda^{j-1-p} A_j \right) \right\|_{\mathcal{L}(\mathcal{Z})} \leq \frac{C_1}{|\lambda|^{\alpha-m+2}},$$

$\alpha - m + 2 > 1$, so, $\mathfrak{L}[D_t^{\alpha-m+n} Z_{m-1}](\mu) = \mu^{n-m} R_\mu$, at $t \in [t_0, t_1]$, $n = 0, 1, \ldots, m-2$,

$$\|D_t^{\alpha-m+n}Z_{m-1}(t)\|_{\mathcal{L}(\mathcal{Z})} \leq \frac{1}{2\pi}\int_\Gamma \|\lambda^{n-m}R_\lambda\|_{\mathcal{L}(\mathcal{Z})}|e^{\lambda t}|ds \leq C_2\int_\delta^\infty r^{n-m}dr + C_3 \leq C_4. \quad (10)$$

At $n = m-1$ we have

$$D_t^{\alpha-1}Z_{m-1}(t) = \frac{1}{2\pi i}\int_\Gamma \frac{R_\lambda}{\lambda}e^{\lambda t}d\lambda$$

$$= I + \frac{1}{2\pi i}\int_\Gamma \lambda^{-1}\left(\sum_{j=1}^{m-1}\lambda^{j-m}A_j + \sum_{l=1}^{n}\lambda^{\alpha_l-\alpha}B_l + \sum_{s=1}^{r}\lambda^{-\beta_s-\alpha}C_s\right)R_\lambda e^{\lambda t}d\lambda,$$

for $\lambda \in \Gamma$

$$\left\|\lambda^{-1}\left(\sum_{j=1}^{m-1}\lambda^{j-m}A_j + \sum_{l=1}^{n}\lambda^{\alpha_l-\alpha}B_l + \sum_{s=1}^{r}\lambda^{-\beta_s-\alpha}C_s\right)R_\lambda\right\|_{\mathcal{L}(\mathcal{Z})} \leq \frac{C_5}{|\lambda|^{1+\delta}},$$

where $\delta = \min\{1, \alpha - \alpha_l : l = 1, 2, \dots, n\}$. Consequently, at $t \in [t_0, t_1]$

$$\|D_t^{\alpha-1}Z_{m-1}(t)\|_{\mathcal{L}(\mathcal{Z})} \leq C_6. \quad (11)$$

Let $\tau > 0$ and $\delta > 0$ be such that $[t_0, t_0+\tau] \times S_\delta(\bar{z}) \subset Z$, where $\bar{z} = (0, 0, \dots, 0, z_{m^*}, z_{m^*+1}, \dots, z_{m-1})$ is constructed using initial data (7). Denote by $\mathcal{S}$ the set of functions $y \in C^{m-1}([t_0, t_0+\tau]; \mathcal{Z})$ such that $\|y^{(q)}(t)\| \leq \delta$, $q = 0, 1, \dots, m^*-1$, $\|y^{(k)}(t) - z_k\| \leq \delta$, $k = m^*, m^*+1, \dots, m-1$ for $t_0 \leq t \leq t_0+\tau$. We define a metric on $\mathcal{S}$

$$d(y, v) := \sum_{k=0}^{m-1}\sup_{t\in[t_0,t_0+\tau]}\|y^{(k)}(t) - v^{(k)}(t)\|_{\mathcal{Z}},$$

then $\mathcal{S}$ is a complete metric space.

Note that

$$J_t^{m-\alpha}\int_{t_0}^{t} Z_{m-1}(t-s)F(s, y(s), y^{(1)}(s), \dots, y^{(m-1)}(s))\,ds$$

$$= \int_{t_0}^{t} J_t^{m-\alpha}Z_{m-1}(t-s)F(s, y(s), y^{(1)}(s), \dots, y^{(m-1)}(s))\,ds.$$

This equality can be proved by changing the order of integration in its left part.

Define for $y \in \mathcal{S}$

$$G(y)(t) := \sum_{p=m^*}^{m-1} J_t^{m-\alpha}Z_p(t-t_0)z_p + \int_{t_0}^{t} J_t^{m-\alpha}Z_{m-1}(t-s)F(s, y(s), y^{(1)}(s), \dots, y^{(m-1)}(s))\,ds$$

for $t \in [t_0, t_0+\tau]$. Let us prove that G maps the metric space $\mathcal{S}$ into itself and it is a contraction operator, if $\tau > 0$ is sufficiently small. Indeed, for $n = 0, 1, \dots, m-1$

$$[G(y)]^{(n)}(t) = \sum_{p=m^*}^{m-1} D_t^{\alpha-m+n}Z_p(t-t_0)z_p$$

$$+ \int_{t_0}^{t} D_t^{\alpha-m+n}Z_{m-1}(t-s)F(s, y(s), y^{(1)}(s), \dots, y^{(m-1)}(s))ds,$$

since $D_t^{\alpha-m+n}Z_{m-1}(0) = 0$, $n = 0,1,\dots,m-2$. By Theorem 2 [14] we have $G(y) \in C^{m-1}([t_0,t_0+\tau];\mathcal{Z})$, $[G(y)]^{(q)}(t_0) = 0$, $q = 0,1,\dots,m^*-1$, $[G(y)]^{(k)}(t_0) = z_k$, $k = m^*,m^*+1,\dots,m-1$. Therefore, for $t \in [t_0,t_0+\tau]$ $\|[G(y)]^{(q)}(t)\|_{\mathcal{Z}} \le \delta, q = 0,1,\dots,m^*-1$, $\|[G(y)]^{(k)}(t) - z_k\|_{\mathcal{Z}} \le \delta$, $k = m^*,m^*+1,\dots,m-1$, for a small enough $\tau > 0$. So, $G : \mathcal{S} \to \mathcal{S}$.

Denote $F(t,\overline{D}y(t)) := F(t,y(t),y^{(1)}(t),\dots,y^{(m-1)}(t))$ for brevity. We have at $n = 0,1,\dots,m-1$, $t \in [t_0,t_0+\tau]$ due to (10), (11)

$$\|[G(y)]^{(n)}(t) - [G(v)]^{(n)}(t)\|_{\mathcal{Z}} = \left\| \int_{t_0}^{t} D_t^{\alpha-m+n}Z_{m-1}(t-s)(F(s,\overline{D}y(s)) - F(s,\overline{D}v(s)))ds \right\|$$

$$\le \tau \sup_{t\in[t_0,t_0+\tau]} \|D_t^{\alpha-m+n}Z_{m-1}(t)\|_{\mathcal{L}(\mathcal{Z})} l \sum_{k=0}^{m-1} \sup_{t\in[t_0,t_0+\tau]} \|y^{(k)}(t) - v^{(k)}(t)\|_{\mathcal{Z}} ds$$

$$\le C_7 \tau d(y,v) \le \frac{d(y,v)}{2m}$$

for small enough τ. Therefore, $d(G(y),G(v)) \le \frac{1}{2}d(y,v)$, the operator G has a unique fixed point $y_0 \in \mathcal{S}$, it is an unique local solution of integro-differential Equation (9). Thus, there exists a unique solution to problem (6), (7) on the segment $[t_0,t_0+\tau]$, it is uniquely defined by the equality $z = D_t^{m-\alpha}y_0$. □

4. A Class of Initial-Boundary Value Problems

Assume given the polynomials

$$P_1(\lambda) = \sum_{p=0}^{\nu} a_p\lambda^p,\ P_2^j(\lambda) = \sum_{p=0}^{\nu} b_p^j\lambda^p,\ P_3^l(\lambda) = \sum_{p=0}^{\nu} c_p^l\lambda^p,\ P_4^s(\lambda) = \sum_{p=0}^{\nu} d_p^s\lambda^p,$$

$a_p, b_p^j, c_p^l, d_p^s \in \mathbb{C}$, $p = 0,1,\dots,\nu \in \mathbb{N}$, $j = 1,2,\dots,m-1$, $l = 1,2,\dots,n$, $s = 1,2,\dots,r$, $a_\nu \ne 0$, $\Omega \subset \mathbb{R}^d$ is a bounded domain with a smooth boundary $\partial\Omega$,

$$(\mathcal{A}u)(\xi) = \sum_{|q|\le 2\rho} a_q(\xi) \frac{\partial^{|q|}u(\xi)}{\partial\xi_1^{q_1}\partial\xi_2^{q_2}\dots\partial\xi_d^{q_d}},\quad a_q \in C^\infty(\overline{\Omega}),$$

$$(\mathcal{B}_l u)(\xi) = \sum_{|q|\le \rho_l} b_{lq}(\xi) \frac{\partial^{|q|}u(\xi)}{\partial\xi_1^{q_1}\partial\xi_2^{q_2}\dots\partial\xi_d^{q_d}},\quad b_{lq} \in C^\infty(\partial\Omega),\ l = 1,2,\dots,\rho,$$

$q = (q_1,q_2,\dots,q_d) \in \mathbb{N}_0^d$, $|q| = q_1 + \dots + q_d$, and the operator pencil $\mathcal{A},\mathcal{B}_1,\mathcal{B}_2,\dots,\mathcal{B}_\rho$ is regularly elliptic [16]. Define the operator $\mathcal{A}_1 \in \mathcal{C}l(L_2(\Omega))$ with the domain

$$D_{\mathcal{A}_1} = H^{2\rho}_{\{\mathcal{B}_l\}}(\Omega) := \{v \in H^{2\rho}(\Omega) : \mathcal{B}_l v(\xi) = 0,\ l = 1,2,\dots,\rho,\ \xi \in \partial\Omega\}$$

by the rule $\mathcal{A}_1 u := \mathcal{A}u$. Suppose that $\mathcal{A}_1$ is a selfadjoint operator; then its spectrum $\sigma(\mathcal{A}_1)$ is real and discrete [16]. Moreover, assume that the spectrum $\sigma(\mathcal{A}_1)$ is bounded from the right and does not contain zero, $\{\varphi_k : k \in \mathbb{N}\}$ is an orthonormal in $L_2(\Omega)$ system of eigenfunctions of $\mathcal{A}_1$ in $L_2(\Omega)$ which is enumerated in nonincreasing order of the corresponding eigenvalues $\{\lambda_k : k \in \mathbb{N}\}$ with their multiplicities counted.

Take $m-1 < \alpha \le m \in \mathbb{N}$, $0 < \alpha_1 < \alpha_2 < \dots < \alpha_n < \alpha$, $m_l - 1 < \alpha_l \le m_l \in \mathbb{N}$, $\alpha_l - m_l \ne \alpha - m$, $l = 1,2,\dots,n$, $\beta_1 > \beta_2 > \dots > \beta_r \ge 0$, $H : \mathbb{R}^{m+1} \to \mathbb{R}$. Denote by m^* the defect of the Cauchy type problem, which is defined by the set of numbers $\alpha_1,\alpha_2,\dots,\alpha_n,\alpha$ (see the second section), and consider the initial-boundary value problem

$$D_t^{\alpha-m+k}u(\xi,0) = u_k(\xi),\ k = m^*,m^*+1,\dots,m-1,\quad \xi \in \Omega, \tag{12}$$

$$\mathcal{B}_l\mathcal{A}^k u(\xi,t)=0,\ \ k=0,1,\dots,\nu-1,\ \ l=1,2,\dots,\rho,\ \ (\xi,t)\in\partial\Omega\times(t_0,t_1], \tag{13}$$

$$\begin{aligned} P_1(\mathcal{A})D_t^\alpha u(\xi,t) &= \sum_{j=1}^{m-1} P_2^j(\mathcal{A})D_t^{\alpha-m+j}u(\xi,t)\\ &+\sum_{l=1}^{n} P_3^l(\mathcal{A})D_t^{\alpha_l}u(\xi,t)+\sum_{s=1}^{r} P_4^s(\mathcal{A})J_t^{\beta_s}u(\xi,t)\\ &+H(\xi,D_t^{\alpha-m}(\xi,t),D_t^{\alpha-m+1}(\xi,t),\dots,D_t^{\alpha-1}(\xi,t)),\ \ (\xi,t)\in\Omega\times(t_0,t_1]. \end{aligned} \tag{14}$$

Put $\rho_0\geq 0$, $\mathcal{X}:=\{v\in H^{2\rho\nu+\rho_0}(\Omega):\mathcal{B}_l\mathcal{A}^k v(\xi)=0, k=0,1,\dots,\nu-1, l=1,2,\dots,\rho,\ \xi\in\partial\Omega\}$, $\mathcal{Y}:=H^{\rho_0}(\Omega)$ is a Sobolev space $W_2^{\rho_0}(\Omega)$ for $\rho_0>0$, or the Lebesgue space $L^{\rho_0}(\Omega)$, if $\rho_0=0$; $L:=P_1(\mathcal{A})\in\mathcal{L}(\mathcal{X};\mathcal{Y})$, $M_j:=P_2^j(\mathcal{A})\in\mathcal{L}(\mathcal{X};\mathcal{Y})$, $j=1,2,\dots,m-1$, $N_l:=P_3^l(\mathcal{A})\in\mathcal{L}(\mathcal{X};\mathcal{Y})$, $l=1,2,\dots,n$, $S_s:=P_4^s(\mathcal{A})\in\mathcal{L}(\mathcal{X};\mathcal{Y})$, $s=1,2,\dots,r$.

If $P_1(\lambda_k)\neq 0$ for all $k\in\mathbb{N}$, then there exists the inverse operator $L^{-1}\in\mathcal{L}(\mathcal{Y};\mathcal{X})$ and (12)–(14) is representable in form (6), (7), where $\mathcal{Z}=\mathcal{X}$, $A_j=L^{-1}M_j\in\mathcal{L}(\mathcal{Z})$, $j=1,2,\dots,m-1$, $B_l=L^{-1}N_l\in\mathcal{L}(\mathcal{Z})$, $l=1,2,\dots,n$, $C_s=L^{-1}S_s\in\mathcal{L}(\mathcal{Z})$, $s=1,2,\dots,r$, $z_k=u_k(\cdot)$, $k=m^*,m^*+1,\dots,m-1$, $F(x_0,x_1,\dots,x_{m-1})=L^{-1}H(\cdot,x_0,x_1,\dots,x_{m-1})$.

Theorem 3. *Let $m-1<\alpha\leq m\in\mathbb{N}$, $0<\alpha_1<\alpha_2<\cdots<\alpha_n<\alpha$, $m_l-1<\alpha_l\leq m_l\in\mathbb{N}$, $\alpha_l-m_l\neq\alpha-m$, $l=1,2,\dots,n$, $\beta_1>\beta_2>\cdots>\beta_r\geq 0$, the spectrum $\sigma(\mathcal{A}_1)$ do not contain the origin and zeros of the polynomial P_1, $4\rho\nu+2\rho_0>d$, $u_k\in\mathcal{X}$, $k=m^*,m^*+1,\dots,m-1$, $H\in C^\infty(\Omega\times\mathbb{R}^n;\mathbb{R})$. Then at some $t_1>t_0$ there exists an unique solution of problem* (12)–(14).

Proof. In this problem the domain of nonlinear operator is $Z=\mathbb{R}\times\mathcal{X}^m$ and due to the inequality $4\rho\nu+2\rho_0>d$ by Proposition 1 ([17], Appendix B) we have

$$H(\cdot,x_0(\cdot),x_1(\cdot),\dots,x_{n-1}(\cdot))\in C^\infty(\mathcal{X}^m;H^{2\rho\nu+\rho_0}(\Omega)),$$

hence, $F(x_0(\cdot),x_1(\cdot),\dots,x_{m-1}(\cdot)):=L^{-1}H(\cdot,x_0(\cdot),x_1(\cdot),\dots,x_{m-1}(\cdot))\in C^\infty(\mathcal{X}^m;\mathcal{X})$. Then by Theorem 2 we obtain the statement of this theorem. □

Example 1. *Take $\alpha=5/2$, $m=3$, $n=1$, $r=1$, $\alpha_1=2/3$, $\beta_1=1/2$, $\nu=2$, $P_1(\lambda)=\lambda^2$, $P_2^1(\lambda)=b_0+b_1\lambda+b_2\lambda^2$, $P_2^2(\lambda)\equiv 0$, $P_3^1(\lambda)=c_0+c_1\lambda+c_2\lambda^2$, $P_4^1(\lambda)=d_0+d_1\lambda+d_2\lambda^2$, $d=1$, $\Omega=(0,\pi)$, $\rho=1$, $\mathcal{A}u=\frac{\partial^2 u}{\partial\xi^2}$, $\mathcal{B}_1=I$. Then $\underline{\alpha}:=\max\varnothing:=0$, $\underline{m}:=\lceil 0\rceil=0$, $\overline{\alpha}:=\max\{2/3\}=2/3$, $\overline{m}:=\lceil 2/3\rceil=1$, $m^*=1$, problem* (12)–(14) *has the form*

$$\begin{aligned} D_t^{5/2}\frac{\partial^4 u}{\partial\xi^4}(\xi,t) &= \left(b_0+b_1\frac{\partial^2}{\partial\xi^2}+b_2\frac{\partial^4}{\partial\xi^4}\right)D_t^{1/2}u(\xi,t)\\ +\left(c_0+c_1\frac{\partial^2}{\partial\xi^2}+c_2\frac{\partial^4}{\partial\xi^4}\right)&D_t^{2/3}u(\xi,t)+\left(d_0+d_1\frac{\partial^2}{\partial\xi^2}+d_2\frac{\partial^4}{\partial\xi^4}\right)J_t^{1/2}u(\xi,t)\\ +F(\xi,J_t^{1/2}u(\xi,t),&D_t^{1/2}u(\xi,t),D_t^{3/2}u(\xi,t)),\ \ (\xi,t)\in(0,\pi)\times(t_0,t_1], \end{aligned}$$

$$u(0,t)=u(\pi,t)=\frac{\partial^2 u}{\partial\xi^2}(0,t)=\frac{\partial^2 u}{\partial\xi^2}(\pi,t)=0,\ \ t\in(t_0,t_1],$$

$$D^{1/2}u(\xi,0)=u_1(\xi),\quad D^{3/2}u(\xi,0)=u_2(\xi)\quad \xi\in(0,\pi).$$

5. Conclusions

The local solvability is shown for the incomplete Cauchy type problem to a solved with respect to a highest derivative multi-term fractional differential equation with bounded operators at Riemann–Liouville derivatives in a Banach space with locally Lipschitzian nonlinear part. The results of the work [14] on inhomogeneous linear multi-term equation are used here for the research of the quasilinear equation, depending on lower order fractional derivatives with orders, which fractional part is equal to the fractional part of the highest fractional derivative. The abstract result was applied to the investigation of initial-boundary value problems to partial differential equations containing polynomials with respect to self-adjoint elliptic differential in spatial variables operator at time-fractional derivatives. Here the highest time-fractional partial derivative acts on the highest spatial derivative.

Our next step is to abandon this condition by allowing unlimited operators in an abstract equation. The linear case of this type has been investigated in [15], the nonlinear one has not yet been studied. Another significant step planned by the authors in the coming papers will be the rejection of conditions for the fractional part of the orders of derivatives on which the nonlinear operator depends (see above).

Author Contributions: Conceptualization, V.E.F. and B.T.K.; methodology, V.E.F. and B.T.K.; software, M.M.T.; validation, V.E.F. and M.M.T.; formal analysis, M.M.T.; investigation, M.M.T.; resources, V.E.F. and B.T.K.; data curation, M.M.T.; writing—original draft preparation, V.E.F. and M.M.T.; writing—review and editing, V.E.F. and B.T.K.; visualization, M.M.T.; supervision, V.E.F.; project administration, V.E.F.; funding acquisition, V.E.F. All authors have read and agreed to the published version of the manuscript.

Funding: The work is supported by the Russian Foundation for Basic Research and Vietnam Academy of Science and Technology, grants numbers 21-51-54003 and QTRU 01-01/21-22, the first two co-authors are funded also by the grant of President of the Russian Federation to support leading scientific schools, project number NSh-2708.2022.1.1.

Data Availability Statement: Not applicable.

Conflicts of Interest: The authors declare no conflict of interest. The funders had no role in the design of the study; in the collection, analyses, or interpretation of data; in the writing of the manuscript, or in the decision to publish the results.

References

1. Samko, S.G.; Kilbas, A.A.; Marichev, O.I. *Fractional Integrals and Derivatives. Theory and Applications*; Gordon and Breach Science Publishers: Philadelphia, PA, USA, 1993.
2. Podlubny, I. *Fractional Differential Equations*; Academic Press: Boston, MA, USA, 1999.
3. Nakhushev, A.M. *Fractional Calculus and Its Applications*; Fizmatlit: Moscow, Russia, 2003. (In Russian)
4. Kilbas, A.A.; Srivastava, H.M.; Trujillo, J.J. *Theory and Applications of Fractional Differential Equations*; Elsevier Science Publishing: Amsterdam, The Netherlands; Boston, MA, USA; Heidelberg, Germany, 2006.
5. Kostić, M. *Abstract Volterra Integro-Differential Equations*; CRC Press: Boca Raton, FL, USA, 2015.
6. Tarasov, V.E. *Fractional Dynamics: Applications of Fractional Calculus to Dynamics of Particles, Fields and Media*; Springer: New York, NY, USA, 2011.
7. Uchaykin, V.V. *Fractional Derivatives for Physicists and Engineers*; Higher Education Press: Beijing, China, 2012.
8. Fedorov, V.E.; Boyko, K.V.; Phuong, T.D. Initial value problems for some classes of linear evolution equations with several fractional derivatives. *Math. Notes NEFU* **2021**, *28*, 85–104.
9. Plekhanova, M.V. Nonlinear equations with degenerate operator at fractional Caputo derivative. *Math. Methods Appl. Sci.* **2016**, *40*, 41–44. [CrossRef]
10. Plekhanova, M.V.; Baybulatova, G.D. Semilinear equations in Banach spaces with lower fractional derivatives. In *International Conference in Nonlinear Analysis and Boundary Value Problems*; Springer: Cham, Switzerland, 2019; Volume 292, pp. 81–93.
11. Fedorov, V.E.; Phuong, T.D.; Kien, B.T.; Boyko, K.V.; Izhberdeeva, E.M. A class of semilinear distributed order equations in Banach spaces. *Chelyabinsk Phys. Math. J.* **2020**, *5*, 343–351.
12. Fedorov, V.E.; Avilovich, A.S. Semilinear fractional-order evolution equations of Sobolev type in the sectorial. *Complex Var. Elliptic Equ.* **2021**, *66*, 1108–1121. [CrossRef]
13. Fedorov, V.E.; Nagumanova, A.V.; Avilovich, A.S. A class of inverse problems for evolution equations with the Riemann–Liouville derivative in the sectorial case. *Math. Methods Appl. Sci.* **2021**, *44*, 11961–11969. [CrossRef]
14. Fedorov, V.E.; Turov, M.M. The defect of a Cauchy type problem for linear equations with several Riemann–Liouville derivatives. *Sib. Math. J.* **2021**, *62*, 925–942. [CrossRef]
15. Fedorov, V.E.; Du, W.-S.; Turov, M.M. On the unique solvability of incomplete Cauchy type problems for a class of multi-term equations with the Riemann–Liouville derivatives. *Symmetry* **2022**, *14*, 75. [CrossRef]
16. Triebel, H. Interpolation Theory. *Function Spaces. Differential Operators*; Deutscher Verlag des Wissenschaften: Berlin, Germany, 1978.
17. Hassard, B.D.; Kazarinoff, N.D.; Wan, Y.-H. *Theory and Applications of Hopf Bifurcation*; Cambridge University Press: Cambridge, UK, 1981.

axioms

MDPI

Article

Positive Solutions for a System of Fractional Boundary Value Problems with r-Laplacian Operators, Uncoupled Nonlocal Conditions and Positive Parameters

Alexandru Tudorache [1] **and Rodica Luca** [2,*]

1 Department of Computer Science and Engineering, Gh. Asachi Technical University, 700050 Iasi, Romania; alexandru-gabriel.tudorache@academic.tuiasi.ro
2 Department of Mathematics, Gh. Asachi Technical University, 700506 Iasi, Romania
* Correspondence: rluca@math.tuiasi.ro

Abstract: In this paper, we investigate the existence and nonexistence of positive solutions for a system of Riemann–Liouville fractional differential equations with r-Laplacian operators, subject to nonlocal uncoupled boundary conditions that contain Riemann–Stieltjes integrals, various fractional derivatives and positive parameters. We first change the unknown functions such that the new boundary conditions have no positive parameters, and then, by using the corresponding Green functions, we equivalently write this new problem as a system of nonlinear integral equations. By constructing an appropriate operator $\mathcal{A}$, the solutions of the integral system are the fixed points of $\mathcal{A}$. Following some assumptions regarding the nonlinearities of the system, we show (by applying the Schauder fixed-point theorem) that operator $\mathcal{A}$ has at least one fixed point, which is a positive solution of our problem, when the positive parameters belong to some intervals. Then, we present intervals for the parameters for which our problem has no positive solution.

Keywords: Riemann–Liouville fractional differential equations; nonlocal boundary conditions; positive parameters; positive solutions; existence; nonexistence

MSC: 34A08; 34B10; 34B18

Citation: Tudorache, A.; Luca, R. Positive Solutions for a System of Fractional Boundary Value Problems with r-Laplacian Operators, Uncoupled Nonlocal Conditions and Positive Parameters. *Axioms* **2022**, *11*, 164. https://doi.org/10.3390/axioms11040164

Academic Editor: Hans J. Haubold

Received: 20 February 2022
Accepted: 4 April 2022
Published: 6 April 2022

Publisher's Note: MDPI stays neutral with regard to jurisdictional claims in published maps and institutional affiliations.

Copyright: © 2022 by the authors. Licensee MDPI, Basel, Switzerland. This article is an open access article distributed under the terms and conditions of the Creative Commons Attribution (CC BY) license (https://creativecommons.org/licenses/by/4.0/).

1. Introduction

We consider the system of fractional differential equations with r_1-Laplacian and r_2-Laplacian operators

$$\begin{cases} D_{0+}^{\alpha_1}(\varphi_{r_1}(D_{0+}^{\beta_1}u(t))) + \mathfrak{a}(t)\mathfrak{f}(v(t)) = 0, & t \in (0,1), \\ D_{0+}^{\alpha_2}(\varphi_{r_2}(D_{0+}^{\beta_2}v(t))) + \mathfrak{b}(t)\mathfrak{g}(u(t)) = 0, & t \in (0,1), \end{cases} \tag{1}$$

supplemented with the uncoupled nonlocal boundary conditions

$$\begin{cases} u^{(j)}(0) = 0,\ j = 0,\dots,n-2;\ D_{0+}^{\beta_1}u(0) = 0,\ D_{0+}^{\gamma_0}u(1) = \sum_{j=1}^{p}\int_0^1 D_{0+}^{\gamma_j}u(\tau)\,d\mathfrak{H}_j(\tau) + \mathfrak{a}_0, \\ v^{(j)}(0) = 0,\ j = 0,\dots,m-2;\ D_{0+}^{\beta_2}v(0) = 0,\ D_{0+}^{\delta_0}v(1) = \sum_{j=1}^{q}\int_0^1 D_{0+}^{\delta_j}v(\tau)\,d\mathfrak{K}_j(\tau) + \mathfrak{b}_0, \end{cases} \tag{2}$$

where $\alpha_1, \alpha_2 \in (0,1]$, $\beta_1 \in (n-1,n]$, $\beta_2 \in (m-1,m]$, $n, m \in \mathbb{N}$, $n, m \geq 3$, $p, q \in \mathbb{N}$, $\gamma_j \in \mathbb{R}$ for all $j = 0,1,\dots,p$, $0 \leq \gamma_1 < \gamma_2 < \cdots < \gamma_p \leq \gamma_0 < \beta_1 - 1$, $\gamma_0 \geq 1$, $\delta_j \in \mathbb{R}$ for all $j = 0,1,\dots,q$, $0 \leq \delta_1 < \delta_2 < \cdots < \delta_q \leq \delta_0 < \beta_2 - 1$, $\delta_0 \geq 1$, $r_1, r_2 > 1$, $\varphi_{r_j}(\zeta) = |\zeta|^{r_j-2}\zeta$, $\varphi_{r_j}^{-1} = \varphi_{\varrho_j}$, $\varrho_j = \frac{r_j}{r_j-1}$, $j = 1,2$, $\mathfrak{a}_0$ and $\mathfrak{b}_0$ are positive parameters, the functions $\mathfrak{a}$, $\mathfrak{b} : [0,1] \to \mathbb{R}_+$ and $\mathfrak{f}$, $\mathfrak{g} : \mathbb{R}_+ \to \mathbb{R}_+$ are continuous, ($\mathbb{R}_+ = [0,\infty)$), the integrals from (2) are Riemann–Stieltjes integrals with $\mathfrak{H}_i$, $i = 1,\dots,p$ and $\mathfrak{K}_j$, $j = 1,\cdots,q$ functions of bounded variation, and D_{0+}^{κ} denotes the Riemann–Liouville derivative of order κ (for $\kappa = \alpha_1, \beta_1, \alpha_2, \beta_2, \gamma_i$ for $i = 0, 1, \ldots, p$, δ_j for $j = 0, 1, \ldots, q$). This paper is motivated by the applications of r-Laplacian operators in various fields such as fluid flow through porous media, nonlinear elasticity, nonlinear electrorheological fluid and glaciology, (for details, see [1] and its references).

In this paper, we provide sufficient conditions for the functions $\mathfrak{f}$ and $\mathfrak{g}$, and intervals for the parameters $\mathfrak{a}_0$ and $\mathfrak{b}_0$ such that problem (1), (2) has at least one positive solution or no positive solution. For the proof of the main existence result, we use the Schauder fixed-point theorem. Using a positive solution of (1), (2) we understand a pair of functions $(u, v) \in (C([0,1]; \mathbb{R}_+))^2$, satisfying the system (1) and the boundary conditions (2), with $u(t) > 0$ and $v(t) > 0$ for all $t \in (0,1]$. The method for studying problem (1), (2) consists of the following stages. First, we make a change in the unknown functions such that the new boundary conditions have no positive parameters, and then, by using the corresponding Green functions, we equivalently write this new problem as a system of nonlinear integral equations. By constructing an appropriate operator $\mathcal{A}$, the solutions of the integral system are the fixed points of $\mathcal{A}$. Following some assumptions regarding the nonlinearities of the system, we show that operator $\mathcal{A}$ has at least one fixed point, which is a positive solution of our problem, when the positive parameters belong to certain intervals. Then, we provide intervals for the parameters for which problem (1), (2) has no positive solution. We now present some recent results related to our problem. In [2], by using Guo-Krasnosel'skii fixed-point theorem, the author studied the system of fractional differential equations

$$\begin{cases} D_{0+}^{\alpha_1}(\varphi_{r_1}(D_{0+}^{\beta_1}u(t))) + \lambda f(t, u(t), v(t)) = 0, & t \in (0,1), \\ D_{0+}^{\alpha_2}(\varphi_{r_2}(D_{0+}^{\beta_2}v(t))) + \mu g(t, u(t), v(t)) = 0, & t \in (0,1), \end{cases} \tag{3}$$

subject to the boundary conditions (2) with $\mathfrak{a}_0 = \mathfrak{b}_0 = 0$, where $f, g \in C([0,1] \times \mathbb{R}_+ \times \mathbb{R}_+, \mathbb{R}_+)$, and λ, μ are positive parameters. The author presented various intervals for λ and μ, such that problem (3), (2) with $\mathfrak{a}_0 = \mathfrak{b}_0 = 0$ has at least one positive solution ($u(t) > 0$ for all $t \in (0,1]$, or $v(t) > 0$ for all $t \in (0,1]$). The author also investigated the nonexistence of positive solutions. In [3], the authors studied the existence and nonexistence of positive solutions for the system (3) with the coupled boundary conditions

$$\begin{cases} u^{(j)}(0) = 0,\ j = 0, \ldots, n-2;\ D_{0+}^{\beta_1}u(0) = 0,\ D_{0+}^{\gamma_0}u(1) = \sum_{j=1}^{p} \int_0^1 D_{0+}^{\gamma_j}v(\tau)\, d\mathfrak{H}_j(\tau), \\ v^{(j)}(0) = 0,\ j = 0, \ldots, m-2;\ D_{0+}^{\beta_2}v(0) = 0,\ D_{0+}^{\delta_0}v(1) = \sum_{j=1}^{q} \int_0^1 D_{0+}^{\delta_j}u(\tau)\, d\mathfrak{K}_j(\tau), \end{cases}$$

where $\gamma_j \in \mathbb{R}$ for all $j = 0, 1, \ldots, p$, $0 \leq \gamma_1 < \gamma_2 < \cdots < \gamma_p \leq \delta_0 < \beta_2 - 1$, $\delta_0 \geq 1$, $\delta_j \in \mathbb{R}$ for all $j = 0, 1, \ldots, q$, $0 \leq \delta_1 < \delta_2 < \cdots < \delta_q \leq \gamma_0 < \beta_1 - 1$, $\gamma_0 \geq 1$, $\mathfrak{H}_i$, $i = 1, \ldots, p$ and $\mathfrak{K}_j$, $j = 1, \ldots, q$ are functions of bounded variation. In [4], the authors investigated the positive solutions for the system of fractional differential equations

$$\begin{cases} D_{0+}^{\alpha}u(t) + \mathfrak{a}(t)\mathfrak{f}(v(t)) = 0, & t \in (0,1), \\ D_{0+}^{\beta}v(t) + \mathfrak{b}(t)\mathfrak{g}(u(t)) = 0, & t \in (0,1), \end{cases}$$

supplemented with the integral boundary conditions

$$\begin{cases} u(0) = u'(0) = \cdots = u^{(n-2)}(0) = 0,\ u(1) = \int_0^1 u(\tau)d\mathfrak{H}(\tau) + \mathfrak{a}_0, \\ v(0) = v'(0) = \cdots = v^{(m-2)}(0) = 0,\ v(1) = \int_0^1 v(\tau)d\mathfrak{K}(\tau) + \mathfrak{b}_0, \end{cases}$$

where $n - 1 < \alpha \leq n$, $m - 1 < \beta \leq m$, $n, m \in \mathbb{N}$, $n, m \geq 3$, $\mathfrak{a}, \mathfrak{b}, \mathfrak{f}, \mathfrak{g}$ are nonnegative continuous functions, $\mathfrak{H}$ and $\mathfrak{K}$ are bounded variation functions, and $\mathfrak{a}_0$, $\mathfrak{b}_0$ are positive parameters. Other recent research regarding fractional differential equations and systems of fractional differential equations with or without Laplacian operators and their applications can be found in the papers [5–9], and in the monographs [10–12]. In comparison with other papers, the novelty of our work consists of the combination between the system of fractional differential equations (1), in which sequential fractional derivatives with r-Laplacian operators are considered, and the existence of positive parameters in the general integro-differential boundary conditions (2).

The paper has the following structure. In Section 2, we provide some preliminary results, including the Green functions associated with our problem (1), (2) and their properties. In Section 3, we present the main theorems for the existence and nonexistence of positive solutions for (1), (2). Section 4 contains an example to illustrate our results, and in Section 5, we provide our conclusions.

2. Auxiliary Results

In this section, we present some auxiliary results related to our problem (1), (2) from [2]. We first consider the fractional differential equation

$$D_{0+}^{\alpha_1}(\varphi_{r_1}(D_{0+}^{\beta_1}u(t)))+\mathfrak{h}(t)=0,\ \ t\in(0,1), \tag{4}$$

with the boundary conditions

$$u^{(j)}(0)=0,\ j=0,\ldots,n-2;\ D_{0+}^{\beta_1}u(0)=0,\ \ D_{0+}^{\gamma_0}u(1)=\sum_{j=1}^{p}\int_0^1 D_{0+}^{\gamma_j}u(\tau)\,d\mathfrak{H}_j(\tau), \tag{5}$$

where $\alpha_1\in(0,1]$, $\beta_1\in(n-1,n]$, $n\in\mathbb{N}$, $n\geq 3$, $p\in\mathbb{N}$, $\gamma_j\in\mathbb{R}$ for all $j=0,1,\ldots,p$, $0\leq\gamma_1<\gamma_2<\cdots<\gamma_p\leq\gamma_0<\beta_1-1$, $\gamma_0\geq 1$, $\mathfrak{H}_j$, $j=1,\ldots,p$ are bounded variation functions, and $\mathfrak{h}\in C[0,1]$. We denote using

$$\Delta_1=\frac{\Gamma(\beta_1)}{\Gamma(\beta_1-\gamma_0)}-\sum_{j=1}^{p}\frac{\Gamma(\beta_1)}{\Gamma(\beta_1-\gamma_j)}\int_0^1\zeta^{\beta_1-\gamma_j-1}\,d\mathfrak{H}_j(\zeta).$$

Lemma 1. *If $\Delta_1\neq 0$, then the unique solution $u\in C[0,1]$ of problem (4), (5) is*

$$u(t)=\int_0^1\mathfrak{G}_1(t,s)\varphi_{\varrho_1}(I_{0+}^{\alpha_1}\mathfrak{h}(s))\,ds,\ \ t\in[0,1], \tag{6}$$

where the Green function $\mathfrak{G}_1$ is given by

$$\mathfrak{G}_1(t,s)=\mathfrak{g}_1(t,s)+\frac{t^{\beta_1-1}}{\Delta_1}\sum_{i=1}^{p}\left(\int_0^1\mathfrak{g}_{2i}(\tau,s)\,d\mathfrak{H}_i(\tau)\right),\ \ t,s\in[0,1], \tag{7}$$

with

$$\mathfrak{g}_1(t,s)=\frac{1}{\Gamma(\beta_1)}\begin{cases}t^{\beta_1-1}(1-s)^{\beta_1-\gamma_0-1}-(t-s)^{\beta_1-1}, & 0\leq s\leq t\leq 1,\\ t^{\beta_1-1}(1-s)^{\beta_1-\gamma_0-1}, & 0\leq t\leq s\leq 1,\end{cases}$$

$$\mathfrak{g}_{2i}(\tau,s)=\frac{1}{\Gamma(\beta_1-\gamma_i)}\begin{cases}\tau^{\beta_1-\gamma_i-1}(1-s)^{\beta_1-\gamma_0-1}-(\tau-s)^{\beta_1-\gamma_i-1}, & 0\leq s\leq\tau\leq 1,\\ \tau^{\beta_1-\gamma_i-1}(1-s)^{\beta_1-\gamma_0-1}, & 0\leq\tau\leq s\leq 1,\end{cases}$$

$i=1,\ldots,p.$

Now, we consider the nonlinear fractional differential equation

$$D_{0+}^{\alpha_2}(\varphi_{r_2}(D_{0+}^{\beta_2}v(t)))+\mathfrak{y}(t)=0,\ \ t\in(0,1), \tag{8}$$

with the boundary conditions

$$v^{(j)}(0) = 0,\ j = 0, \ldots, m-2;\ D_{0+}^{\beta_2} v(0) = 0,\ D_{0+}^{\delta_0} v(1) = \sum_{i=1}^{q} \int_0^1 D_{0+}^{\delta_i} v(t)\, d\mathfrak{K}_i(t), \tag{9}$$

where $\alpha_2 \in (0,1]$, $\beta_2 \in (m-1, m]$, $m \in \mathbb{N}$, $m \geq 3$, $q \in \mathbb{N}$, $\delta_i \in \mathbb{R}$ for all $i = 0, \ldots, q$, $0 \leq \delta_1 < \delta_2 < \cdots < \delta_q \leq \delta_0 < \beta_2 - 1$, $\delta_0 \geq 1$, $\mathfrak{K}_i$, $i = 1, \ldots, q$ are bounded variation functions, and $\mathfrak{y} \in C[0,1]$. We denote using

$$\Delta_2 = \frac{\Gamma(\beta_2)}{\Gamma(\beta_2 - \delta_0)} - \sum_{j=1}^{q} \frac{\Gamma(\beta_2)}{\Gamma(\beta_2 - \delta_j)} \int_0^1 \zeta^{\beta_2 - \delta_j - 1}\, d\mathfrak{K}_j(\zeta).$$

Lemma 2. *If $\Delta_2 \neq 0$, then the unique solution $v \in C[0,1]$ of problem (8), (9) is*

$$v(t) = \int_0^1 \mathfrak{G}_2(t,s) \varphi_{\varrho_2}(I_{0+}^{\alpha_2} \mathfrak{y}(s))\, ds,\ t \in [0,1], \tag{10}$$

where the Green function $\mathfrak{G}_2$ is given by

$$\mathfrak{G}_2(t,s) = \mathfrak{g}_3(t,s) + \frac{t^{\beta_2 - 1}}{\Delta_2} \sum_{i=1}^{q} \left(\int_0^1 \mathfrak{g}_{4i}(\tau, s)\, d\mathfrak{K}_i(\tau) \right),\ t, s \in [0,1], \tag{11}$$

with

$$\mathfrak{g}_3(t,s) = \frac{1}{\Gamma(\beta_2)} \begin{cases} t^{\beta_2-1}(1-s)^{\beta_2-\delta_0-1} - (t-s)^{\beta_2-1}, & 0 \leq s \leq t \leq 1, \\ t^{\beta_2-1}(1-s)^{\beta_2-\delta_0-1}, & 0 \leq t \leq s \leq 1, \end{cases}$$
$$\mathfrak{g}_{4i}(\tau,s) = \frac{1}{\Gamma(\beta_2 - \delta_i)} \begin{cases} \tau^{\beta_2-\delta_i-1}(1-s)^{\beta_2-\delta_0-1} - (\tau-s)^{\beta_2-\delta_i-1}, & 0 \leq s \leq \tau \leq 1, \\ \tau^{\beta_2-\delta_i-1}(1-s)^{\beta_2-\delta_0-1}, & 0 \leq \tau \leq s \leq 1, \end{cases}$$
$$i = 1, \ldots, q.$$

Lemma 3. *Assume that $\mathfrak{H}_i : [0,1] \to \mathbb{R}$, $i = 1, \ldots, p$, and $\mathfrak{K}_i : [0,1] \to \mathbb{R}$, $i = 1, \ldots, q$ are nondecreasing functions and $\Delta_1 > 0$, $\Delta_2 > 0$. Then, the Green functions $\mathfrak{G}_1$ and $\mathfrak{G}_2$ given by (7) and (11) have the following properties:*

(a) $\mathfrak{G}_1, \mathfrak{G}_2 : [0,1] \times [0,1] \to \mathbb{R}_+$ are continuous functions;

(b) $\mathfrak{G}_1(t,s) \leq \mathfrak{J}_1(s)$ for all $t, s \in [0,1]$, where $\mathfrak{J}_1(s) = \mathfrak{h}_1(s) + \frac{1}{\Delta_1} \sum_{i=1}^{p} \int_0^1 \mathfrak{g}_{2i}(\tau,s)\, d\mathfrak{H}_i(\tau)$, and $\mathfrak{h}_1(s) = \frac{1}{\Gamma(\beta_1)}[(1-s)^{\beta_1-\gamma_0-1} - (1-s)^{\beta_1-1}]$, $s \in [0,1]$;

(c) $\mathfrak{G}_1(t,s) \geq t^{\beta_1-1}\mathfrak{J}_1(s)$ for all $t, s \in [0,1]$;

(d) $\mathfrak{G}_2(t,s) \leq \mathfrak{J}_2(s)$ for all $t, s \in [0,1]$, where $\mathfrak{J}_2(s) = \mathfrak{h}_2(s) + \frac{1}{\Delta_2} \sum_{i=1}^{q} \int_0^1 \mathfrak{g}_{4i}(\tau,s)\, d\mathfrak{K}_i(\tau)$, and $\mathfrak{h}_2(s) = \frac{1}{\Gamma(\beta_2)}[(1-s)^{\beta_2-\delta_0-1} - (1-s)^{\beta_2-1}]$, $s \in [0,1]$;

(e) $\mathfrak{G}_2(t,s) \geq t^{\beta_2-1}\mathfrak{J}_2(s)$ for all $t, s \in [0,1]$.

Lemma 4. *Assume that $\mathfrak{H}_i : [0,1] \to \mathbb{R}$, $i = 1, \ldots, p$, and $\mathfrak{K}_i : [0,1] \to \mathbb{R}$, $i = 1, \ldots, q$ are nondecreasing functions, $\Delta_1 > 0$, $\Delta_2 > 0$, and $\mathfrak{h}, \mathfrak{y} \in C([0,1], \mathbb{R}_+)$. Then, the solutions u and v of problems (4), (5) and (8), (9), respectively, given by (6) and (10) satisfy the inequalities $u(t) \geq 0$, $v(t) \geq 0$, $u(t) \geq t^{\beta_1-1}u(\tau)$, $v(t) \geq t^{\beta_2-1}v(\tau)$ for all $t, \tau \in [0,1]$.*

3. Main Results

In this section, we study the existence and nonexistence of positive solutions for problem (1), (2) by imposing various conditions on the functions $\mathfrak{a}$, $\mathfrak{b}$, $\mathfrak{f}$ and $\mathfrak{g}$. We present the assumptions that we will use in the sequel.

(I1) $\alpha_1, \alpha_2 \in (0,1]$, $\beta_1 \in (n-1, n]$, $\beta_2 \in (m-1, m]$, $n, m \in \mathbb{N}$, $n, m \geq 3$, $p, q \in \mathbb{N}$, $\gamma_j \in \mathbb{R}$ for all $j = 0, 1, \ldots, p$, $0 \leq \gamma_1 < \gamma_2 < \cdots < \gamma_p \leq \gamma_0 < \beta_1 - 1$, $\gamma_0 \geq 1$, $\delta_j \in \mathbb{R}$ for all $j = 0, 1, \ldots, q$, $0 \leq \delta_1 < \delta_2 < \cdots < \delta_q \leq \delta_0 < \beta_2 - 1$, $\delta_0 \geq 1$, $r_1, r_2 > 1$,

$\varphi_{r_j}(\zeta) = |\zeta|^{r_j-2}\zeta$, $\varphi_{r_j}^{-1} = \varphi_{\varrho_j}$, $\varrho_j = \frac{r_j}{r_j-1}$, $j = 1,2$, $\mathfrak{a}_0 > 0$, $\mathfrak{b}_0 > 0$, $\mathfrak{H}_i$, $i = 1,\dots,p$ and $\mathfrak{K}_j$, $j = 1,\dots,q$ are nondecreasing functions, $\Delta_1 > 0$ and $\Delta_2 > 0$.

$(I2)$ The functions $\mathfrak{a}$, $\mathfrak{b} : [0,1] \to \mathbb{R}_+$ are continuous and there exist $t_1, t_2 \in (0,1)$ such that $\mathfrak{a}(t_1) > 0$, $\mathfrak{b}(t_2) > 0$.

$(I3)$ The functions $\mathfrak{f}$, $\mathfrak{g} : \mathbb{R}_+ \to \mathbb{R}_+$ are continuous, and there exists $\mathfrak{c}_0 > 0$ such that $\mathfrak{f}(z) < \frac{\mathfrak{c}_0^{r_1-1}}{L}$, $\mathfrak{g}(z) < \frac{\mathfrak{c}_0^{r_2-1}}{L}$ for all $z \in [0,\mathfrak{c}_0]$, where

$$L = \max\left\{ \frac{\Lambda_i}{\Gamma(\alpha_i+1)} \left(\int_0^1 s^{\alpha_i(\varrho_i-1)} \mathfrak{J}_i(s)\, ds \right)^{r_i-1},\ i = 1,2 \right\},$$

with $\Lambda_1 = \sup_{t\in[0,1]} \mathfrak{a}(t)$, $\Lambda_2 = \sup_{t\in[0,1]} \mathfrak{b}(t)$.

$(I4)$ The functions $\mathfrak{f}$, $\mathfrak{g} : \mathbb{R}_+ \to \mathbb{R}_+$ are continuous and satisfy the conditions $\lim_{z\to\infty} \frac{\mathfrak{f}(z)}{z^{r_1-1}} = \infty$, $\lim_{z\to\infty} \frac{\mathfrak{g}(z)}{z^{r_2-1}} = \infty$.

Using $(I1)$, $(I2)$ and Lemma 3, we deduce that the constant L from assumption $(I3)$ is positive.

We consider the problems

$$\begin{cases} D_{0+}^{\alpha_1}\left(\varphi_{r_1}(D_{0+}^{\beta_1}x(t))\right) = 0,\ t \in (0,1), \\ x^{(j)}(0) = 0,\ j = 0,\dots,n-2,\ D_{0+}^{\beta_1}x(0) = 0,\ D_{0+}^{\gamma_0}x(1) = \sum_{i=1}^{p} \int_0^1 D_{0+}^{\gamma_i}x(\tau)\, d\mathfrak{H}_i(\tau) + 1, \end{cases} \tag{12}$$

$$\begin{cases} D_{0+}^{\alpha_2}\left(\varphi_{r_2}(D_{0+}^{\beta_2}y(t))\right) = 0,\ t \in (0,1), \\ y^{(j)}(0) = 0,\ j = 0,\dots,m-2,\ D_{0+}^{\beta_2}y(0) = 0,\ D_{0+}^{\delta_0}y(1) = \sum_{i=1}^{q} \int_0^1 D_{0+}^{\delta_i}y(\tau)\, d\mathfrak{K}_i(\tau) + 1, \end{cases} \tag{13}$$

The aforementioned problems (12) and (13) have the solutions $x(t) = \frac{t^{\beta_1-1}}{\Delta_1}$ and $y(t) = \frac{t^{\beta_2-1}}{\Delta_2}$, $t \in [0,1]$, respectively. Using $(I1)$, we have $x(t) > 0$ and $y(t) > 0$ for all $t \in (0,1]$. For a solution (u,v) of problem (1), (2), we define the functions $h(t) = u(t) - \mathfrak{a}_0 x(t) = u(t) - \frac{\mathfrak{a}_0 t^{\beta_1-1}}{\Delta_1}$, and $k(t) = v(t) - \mathfrak{b}_0 y(t) = v(t) - \frac{\mathfrak{b}_0 t^{\beta_2-1}}{\Delta_2}$, for $t \in [0,1]$. Then (1), (2) can equivalently be written as the system of fractional differential equations

$$\begin{cases} D_{0+}^{\alpha_1}\left(\varphi_{r_1}(D_{0+}^{\beta_1}h(t))\right) + \mathfrak{a}(t)\mathfrak{f}(k(t) + \mathfrak{b}_0 y(t)) = 0,\ t \in (0,1), \\ D_{0+}^{\alpha_2}\left(\varphi_{r_2}(D_{0+}^{\beta_2}k(t))\right) + \mathfrak{b}(t)\mathfrak{g}(h(t) + \mathfrak{a}_0 x(t)) = 0,\ t \in (0,1), \end{cases} \tag{14}$$

with the boundary conditions

$$\begin{cases} h^{(j)}(0) = 0,\ j = 0,\dots,n-2;\ D_{0+}^{\beta_1}h(0) = 0,\ D_{0+}^{\gamma_0}h(1) = \sum_{j=1}^{p} \int_0^1 D_{0+}^{\gamma_j}h(\tau)\, d\mathfrak{H}_j(\tau), \\ k^{(j)}(0) = 0,\ j = 0,\dots,m-2;\ D_{0+}^{\beta_2}k(0) = 0,\ D_{0+}^{\delta_0}k(1) = \sum_{j=1}^{q} \int_0^1 D_{0+}^{\delta_j}k(\tau)\, d\mathfrak{K}_j(\tau). \end{cases} \tag{15}$$

Using the Green functions $\mathfrak{G}_1$ and $\mathfrak{G}_2$, Lemmas 1 and 2, a pair of functions (h,k) is a solution of problem (14), (15) if, and only if, (h,k) is a solution of the system of nonlinear integral equations

$$\begin{cases} h(t) = \int_0^1 \mathfrak{G}_1(t,s)\varphi_{\varrho_1}(I_{0+}^{\alpha_1}(\mathfrak{a}(s)\mathfrak{f}(k(s) + \mathfrak{b}_0 y(s))))\, ds,\ t \in [0,1], \\ k(t) = \int_0^1 \mathfrak{G}_2(t,s)\varphi_{\varrho_2}(I_{0+}^{\alpha_2}(\mathfrak{b}(s)\mathfrak{g}(h(s) + \mathfrak{a}_0 x(s))))\, ds,\ t \in [0,1]. \end{cases} \tag{16}$$

We consider the Banach space $\mathcal{X} = C[0,1]$ with the supremum norm $\|h\| = \sup_{\tau\in[0,1]} |h(\tau)|$ for $h \in \mathcal{X}$, and the Banach space $\mathcal{Y} = \mathcal{X} \times \mathcal{X}$ with the norm $\|(h,k)\|_{\mathcal{Y}} = \max\{\|h\|, \|k\|\}$ for $(h,k) \in \mathcal{Y}$. We define the set $\mathcal{E} = \{(h,k) \in \mathcal{Y},\ 0 \le h(t) \le \mathfrak{c}_0,\ 0 \le k(t) \le \mathfrak{c}_0\}$. We also define the operator $\mathcal{A} : \mathcal{E} \to \mathcal{Y}$, $\mathcal{A} = (\mathcal{A}_1, \mathcal{A}_2)$,

$$\mathcal{A}_1(h,k)(t) = \int_0^1 \mathfrak{G}_1(t,s)\varphi_{\varrho_1}(I_{0+}^{\alpha_1}(\mathfrak{a}(s)\mathfrak{f}(k(s)+\mathfrak{b}_0 y(s))))\,ds,\ t \in [0,1],$$
$$\mathcal{A}_2(h,k)(t) = \int_0^1 \mathfrak{G}_2(t,s)\varphi_{\varrho_2}(I_{0+}^{\alpha_2}(\mathfrak{b}(s)\mathfrak{g}(h(s)+\mathfrak{a}_0 x(s))))\,ds,\ t \in [0,1],$$

for $(h,k) \in \mathcal{E}$. We remark that (h,k) is a solution of system (16) if and only if (h,k) is a fixed point of operator $\mathcal{A}$.

Our Theorem 1 is the following existence result for problem (1), (2).

Theorem 1. *We suppose that assumptions $(I1)-(I3)$ are satisfied. Then, there exist $\mathfrak{a}_1 > 0$ and $\mathfrak{b}_1 > 0$ such that for any $\mathfrak{a}_0 \in (0, \mathfrak{a}_1]$ and $\mathfrak{b}_0 \in (0, \mathfrak{b}_1]$, the problem (1), (2) has at least one positive solution.*

Proof. By assumption $(I3)$, we find that there exist $p_0 > 0$ and $q_0 > 0$ such that $\mathfrak{f}(z) \le \frac{\mathfrak{c}_0^{r_1-1}}{L}$ for all $z \in [0, \mathfrak{c}_0 + p_0]$, and $\mathfrak{g}(z) \le \frac{\mathfrak{c}_0^{r_2-1}}{L}$ for all $z \in [0, \mathfrak{c}_0 + q_0]$. We define $\mathfrak{a}_1 = q_0\Delta_1$ and $\mathfrak{b}_1 = p_0\Delta_2$. Let $\mathfrak{a}_0 \in (0, \mathfrak{a}_1]$ and $\mathfrak{b}_0 \in (0, \mathfrak{b}_1]$. Then, we obtain

$$\mathfrak{f}(k(s)+\mathfrak{b}_0 y(s)) \le \frac{\mathfrak{c}_0^{r_1-1}}{L},\ \ \mathfrak{g}(h(s)+\mathfrak{a}_0 x(s)) \le \frac{\mathfrak{c}_0^{r_2-1}}{L},$$

for all $s \in [0,1]$ and $(h,k) \in \mathcal{E}$. Hence, by using Lemma 4, we deduce that $\mathcal{A}_i(h,k)(t) \ge 0$, $i = 1,2$, for all $t \in [0,1]$ and $(h,k) \in \mathcal{E}$. By Lemma 3, for all $(h,k) \in \mathcal{E}$, we obtain

$$\begin{aligned}
I_{0+}^{\alpha_1}(\mathfrak{a}(s)\mathfrak{f}(k(s)+\mathfrak{b}_0 y(s))) &= \frac{1}{\Gamma(\alpha_1)}\int_0^s (s-\tau)^{\alpha_1-1}\mathfrak{a}(\tau)\mathfrak{f}(k(\tau)+\mathfrak{b}_0 y(\tau))\,d\tau \\
&\le \frac{\mathfrak{c}_0^{r_1-1}}{L\Gamma(\alpha_1)}\int_0^s (s-\tau)^{\alpha_1-1}\mathfrak{a}(\tau)\,d\tau \le \frac{\Lambda_1\mathfrak{c}_0^{r_1-1}}{L\Gamma(\alpha_1)}\int_0^s (s-\tau)^{\alpha_1-1}\,d\tau = \frac{\Lambda_1\mathfrak{c}_0^{r_1-1}s^{\alpha_1}}{L\Gamma(\alpha_1+1)},\ \ \forall s \in [0,1],
\end{aligned}$$

and then,

$$\begin{aligned}
\mathcal{A}_1(h,k)(t) &\le \int_0^1 \mathfrak{J}_1(s)\varphi_{\varrho_1}\left(\frac{\Lambda_1\mathfrak{c}_0^{r_1-1}s^{\alpha_1}}{L\Gamma(\alpha_1+1)}\right)ds \\
&= \left(\frac{\Lambda_1\mathfrak{c}_0^{r_1-1}}{L\Gamma(\alpha_1+1)}\right)^{\varrho_1-1}\int_0^1 \mathfrak{J}_1(s)s^{\alpha_1(\varrho_1-1)}\,ds \le \mathfrak{c}_0,\ \ \forall t \in [0,1].
\end{aligned}$$

In a similar manner, for all $(h,k) \in \mathcal{E}$ we have

$$I_{0+}^{\alpha_2}(\mathfrak{b}(s)\mathfrak{g}(h(s)+\mathfrak{a}_0 x(s))) \le \frac{\Lambda_2\mathfrak{c}_0^{r_2-1}s^{\alpha_2}}{L\Gamma(\alpha_2+1)},\ \ \forall s \in [0,1],$$

and

$$\mathcal{A}_2(h,k)(t) \le \int_0^1 \mathfrak{J}_2(s)\varphi_{\varrho_2}\left(\frac{\Lambda_2\mathfrak{c}_0^{r_2-1}s^{\alpha_2}}{L\Gamma(\alpha_2+1)}\right)ds \le \mathfrak{c}_0,\ \ \forall t \in [0,1].$$

Therefore, we find that $\mathcal{A}(\mathcal{E}) \subset \mathcal{E}$. By using standard arguments, we deduce that $\mathcal{A}$ is a completely continuous operator. Therefore, using the Schauder fixed-point theorem, we conclude that $\mathcal{A}$ has a fixed point $(h,k) \in \mathcal{E}$, which is a nonnegative solution for problem (16) or, equivalently, for problem (14), (15). Therefore, (u,v), where $u(t) = h(t) + \mathfrak{a}_0 x(t) = h(t) + \mathfrak{a}_0\frac{t^{\beta_1-1}}{\Delta_1}$, $v(t) = k(t) + \mathfrak{b}_0 y(t) = k(t) + \mathfrak{b}_0\frac{t^{\beta_2-1}}{\Delta_2}$ for $t \in [0,1]$, is a positive solution of problem (1), (2). This solution (u,v) satisfies the conditions $\frac{\mathfrak{a}_0 t^{\beta_1-1}}{\Delta_1} \leq u(t) \leq \frac{\mathfrak{a}_0 t^{\beta_1-1}}{\Delta_1} + \mathfrak{c}_0$ and $\frac{\mathfrak{b}_0 t^{\beta_2-1}}{\Delta_2} \leq v(t) \leq \frac{\mathfrak{b}_0 t^{\beta_2-1}}{\Delta_2} + \mathfrak{c}_0$ for all $t \in [0,1]$. □

Theorem 2 is the following nonexistence result for problem (1), (2).

Theorem 2. *We suppose that assumptions $(I1)$, $(I2)$ and $(I4)$ are satisfied. Then, there exist $\mathfrak{a}_2 > 0$ and $\mathfrak{b}_2 > 0$ such that, for any $\mathfrak{a}_0 \geq \mathfrak{a}_2$ and $\mathfrak{b}_0 \geq \mathfrak{b}_2$, the problem (1), (2) has no positive solution.*

Proof. By $(I2)$, there exist $[\theta_1,\theta_2] \subset (0,1)$, $\theta_1 < \theta_2$ such that $t_1, t_2 \in (\theta_1,\theta_2)$, and then

$$\Xi_1 := \int_{\theta_1}^{\theta_2} \mathfrak{J}_1(s)\left(\int_{\theta_1}^{s} \mathfrak{a}(\tau)(s-\tau)^{\alpha_1-1}\,d\tau\right)^{\varrho_1-1} ds > 0,$$
$$\Xi_2 := \int_{\theta_1}^{\theta_2} \mathfrak{J}_2(s)\left(\int_{\theta_1}^{s} \mathfrak{b}(\tau)(s-\tau)^{\alpha_2-1}\,d\tau\right)^{\varrho_2-1} ds > 0.$$

We consider

$$R = \max\left\{2^{r_i-1}\Gamma(\alpha_i)\left(\Xi_i \theta_1^{\beta_1+\beta_2-1}\right)^{1-r_i},\ i=1,2\right\}.$$

By using $(I4)$, for teh R defined above, we deduce that there exists $M_0 > 0$ such that $\mathfrak{f}(z) \geq R z^{r_1-1}$, $\mathfrak{g}(z) \geq R z^{r_2-1}$ for all $z \geq M_0$. We define $\mathfrak{a}_2 = \frac{M_0 \Delta_1}{\theta_1^{\beta_1}}$ and $\mathfrak{b}_2 = \frac{M_0 \Delta_2}{\theta_1^{\beta_2-1}}$. Let $\mathfrak{a}_0 \geq \mathfrak{a}_2$ and $\mathfrak{b}_0 \geq \mathfrak{b}_2$. We assume that (u,v) is a positive solution of (1), (2). Then, (h,k) where $h(t) = u(t) - \mathfrak{a}_0 x(t) = u(t) - \mathfrak{a}_0 \frac{t^{\beta_1-1}}{\Delta_1}$, $k(t) = v(t) - \mathfrak{b}_0 y(t) = v(t) - \mathfrak{b}_0 \frac{t^{\beta_2-1}}{\Delta_2}$ for $t \in [0,1]$, is a solution for (14), (15) or equivalently for (16). By using Lemma 4, we have $h(t) \geq t^{\beta_1-1}\|h\|$, $k(t) \geq t^{\beta_2-1}\|k\|$ for all $t \in [0,1]$. Then, $\inf_{t\in[\theta_1,\theta_2]} h(t) \geq \theta_1^{\beta_1-1}\|h\|$, $\inf_{t\in[\theta_1,\theta_2]} k(t) \geq \theta_1^{\beta_2-1}\|k\|$. Using the definition of x and y, we obtain $\inf_{t\in[\theta_1,\theta_2]} x(t) = \frac{\theta_1^{\beta_1-1}}{\Delta_1} = \theta_1^{\beta_1-1}\|x\|$ and $\inf_{t\in[\theta_1,\theta_2]} y(t) = \frac{\theta_1^{\beta_2-1}}{\Delta_2} = \theta_1^{\beta_2-1}\|y\|$. Therefore, we find

$$\begin{aligned}
&\inf_{t\in[\theta_1,\theta_2]}(h(t)+\mathfrak{a}_0 x(t)) \geq \inf_{t\in[\theta_1,\theta_2]} h(t) + \mathfrak{a}_0 \inf_{t\in[\theta_1,\theta_2]} x(t)\\
&\geq \theta_1^{\beta_1-1}\|h\| + \mathfrak{a}_0\theta_1^{\beta_1-1}\|x\| \geq \theta_1^{\beta_1-1}\|h+\mathfrak{a}_0 x\|,\\
&\inf_{t\in[\theta_1,\theta_2]}(k(t)+\mathfrak{b}_0 y(t)) \geq \inf_{t\in[\theta_1,\theta_2]} k(t) + \mathfrak{b}_0 \inf_{t\in[\theta_1,\theta_2]} y(t)\\
&\geq \theta_1^{\beta_2-1}\|k\| + \mathfrak{b}_0\theta_1^{\beta_2-1|}\|y\| \geq \theta_1^{\beta_2-1}\|k+\mathfrak{b}_0 y\|.
\end{aligned}$$

In addition, we have

$$\inf_{t\in[\theta_1,\theta_2]}(h(t)+\mathfrak{a}_0 x(t)) \geq \theta_1^{\beta_1-1}\|h\| + \frac{\mathfrak{a}_0\theta_1^{\beta_1-1}}{\Delta_1} \geq \frac{\mathfrak{a}_0\theta_1^{\beta_1-1}}{\Delta_1} \geq \frac{\mathfrak{a}_2\theta_1^{\beta_1-1}}{\Delta_1} = M_0,$$
$$\inf_{t\in[\theta_1,\theta_2]}(k(t)+\mathfrak{a}_0 y(t)) \geq \theta_1^{\beta_2-1}\|k\| + \frac{\mathfrak{b}_0\theta_1^{\beta_2-1}}{\Delta_2} \geq \frac{\mathfrak{b}_0\theta_1^{\beta_2-1}}{\Delta_2} \geq \frac{\mathfrak{b}_2\theta_1^{\beta_2-1}}{\Delta_2} = M_0.$$

Now, by using Lemma 4 and the above inequalities, we obtain

$$
\begin{aligned}
&I_{0+}^{\alpha_1}(\mathfrak{a}(s)\mathfrak{f}(k(s)+\mathfrak{b}_0 y(s)))\\
&\geq \frac{1}{\Gamma(\alpha_1)}\int_{\theta_1}^{s}(s-\tau)^{\alpha_1-1}\mathfrak{a}(\tau)\mathfrak{f}(k(\tau)+\mathfrak{b}_0 y(\tau))\,d\tau\\
&\geq \frac{R}{\Gamma(\alpha_1)}\int_{\theta_1}^{s}(s-\tau)^{\alpha_1-1}\mathfrak{a}(\tau)(k(\tau)+\mathfrak{b}_0 y(\tau))^{r_1-1}\,d\tau\\
&\geq \frac{R}{\Gamma(\alpha_1)}\int_{\theta_1}^{s}(s-\tau)^{\alpha_1-1}\mathfrak{a}(\tau)\left(\inf_{\tau\in[\theta_1,\theta_2]}(k(\tau)+\mathfrak{b}_0 y(\tau))\right)^{r_1-1}d\tau\\
&\geq \frac{RM_0^{r_1-1}}{\Gamma(\alpha_1)}\int_{\theta_1}^{s}(s-\tau)^{\alpha_1-1}\mathfrak{a}(\tau)\,d\tau,\ \ \forall s\in[\theta_1,\theta_2],
\end{aligned}
$$

and then

$$
\begin{aligned}
h(\theta_1) &\geq \int_0^1 \theta_1^{\beta_1-1}\mathfrak{J}_1(s)\varphi_{\varrho_1}(I_{0+}^{\alpha_1}(\mathfrak{a}(s)\mathfrak{f}(k(s)+\mathfrak{b}_0 y(s))))ds\\
&\geq \int_{\theta_1}^{\theta_2}\theta_1^{\beta_1-1}\mathfrak{J}_1(s)\varphi_{\varrho_1}\left(\frac{RM_0^{r_1-1}}{\Gamma(\alpha_1)}\int_{\theta_1}^{s}(s-\tau)^{\alpha_1-1}\mathfrak{a}(\tau)\,d\tau\right)ds\\
&= \frac{R^{\varrho_1-1}M_0\theta_1^{\beta_1-1}}{(\Gamma(\alpha_1))^{\varrho_1-1}}\int_{\theta_1}^{\theta_2}\mathfrak{J}_1(s)\left(\int_{\theta_1}^{s}(s-\tau)^{\alpha_1-1}\mathfrak{a}(\tau)\,d\tau\right)^{\varrho_1-1}ds\\
&= \frac{R^{\varrho_1-1}M_0\theta_1^{\beta_1-1}\Xi_1}{(\Gamma(\alpha_1))^{\varrho_1-1}} > 0.
\end{aligned}
$$

We deduce that $\|h\| \geq h(\theta_1) > 0$. In a similar manner, we find

$$
\begin{aligned}
&I_{0+}^{\alpha_2}(\mathfrak{b}(s)\mathfrak{g}(h(s)+\mathfrak{a}_0 x(s)))\\
&\geq \frac{R}{\Gamma(\alpha_2)}\int_{\theta_1}^{s}(s-\tau)^{\alpha_2-1}\mathfrak{b}(\tau)\left(\inf_{\tau\in[\theta_1,\theta_2]}(h(\tau)+\mathfrak{a}_0 x(\tau))\right)^{r_2-1}d\tau\\
&\geq \frac{RM_0^{r_2-1}}{\Gamma(\alpha_2)}\int_{\theta_1}^{s}(s-\tau)^{\alpha_2-1}\mathfrak{b}(\tau)\,d\tau,\ \ \forall s\in[\theta_1,\theta_2],
\end{aligned}
$$

and so

$$
k(\theta_1) \geq \frac{R^{\varrho_2-1}M_0\theta_1^{\beta_2-1}\Xi_2}{(\Gamma(\alpha_2))^{\varrho_2-1}} > 0.
$$

Therefore, $\|k\| \geq k(\theta_1) > 0$.

Besides, from the above inequalities, we obtain

$$
\begin{aligned}
&I_{0+}^{\alpha_1}(\mathfrak{a}(s)\mathfrak{f}(k(s)+\mathfrak{b}_0 y(s)))\\
&\geq \frac{R}{\Gamma(\alpha_1)}\int_{\theta_1}^{s}(s-\tau)^{\alpha_1-1}\mathfrak{a}(\tau)\left(\inf_{\tau\in[\theta_1,\theta_2]}(k(\tau)+\mathfrak{b}_0 y(\tau))\right)^{r_1-1}d\tau\\
&\geq \frac{R\theta_1^{(\beta_2-1)(r_1-1)}}{\Gamma(\alpha_1)}\|k+\mathfrak{b}_0 y\|^{r_1-1}\int_{\theta_1}^{s}(s-\tau)^{\alpha_1-1}\mathfrak{a}(\tau)\,d\tau,\ \ \forall s\in[\theta_1,\theta_2],
\end{aligned}
$$

and then

$$
\begin{aligned}
h(\theta_1) &\geq \int_{\theta_1}^{\theta_2}\theta_1^{\beta_1-1}\mathfrak{J}_1(s)\left(\frac{R\theta_1^{(\beta_2-1)(r_1-1)}}{\Gamma(\alpha_1)}\right)^{\varrho_1-1}\|k+\mathfrak{b}_0 y\|\left(\int_{\theta_1}^{s}(s-\tau)^{\alpha_1-1}\mathfrak{a}(\tau)\,d\tau\right)^{\varrho_1-1}ds\\
&= \frac{\theta_1^{\beta_1+\beta_2-2}R^{\varrho_1-1}}{(\Gamma(\alpha_1))^{\varrho_1-1}}\|k+\mathfrak{b}_0 y\|\int_{\theta_1}^{\theta_2}\mathfrak{J}_1(s)\left(\int_{\theta_1}^{s}(s-\tau)^{\alpha_1-1}\mathfrak{a}(\tau)\,d\tau\right)^{\varrho_1-1}ds\\
&= \frac{\theta_1^{\beta_1+\beta_2-2}R^{\varrho_1-1}}{(\Gamma(\alpha_1))^{\varrho_1-1}}\Xi_1\|k+\mathfrak{b}_0 y\| \geq 2\|k+\mathfrak{b}_0 y\| \geq 2\|k\|.
\end{aligned}
$$

Hence,

$$
\|k\| \leq \frac{h(\theta_1)}{2} \leq \frac{\|h\|}{2}. \tag{17}
$$

In a similar manner, we deduce

$$\begin{aligned} & I_{0+}^{\alpha_2}(\mathfrak{b}(s)\mathfrak{g}(h(s)+\mathfrak{a}_0x(s))) \\ & \geq \frac{R}{\Gamma(\alpha_2)}\int_{\theta_1}^{s}(s-\tau)^{\alpha_2-1}\mathfrak{b}(\tau)\left(\inf_{\tau\in[\theta_1,\theta_2]}(h(\tau)+\mathfrak{a}_0x(\tau))\right)^{r_2-1}d\tau \\ & \geq \frac{R\theta_1^{(\beta_1-1)(r_2-1)}}{\Gamma(\alpha_2)}\|h+\mathfrak{a}_0x\|^{r_2-1}\int_{\theta_1}^{s}(s-\tau)^{\alpha_2-1}\mathfrak{b}(\tau)\,d\tau,\ \ \forall s\in[\theta_1,\theta_2], \end{aligned}$$

and so

$$\begin{aligned} k(\theta_1) & \geq \int_{\theta_1}^{\theta_2}\theta_1^{\beta_2-1}\mathfrak{J}_2(s)\left(\frac{R\theta_1^{(\beta_1-1)(r_2-1)}}{\Gamma(\alpha_2)}\right)^{\varrho_2-1}\|h+\mathfrak{a}_0x\|\left(\int_{\theta_1}^{s}(s-\tau)^{\alpha_2-1}\mathfrak{b}(\tau)\,d\tau\right)^{\varrho_2-1}ds \\ & = \frac{\theta_1^{\beta_1+\beta_2-2}R^{\varrho_2-1}}{(\Gamma(\alpha_2))^{\varrho_2-1}}\|h+\mathfrak{a}_0x\|\int_{\theta_1}^{\theta_2}\mathfrak{J}_2(s)\left(\int_{\theta_1}^{s}(s-\tau)^{\alpha_2-1}\mathfrak{b}(\tau)\,d\tau\right)^{\varrho_2-1}ds \\ & = \frac{\theta_1^{\beta_1+\beta_2-2}R^{\varrho_2-1}}{(\Gamma(\alpha_2))^{\varrho_2-1}}\Xi_2\|h+\mathfrak{a}_0x\| \geq 2\|h+\mathfrak{a}_0x\| \geq 2\|h\|. \end{aligned}$$

Hence,

$$\|h\| \leq \frac{k(\theta_1)}{2} \leq \frac{\|k\|}{2}. \tag{18}$$

Therefore, using (17) and (18), we conclude that $\|h\| \leq \frac{\|k\|}{2} \leq \frac{\|h\|}{4}$, which contradicts the inequality $\|h\| > 0$. Then, problem (1), (2) has no positive solution. □

4. An Example

We consider $\alpha_1 = 1/2$, $\alpha_2 = 1/3$, $\beta_1 = 9/4$, $n = 3$, $\beta_2 = 17/5$, $m = 4$, $p = 2$, $q = 1$, $\gamma_0 = 8/7$, $\gamma_1 = 1/5$, $\gamma_2 = 2/3$, $\delta_0 = 11/6$, $\delta_1 = 3/4$, $r_1 = 21/5$, $\varrho_1 = 21/16$, $r_2 = 11/2$, $\varrho_2 = 11/9$, $\mathfrak{a}(t) = 1$, $\mathfrak{b}(t) = 1$ for all $t \in [0,1]$, $\mathfrak{H}_1(t) = 7t/6$ for all $t \in [0,1]$, $\mathfrak{H}_2(t) = \{1/2,\ t \in [0,11/23);\ 17/18,\ t \in [11/23,1]\}$, $\mathfrak{K}_1(t) = \{2,\ t \in [0,2/5);\ 61/21,\ t \in [2/5,1]\}$. We also consider the functions $\mathfrak{f}(z) = \frac{\sigma_1 z^{\omega_1}}{z^{\omega_2}+\sigma_2}$, $\mathfrak{g}(z) = \frac{\sigma_3 z^{\omega_3}}{z^{\omega_4}+\sigma_4}$ for all $z \in \mathbb{R}_+$, with $\sigma_i > 0$, $\omega_i > 0$, $i = 1,\ldots,4$, $\omega_1 > \omega_2 + 16/5$, $\omega_3 > \omega_4 + 9/2$. We have $\lim_{z\to\infty}\frac{\mathfrak{f}(z)}{z^{r_1-1}} = \infty$ and $\lim_{z\to\infty}\frac{\mathfrak{g}(z)}{z^{r_2-1}} = \infty$.

Hence, we consider the system of Riemann–Liouville fractional differential equations

$$\begin{cases} D_{0+}^{1/2}\left(\varphi_{21/5}(D_{0+}^{9/4}u(t))\right) + \dfrac{\sigma_1(v(t))^{\omega_1}}{(v(t))^{\omega_2}+\sigma_2} = 0, & t \in (0,1), \\ D_{0+}^{1/3}\left(\varphi_{11/2}(D_{0+}^{17/5}v(t))\right) + \dfrac{\sigma_3(u(t))^{\omega_3}}{(u(t))^{\omega_4}+\sigma_4} = 0, & t \in (0,1), \end{cases} \tag{19}$$

with the boundary conditions

$$\begin{cases} u(0) = u'(0) = 0,\ D_{0+}^{9/4}u(0) = 0,\ D_{0+}^{8/7}u(1) = \dfrac{7}{6}\displaystyle\int_0^1 D_{0+}^{1/5}u(\tau)\,d\tau + \dfrac{4}{9}D_{0+}^{2/3}u\left(\dfrac{11}{23}\right) + \mathfrak{a}_0, \\ v(0) = v'(0) = v''(0) = 0,\ D_{0+}^{17/5}v(0) = 0,\ D_{0+}^{11/6}v(1) = \dfrac{19}{21}D_{0+}^{3/4}v\left(\dfrac{2}{5}\right) + \mathfrak{b}_0. \end{cases} \tag{20}$$

We obtain $\Delta_1 \approx 0.19646507 > 0$, $\Delta_2 \approx 2.94848267 > 0$. Therefore, assumptions $(I1)$, $(I2)$ and $(I4)$ are satisfied. In addition, we find

$$
\begin{aligned}
\mathfrak{g}_1(t,s) &= \frac{1}{\Gamma(9/4)}\begin{cases} t^{5/4}(1-s)^{3/28}-(t-s)^{5/4}, & 0\le s\le t\le 1,\\ t^{5/4}(1-s)^{3/28}, & 0\le t\le s\le 1,\end{cases}\\
\mathfrak{g}_{21}(t,s) &= \frac{1}{\Gamma(41/20)}\begin{cases} t^{21/20}(1-s)^{3/28}-(t-s)^{21/20}, & 0\le s\le t\le 1,\\ t^{21/20}(1-s)^{3/28}, & 0\le t\le s\le 1,\end{cases}\\
\mathfrak{g}_{22}(t,s) &= \frac{1}{\Gamma(19/12)}\begin{cases} t^{7/12}(1-s)^{3/28}-(t-s)^{7/12}, & 0\le s\le t\le 1,\\ t^{7/12}(1-s)^{3/28}, & 0\le t\le s\le 1,\end{cases}\\
\mathfrak{g}_3(t,s) &= \frac{1}{\Gamma(17/5)}\begin{cases} t^{12/5}(1-s)^{17/30}-(t-s)^{12/5}, & 0\le s\le t\le 1,\\ t^{12/5}(1-s)^{17/30}, & 0\le t\le s\le 1,\end{cases}\\
\mathfrak{g}_{41}(t,s) &= \frac{1}{\Gamma(53/20)}\begin{cases} t^{33/20}(1-s)^{17/30}-(t-s)^{33/20}, & 0\le s\le t\le 1,\\ t^{33/20}(1-s)^{17/30}, & 0\le t\le s\le 1,\end{cases}\\
\mathfrak{G}_1(t,s) &= \mathfrak{g}_1(t,s)+\frac{t^{5/4}}{\Delta_1}\left[\frac{7}{6}\int_0^1 \mathfrak{g}_{21}(\tau,s)\,d\tau+\frac{4}{9}\mathfrak{g}_{22}\left(\frac{11}{23},s\right)\right],\\
\mathfrak{G}_2(t,s) &= \mathfrak{g}_3(t,s)+\frac{19t^{12/5}}{21\Delta_2}\mathfrak{g}_{41}\left(\frac{2}{5},s\right),\\
\mathfrak{h}_1(s) &= \frac{1}{\Gamma(9/4)}\left[(1-s)^{3/28}-(1-s)^{5/4}\right],\\
\mathfrak{h}_2(s) &= \frac{1}{\Gamma(17/5)}\left[(1-s)^{17/30}-(1-s)^{12/5}\right],
\end{aligned}
$$

for all $t,s\in[0,1]$. In addition, we deduce

$$
\mathfrak{J}_1(s)=\begin{cases} \mathfrak{h}_1(s)+\dfrac{1}{\Delta_1}\left\{\dfrac{7}{6\Gamma(61/20)}(1-s)^{3/28}-\dfrac{7}{6\Gamma(61/20)}(1-s)^{41/20}\right.\\ \quad\left.+\dfrac{4}{9\Gamma(19/12)}\left[\left(\dfrac{11}{23}\right)^{7/12}(1-s)^{3/28}-\left(\dfrac{11}{23}-s\right)^{7/12}\right]\right\}, & 0\le s<\dfrac{11}{23},\\ \mathfrak{h}_1(s)+\dfrac{1}{\Delta_1}\left\{\dfrac{7}{6\Gamma(61/20)}(1-s)^{3/28}-\dfrac{7}{6\Gamma(61/20)}(1-s)^{41/20}\right.\\ \quad\left.+\dfrac{4}{9\Gamma(19/12)}\left(\dfrac{11}{23}\right)^{7/12}(1-s)^{3/28}\right], & \dfrac{11}{23}\le s\le 1,\end{cases}
$$

$$
\mathfrak{J}_2(s)=\begin{cases} \mathfrak{h}_2(s)+\dfrac{19}{21\Delta_2\Gamma(53/20)}\left[\left(\dfrac{2}{5}\right)^{33/20}(1-s)^{17/30}-\left(\dfrac{2}{5}-s\right)^{33/20}\right], & 0\le s<\dfrac{2}{5},\\ \mathfrak{h}_2(s)+\dfrac{19}{21\Delta_2\Gamma(53/20)}\left(\dfrac{2}{5}\right)^{33/20}(1-s)^{17/30}, & \dfrac{2}{5}\le s\le 1,\end{cases}
$$

After some computations, we obtain $\int_0^1 s^{5/32}\mathfrak{J}_1(s)\,ds\approx 2.7671383$, $\int_0^1 s^{2/27}\mathfrak{J}_2(s)\,ds\approx 0.12990129$, $\Lambda_1=1$, $\Lambda_2=1$ and $L\approx 29.30581677$. We take $\mathfrak{c}_0=1$ and, if we choose σ_i, $i=1,\dots,4$ which satisfy the conditions $\sigma_1<\frac{1+\sigma_2}{L}$ and $\sigma_3<\frac{1+\sigma_4}{L}$, then we deduce $\mathfrak{f}(z)\le\frac{\sigma_1}{1+\sigma_2}<\frac{1}{L}$ and $\mathfrak{g}(z)\le\frac{\sigma_3}{1+\sigma_4}<\frac{1}{L}$ for all $z\in[0,1]$. For example, if $\sigma_2=1$ and $\sigma_4=2$, then for $\sigma_1\le 0.068$ and $\sigma_3\le 0.102$, the above conditions for $\mathfrak{f}$ and $\mathfrak{g}$ are satisfied. Hence, assumption $(I3)$ is also satisfied. Using Theorems 1 and 2, we conclude that there exist $\mathfrak{a}_1$, $\mathfrak{b}_1$, $\mathfrak{a}_2$, $\mathfrak{b}_2$ such that, for any $\mathfrak{a}_0\in(0,\mathfrak{a}_1]$ and $\mathfrak{b}_0\in(0,\mathfrak{b}_1]$ there exists at least one positive solution of problem (19), (20), and, for any $\mathfrak{a}_0\ge\mathfrak{a}_2$ and $\mathfrak{b}_0\ge\mathfrak{b}_2$, there exists no positive solution of (19), (20).

5. Conclusions

In this paper, we studied the system of Riemann–Liouville fractional differential Equation (1) with r_1-Laplacian and r_2-Laplacian operators, supplemented with the nonlocal uncoupled boundary conditions (2), which contain fractional derivatives of various orders, Riemann–Stieltjes integrals, and two positive parameters. The functions $\mathfrak{a}$, $\mathfrak{b}$, $\mathfrak{f}$ and $\mathfrak{g}$ from the system are continuous ones and satisfy some additional assumptions. We presented some auxiliary results, including the associated Green functions with their properties. Then, we investigated problem (1), (2) in some stages. First, we made a change in the unknown functions, such that the new boundary conditions have no positive parameters, and then, by using the Green functions, we equivalently wrote this new problem as the system of nonlinear integral equations (16). By constructing an appropriate operator $\mathcal{A}$, the solutions of the integral system are the fixed points of $\mathcal{A}$. By applying the Schauder fixed-point theorem, we showed that the operator $\mathcal{A}$ has at least one fixed point, which is a positive solution of our problem, when the positive parameters belong to some intervals. Then, we provided intervals for the parameters for which problem (1), (2) has no positive solution. We also presented an example to illustrate our obtained results.

Author Contributions: Conceptualization, R.L.; formal analysis, A.T. and R.L.; methodology, A.T. and R.L. All authors have read and agreed to the published version of the manuscript.

Funding: This research received no external funding.

Institutional Review Board Statement: Not applicable.

Informed Consent Statement: Not applicable.

Data Availability Statement: Not applicable.

Acknowledgments: The authors thank the referees for their valuable comments and suggestions.

Conflicts of Interest: The authors declare no conflict of interest.

References

1. Wang, G.; Ren, X.; Zhang, L.; Ahmad, B. Explicit iteration and unique positive solution for a Caputo-Hadamard fractional turbulent flow model. *IEEE Access* **2019**, *7*, 109833–109839. [CrossRef]
2. Luca, R. On a system of fractional differential equations with p-Laplacian operators and integral boundary conditions. *Revue Roum. Math. Pures Appl.* **2021**, *66*, 749–766.
3. Tudorache, A.; Luca, R. Positive solutions for a system of Riemann–Liouville fractional boundary value problems with p-Laplacian operators. *Adv. Differ. Equ.* **2020**, *292*, 1–30. [CrossRef]
4. Henderson, J.; Luca, R.; Tudorache, A. Existence of positive solutions for a system of fractional boundary value problems. In *Proceedings of the "Differential and Difference Equations with Applications", ICDDEA, Amadora, Portugal, May 2015, Selected Contributions*; Pinelas, S., Dosla, Z., Dosly, O., Kloeden, P.E., Eds.; Springer: Berlin/Heidelberg, Germany, 2016; pp. 349–357.
5. Tan, J.; Li, M. Solutions of fractional differential equations with p-Laplacian operator in Banach spaces. *Bound. Value Prob.* **2018**, *15*, 1–13. [CrossRef]
6. Tang, X.; Wang, X.; Wang, Z.; Ouyang, P. The existence of solutions for mixed fractional resonant boundary value problem with $p(t)$-Laplacian operator. *J. Appl. Math. Comput.* **2019**, *61*, 559–572. [CrossRef]
7. Tian, Y.; Sun, S.; Bai, Z. Positive Solutions of Fractional Differential Equations with p-Laplacian. *J. Funct. Spaces* **2017**, *2017*, 3187492. [CrossRef]
8. Wang, H.; Jiang, J. Existence and multiplicity of positive solutions for a system of nonlinear fractional multi-point boundary value problems with p-Laplacian operator. *J. Appl. Anal. Comput.* **2021**, *11*, 351–366.
9. Wang, Y.; Liu, S.; Han, Z. Eigenvalue problems for fractional differential equationswith mixed derivatives and generalized p-Laplacian. *Nonlinear Anal. Model. Control* **2018**, *23*, 830–850. [CrossRef]
10. Ahmad, A.B.; Alsaedi, S.K.; Ntouyas, J. *Tariboon, Hadamard-Type Fractional Differential Equations, Inclusions and Inequalities*; Springer: Cham, Switzerland, 2017.
11. Ahmad, B.; Ntouyas, S.K. *Nonlocal Nonlinear Fractional-Order Boundary Value Problems*; World Scientific: Hackensack, NJ, USA, 2021.
12. Zhou, Y.; Wang, J.R.; Zhang, L. *Basic Theory of Fractional Differential Equations*, 2nd ed.; World Scientific: Singapore, 2016.

axioms MDPI

Article

On Fractional Inequalities Using Generalized Proportional Hadamard Fractional Integral Operator

Vaijanath L. Chinchane [1], Asha B. Nale [2], Satish K. Panchal [2], Christophe Chesneau [3,*] and Amol D. Khandagale [2]

[1] Department of Mathematics, Deogiri Institute of Engineering and Management, Aurangabad 431005, India; chinchane85@gmail.com
[2] Department of Mathematics, Babasaheb Ambedkar Marathwada University, Aurangabad 431004, India; ashabnale@gmail.com (A.B.N.); drpanchalsk@gmail.com (S.K.P.); kamoldsk@gmail.com (A.D.K.)
[3] Department of Mathematics, University of Caen-Normandie, 14000 Caen, France
* Correspondence: christophe.chesneau@unicaen.fr

Abstract: The main objective of this paper is to use the generalized proportional Hadamard fractional integral operator to establish some new fractional integral inequalities for extended Chebyshev functionals. In addition, we investigate some fractional integral inequalities for positive continuous functions by employing a generalized proportional Hadamard fractional integral operator. The findings of this study are theoretical but have the potential to help solve additional practical problems in mathematical physics, statistics, and approximation theory.

Keywords: extended Chebyshev functional; generalized proportional Hadamard fractional integral operator

MSC: 26D10; 26A33; 05A30; 26D53

Citation: Chinchane, V.L.; Nale, A.B.; Panchal, S.K.; Chesneau, C.; Khandagale, A.D. On Fractional Inequalities Using Generalized Proportional Hadamard Fractional Integral Operator. *Axioms* **2022**, *11*, 266. https://doi.org/10.3390/axioms11060266

Academic Editor: Hans J. Haubold

Received: 20 April 2022
Accepted: 30 May 2022
Published: 1 June 2022

Publisher's Note: MDPI stays neutral with regard to jurisdictional claims in published maps and institutional affiliations.

Copyright: © 2022 by the authors. Licensee MDPI, Basel, Switzerland. This article is an open access article distributed under the terms and conditions of the Creative Commons Attribution (CC BY) license (https://creativecommons.org/licenses/by/4.0/).

1. Introduction

Fractional calculus has a new orientation not only with respect to mathematics but also to physics, statistics, engineering, and other applied sciences. Its birth dates back to ancient times, and its development has been rapid in recent years, gaining new momentum, especially with the definition of new fractional integral and derivative operators. New fractional operators lead to useful applications and generalizations in the field, and their kernel structures and properties give them an advantage over classical derivative and integral operators.

Recently, many mathematicians have worked with slightly different fractional integral formulas. For example, see [1–7] for Riemann–Liouville fractional integral operators, [8] for Hadamard fractional integral operators, [9–12] for Saigo fractional integral operators, [13–15] for conformable fractional integral operators, [16–18] for generalized Katugampola fractional operators, and [19–22] for k-generalized (in terms of hypergeometric function) fractional integral operators. In [3,20], the authors investigated fractional integral inequalities for extended Chebyshev functionals by employing Riemann–Liouville and generalized k-fractional integral fractional integrals, respectively. Recently, many mathematicians have examined several kinds of fractional integral and derivative operators with different types of kernels, such as logarithmic kernels, non-singular exponential kernels, etc. During the past few years, numerous analyses of real-world problems, mathematical models, and numerical methods have been resolved by fractional derivatives and integrals [13,15,23–33]. Anber et al. [34] presented some fractional integral inequalities similar to the Minkowski fractional integral inequality, using the Riemann–Liouville fractional integral. In [35], Panchal et al. studied weighted fractional integral inequalities using a generalized Katugampola fractional integral operator. In [36], Andric et al. proposed the reverse fractional Minkowski integral inequality using the extended Mittag-Leffler function with the corresponding fractional integral operator, which was proved together with several related Minkowski-type inequalities. Rahman et al. [37–39] investigated the Minkowski inequality and some other fractional inequalities for convex functions by employing fractional proportional integral operators. Atangana and Baleanu proposed a new fractional derivative operator with a non-local and non-singular kernel [40]. In [41], Jarad et al. proposed fractional conformable integral and derivative operators. In [42–44], Jarad et al. and Rahman presented the concepts of non-local fractional proportional and generalized Hadamard proportional integrals involving exponential functions in their kernels. In [14,45–47], the authors explored various integral inequalities by employing conformable and generalized conformable fractional integrals. Caputo and Fabrizio [48] introduced new fractional derivatives and integrals without singular kernels. Later, Lasada and Niteto proposed certain properties of fractional derivatives without a singular kernel [49]. Nale et al. and Rahaman et al. [44,50] investigated some Minkowski-type inequalities and other integral inequalities by considering the generalized proportional Hadamard fractional integral operator. Kukushkin [51] examined the final terms of a differential operator with a fractional integro-differential operator composition on a bounded domain of n-dimensional Euclidean space, as well as on the real axis. One of the central points was the relation connecting fractional powers of m-accretive operators and fractional derivatives in the most general sense. By virtue of such an approach, we express fractional derivatives in terms of infinitesimal generators. In this regard, operators such as the Kipriyanov operator, Riesz potential, and difference operator are considered. In addition, in [52], Yosida studied the semigroup that generates the involved fractional integro-differential operators due to the Balakrishnyan formula. We think that it would be interesting to illustrate the relevance of the topic by presenting and comparing the well-known Chebyshev inequality in L_1. In [53], the Chebyshev functional for two integrable functions u and v on $[a,b]$ is defined as follows:

$$T[u(x),v(x)] = \frac{1}{b-a}\int_a^b u(x)v(x)dx - \frac{1}{b-a}\left(\int_a^b u(x)dx\right)\frac{1}{b-a}\left(\int_a^b v(x)dx\right). \quad (1)$$

Many applications and several inequalities related to Chebyshev functionals can be found in [6,54–56]. Let us now consider the following extended Chebyshev functional [3]:

$$\begin{aligned} T[u(x),v(x),p(x),q(x)] &= \int_a^b q(x)dx \int_a^b p(x)u(x)v(x)dx + \int_a^b p(x)dx \int_a^b q(x)u(x)v(x)dx \\ &- \left(\int_a^b p(x)u(x)dx\right)\left(\int_a^b q(x)v(x)dx\right) \\ &- \left(\int_a^b q(x)u(x)dx\right)\left(\int_a^b p(x)v(x)dx\right), \end{aligned} \quad (2)$$

where u and v are two integrable functions on $[a,b]$, and p and q are positive integrable functions on $[a,b]$. In order to present a famous inequality for this function, let us now introduce the concept of synchronous (asynchronous) functions.

Definition 1. *Two functions u and v are called synchronous (asynchronous) functions on $[a,b]$ if*

$$\Big(u(\tau)-u(\sigma)\Big)\Big(v(\tau)-v(\sigma)\Big) \geq (\leq) 0, \quad \tau,\sigma \in [a,b]. \quad (3)$$

Hence, if u an v are synchronous on $[a,b]$, then $T[u(x),v(x),p(x),q(x)] \geq 0$.

Motivated by [3,19,34,38,39,43,44], our purpose in this paper is to obtain fractional integral inequalities for the extended Chebyshev functional and other fractional inequalities, using the generalized Hadamard proportional integral. The assumption of synchronous (asynchronous) functions will sometimes be made.

The paper has been organized as follows. in Section 2, we recall basic definitions, remarks, and lemmas related to generalized Hadamard proportional integrals. In Section 3, we obtain fractional integral inequalities for extended Chebyshev functionals using generalized Hadamard proportional integrals. In Section 4, we present some other fractional integral inequalities using generalized Hadamard proportional integrals. In Section 5, we give the concluding remarks.

2. Preliminary

Here, we present some important definitions, remarks, and lemmas of the generalized proportional Hadamard fractional integral operator, which will be used throughout this paper. Recently, Rahman et al. [43] presented the left and right-sided generalized proportional integral operators as follows:

Definition 2. *The left- and right-sided generalized proportional fractional integrals are, respectively, defined by*

$$ {}_{a}\mathfrak{I}^{\alpha,\beta}[z(x)](x) = \frac{1}{\beta^{\alpha}\Gamma(\alpha)} \int_{a}^{x} e^{\left[\frac{\beta-1}{\beta}(x-t)\right]} (x-t)^{\alpha-1} z(t)dt, \quad a < x \tag{4} $$

(here, the first x between the brackets refers to the variable of the function $z(x)$, and the second x between the brackets refers to the integral upper bound; other notations are possible), and

$$ \mathfrak{I}_{b}^{\alpha,\beta}[z(x)](x) = \frac{1}{\beta^{\alpha}\Gamma(\alpha)} \int_{x}^{b} e^{\left[\frac{\beta-1}{\beta}(t-x)\right]} (t-x)^{\alpha-1} z(t)dt, \quad x < b, \tag{5} $$

where the proportionality index is $\beta \in (0,1]$, $\alpha \in \mathbb{C}$ with $\Re(\alpha) > 0$, and $\Gamma(\alpha)$ is the classical well-known gamma function.

Remark 1. *If we consider $\beta = 1$ in Equations (4) and (5), then we obtain the well-known left- and right-sided Riemann–Liouville integrals, which are, respectively, defined by*

$$ {}_{a}\mathfrak{I}^{\alpha}[z(x)](x) = \frac{1}{\Gamma(\alpha)} \int_{a}^{x} (x-t)^{\alpha-1} z(t)dt, \quad a < x \tag{6} $$

and

$$ \mathfrak{I}_{b}^{\alpha}[z(x)](x) = \frac{1}{\Gamma(\alpha)} \int_{x}^{b} (t-x)^{\alpha-1} z(t)dt, \quad x < b, \tag{7} $$

where $\alpha \in \mathbb{C}$ with $\Re(\alpha) > 0$.

On the other hand, recently, Rahman et al. [44] proposed the following generalized Hadamard proportional fractional integrals.

Definition 3. *The left-sided generalized Hadamard proportional fractional integral of order $\alpha > 0$ and proportional index $\beta \in (0,1]$ is defined by*

$$ {}_{a}\mathcal{H}^{\alpha,\beta}[z(x)](x) = \frac{1}{\beta^{\alpha}\Gamma(\alpha)} \int_{a}^{x} e^{\left[\frac{\beta-1}{\beta}(\ln x - \ln t)\right]} (\ln x - \ln t)^{\alpha-1} \frac{z(t)}{t} dt, \quad a < x. \tag{8} $$

Definition 4. *The right-sided generalized Hadamard proportional fractional integral of order $\alpha > 0$ and proportional index $\beta \in (0,1]$ is defined by*

$$ \mathcal{H}_{b}^{\alpha,\beta}[z(x)](x) = \frac{1}{\beta^{\alpha}\Gamma(\alpha)} \int_{x}^{b} e^{\left[\frac{\beta-1}{\beta}(\ln t - \ln x)\right]} (\ln t - \ln x)^{\alpha-1} \frac{z(t)}{t} dt, \quad x < b. \tag{9} $$

Remark 2. *If we consider* $a = 1$ *in Equation* (8), *then we obtain*

$$ {}_1\mathcal{H}^{\alpha,\beta}[z(x)](x) = \frac{1}{\beta^{\alpha}\Gamma(\alpha)} \int_1^x e^{\left[\frac{\beta-1}{\beta}(\ln x - \ln t)\right]} (\ln x - \ln t)^{\alpha-1} \frac{z(t)}{t} dt, \quad x > 1. \tag{10} $$

Hereafter, to lighten the notation, we set $\mathcal{H}^{\alpha,\beta}_{1,x}[z(x)] = {}_1\mathcal{H}^{\alpha,\beta}[z(x)](x)$.

Remark 3. *If we consider* $\beta = 1$, *then Equations* (8)–(10) *will lead to the following well known Hadamard fractional integrals, indicated as*

$$ {}_a\mathcal{H}^{\alpha}[z(x)](x) = \frac{1}{\Gamma(\alpha)} \int_a^x (\ln x - \ln t)^{\alpha-1} \frac{z(t)}{t} dt, \quad a < x, \tag{11} $$

$$ \mathcal{H}^{\alpha}_b[z(x)](x) = \frac{1}{\Gamma(\alpha)} \int_x^b (\ln t - \ln x)^{\alpha-1} \frac{z(t)}{t} dt, \quad x < b, \tag{12} $$

and

$$ \mathcal{H}^{\alpha,\beta}_{1,x}[z(x)] = \frac{1}{\Gamma(\alpha)} \int_1^x (\ln x - \ln t)^{\alpha-1} \frac{z(t)}{t} dt, \quad x > 1. \tag{13} $$

One can easily prove the following results.

Lemma 1. *With the special function:* $z(x) = e^{\left[\frac{\beta-1}{\beta}(\ln x)\right]}(\ln x)^{\lambda-1}$, *we have*

$$ \mathcal{H}^{\alpha,\beta}_{1,x}\left[e^{\left[\frac{\beta-1}{\beta}(\ln x)\right]}(\ln x)^{\lambda-1}\right] = \frac{\Gamma(\lambda)}{\beta^{\alpha}\Gamma(\alpha+\lambda)} e^{\left[\frac{\beta-1}{\beta}(\ln x)\right]}(\ln x)^{\alpha+\lambda-1}, \tag{14} $$

and the following semigroup property holds:

$$ \mathcal{H}^{\alpha,\beta}_{1,x}\left[\mathcal{H}^{\lambda,\beta}_{1,x}[z(x)]\right] = \mathcal{H}^{\alpha+\lambda,\beta}_{1,x}[z(x)]. \tag{15} $$

Remark 4. *If* $\beta = 1$, *then Equation* (14) *reduces to the result of [57] as defined by*

$$ \mathcal{H}^{\alpha}_{1,x}\left[(\ln x)^{\lambda-1}\right] = \frac{\Gamma(\lambda)}{\Gamma(\alpha+\lambda)}(\ln x)^{\alpha+\lambda-1}. \tag{16} $$

3. Fractional Integral Inequalities for Extended Chebyshev Functional

In this section, we establish a fractional integral inequality involving generalized proportional Hadamard fractional integral operators. We now prove the following lemma.

Lemma 2. *Let f and g be two integrable and synchronous functions on* $[1,\infty)$, *and* $u, v : [1,\infty) \to [0,\infty)$. *Then, for all* $x > 1$, $\alpha > 0$ *and* $\beta \in (0,1]$, *we have*

$$ \begin{aligned} &\mathcal{H}^{\alpha,\beta}_{1,x}[u(x)]\mathcal{H}^{\alpha,\beta}_{1,x}[vfg(x)] + \mathcal{H}^{\alpha,\beta}_{1,x}[v(x)]\mathcal{H}^{\alpha,\beta}_{1,x}[ufg(x)] \geq \\ &\mathcal{H}^{\alpha,\beta}_{1,x}[uf(x)]\mathcal{H}^{\alpha,\beta}_{1,x}[vg(x)] + \mathcal{H}^{\alpha,\beta}_{1,x}[vf(x)]\mathcal{H}^{\alpha,\beta}_{1,x}[ug(x)]. \end{aligned} \tag{17} $$

It is understood that, for instance, $vfg(x) = v(x)f(x)g(x)$.

Proof. Since f and g are synchronous functions on $[1,\infty)$, for all $\tau \geq 0$ and $\sigma \geq 0$, the following inequality holds:

$$ \Big(f(\tau) - f(\sigma)\Big)\Big(g(\tau) - g(\sigma)\Big) \geq 0. \tag{18} $$

Then, Equation (18) becomes

$$f(\tau)g(\tau) + f(\sigma)g(\sigma) \geq f(\tau)g(\sigma) + f(\sigma)g(\tau). \tag{19}$$

Let us now consider

$$\psi(x,\tau) = \frac{1}{\beta^{\alpha}\Gamma(\alpha)\tau} e^{\left[\frac{\beta-1}{\beta}(\ln x - \ln \tau)\right]} (\ln x - \ln \tau)^{\alpha-1}. \tag{20}$$

We can clearly state that the function $\psi(x,\tau)u(\tau)$ remains positive, because for all $\tau \in (1,x)$, $(x > 1), \alpha, \beta > 0$. Multiplying both sides of Equation (19) by $\psi(x,\tau)$, then integrating the resulting identity with respect to τ from 1 to x, we obtain

$$\begin{aligned}
&\frac{1}{\beta^{\alpha}\Gamma(\alpha)} \int_1^x e^{\left[\frac{\beta-1}{\beta}(\ln x - \ln \tau)\right]} (\ln x - \ln \tau)^{\alpha-1} u(\tau) f(\tau) g(\tau) \frac{d\tau}{\tau} \\
&+ \frac{1}{\beta^{\alpha}\Gamma(\alpha)} \int_1^x e^{\left[\frac{\beta-1}{\beta}(\ln x - \ln \tau)\right]} (\ln x - \ln \tau)^{\alpha-1} u(\tau) f(\sigma) g(\sigma) \frac{d\tau}{\tau} \\
&\geq \frac{1}{\beta^{\alpha}\Gamma(\alpha)} \int_1^x e^{\left[\frac{\beta-1}{\beta}(\ln x - \ln \tau)\right]} (\ln x - \ln \tau)^{\alpha-1} u(\tau) f(\tau) g(\sigma) \frac{d\tau}{\tau} \\
&+ \frac{1}{\beta^{\alpha}\Gamma(\alpha)} \int_1^x e^{\left[\frac{\beta-1}{\beta}(\ln x - \ln \tau)\right]} (\ln x - \ln \tau)^{\alpha-1} u(\tau) f(\sigma) g(\tau) \frac{d\tau}{\tau}.
\end{aligned} \tag{21}$$

Consequently,

$$\begin{aligned}
&\mathcal{H}_{1,x}^{\alpha,\beta}[ufg(x)] + f(\sigma)g(\sigma)\mathcal{H}_{1,x}^{\alpha,\beta}[u(x)] \\
&\geq g(\sigma)\mathcal{H}_{1,x}^{\alpha,\beta}[uf(x)] + f(\sigma)\mathcal{H}_{1,x}^{\alpha,\beta}[ug(x)].
\end{aligned} \tag{22}$$

Taking both sides of Equation (22) and multiplying them by $\psi(x,\sigma)v(\sigma)$, which remains positive because for all $\sigma \in (1,x)$, $(x > 1), \alpha, \beta > 0$, then integrating the resulting identity with respect to σ from 1 to x, we obtain

$$\begin{aligned}
&\mathcal{H}_{1,x}^{\alpha,\beta}[ufg(x)] \frac{1}{\beta^{\alpha}\Gamma(\alpha)} \int_1^x e^{\left[\frac{\beta-1}{\beta}(\ln x - \ln \sigma)\right]} (\ln x - \ln \sigma)^{\alpha-1} v(\sigma) \frac{d\sigma}{\sigma} \\
&+ \mathcal{H}_{1,x}^{\alpha,\beta}[u(x)] \frac{1}{\beta^{\alpha}\Gamma(\alpha)} \int_1^x e^{\left[\frac{\beta-1}{\beta}(\ln x - \ln \sigma)\right]} (\ln x - \ln \sigma)^{\alpha-1} v(\sigma) f(\sigma) g(\sigma) \frac{d\sigma}{\sigma} \\
&\geq \mathcal{H}_{1,x}^{\alpha,\beta}[uf(x)] \frac{1}{\beta^{\alpha}\Gamma(\alpha)} \int_1^x e^{\left[\frac{\beta-1}{\beta}(\ln x - \ln \sigma)\right]} (\ln x - \ln \sigma)^{\alpha-1} v(\sigma) g(\sigma) \frac{d\sigma}{\sigma} \\
&+ \mathcal{H}_{1,x}^{\alpha,\beta}[ug(x)] \frac{1}{\beta^{\alpha}\Gamma(\alpha)} \int_1^x e^{\left[\frac{\beta-1}{\beta}(\ln x - \ln \sigma)\right]} (\ln x - \ln \sigma)^{\alpha-1} v(\sigma) f(\sigma) \frac{d\sigma}{\sigma}.
\end{aligned} \tag{23}$$

This completes the proof of Inequality (17). □

We present below the major result of the paper.

Theorem 1. *Let f and g be two integrable and synchronous functions on $[1,\infty)$, and $r, p, q : [1,\infty) \to [0,\infty)$ (so are positive). Then, for all $x > 1$, $\alpha > 0$ and $\beta \in (0,1]$, we have*

$$\begin{aligned}
&2\mathcal{H}_{1,x}^{\alpha,\beta}[r(x)]\left[\mathcal{H}_{1,x}^{\alpha,\beta}[p(x)]\mathcal{H}_{1,x}^{\alpha,\beta}[qfg(x)] + \mathcal{H}_{1,x}^{\alpha,\beta}[q(x)]\mathcal{H}_{1,x}^{\alpha,\beta}[pfg(x)]\right] + \\
&2\mathcal{H}_{1,x}^{\alpha,\beta}[p(x)]\mathcal{H}_{1,x}^{\alpha,\beta}[q(x)]\mathcal{H}_{1,x}^{\alpha,\beta}[rfg(x)] \geq \\
&\mathcal{H}_{1,x}^{\alpha,\beta}[r(x)]\left[\mathcal{H}_{1,x}^{\alpha,\beta}[pf(x)]\mathcal{H}_{1,x}^{\alpha,\beta}[qg(x)] + \mathcal{H}_{1,x}^{\alpha,\beta}[qf(x)]\mathcal{H}_{1,x}^{\alpha,\beta}[pg(x)]\right] + \\
&\mathcal{H}_{1,x}^{\alpha,\beta}[p(x)]\left[\mathcal{H}_{1,x}^{\alpha,\beta}[rf(x)]\mathcal{H}_{1,x}^{\alpha,\beta}[qg(x)] + \mathcal{H}_{1,x}^{\alpha,\beta}[qf(x)]\mathcal{H}_{1,x}^{\alpha,\beta}[rg(x)]\right] + \\
&\mathcal{H}_{1,x}^{\alpha,\beta}[q(x)]\left[\mathcal{H}_{1,x}^{\alpha,\beta}[rf(x)]\mathcal{H}_{1,x}^{\alpha,\beta}[pg(x)] + \mathcal{H}_{1,x}^{\alpha,\beta}[pf(x)]\mathcal{H}_{1,x}^{\alpha,\beta}[rg(x)]\right].
\end{aligned} \tag{24}$$

Proof. To prove this theorem, we put $u = p$ and $v = q$ into Lemma 2, and we obtain

$$\begin{aligned} &\mathcal{H}_{1,x}^{\alpha,\beta}[p(x)]\mathcal{H}_{1,x}^{\alpha,\beta}[qfg(x)] + \mathcal{H}_{1,x}^{\alpha,\beta}[q(x)]\mathcal{H}_{1,x}^{\alpha,\beta}[pfg(x)] \geq \\ &\mathcal{H}_{1,x}^{\alpha,\beta}[pf(x)]\mathcal{H}_{1,x}^{\alpha,\beta}[qg(x)] + \mathcal{H}_{1,x}^{\alpha,\beta}[qf(x)]\mathcal{H}_{1,x}^{\alpha,\beta}[pg(x)]. \end{aligned} \tag{25}$$

Now, multiplying both sides of Equation (25) by $\mathcal{H}_{1,x}^{\alpha,\beta}[r(x)]$, we have

$$\begin{aligned} &\mathcal{H}_{1,x}^{\alpha,\beta}[r(x)]\Big[\mathcal{H}_{1,x}^{\alpha,\beta}[p(x)]\mathcal{H}_{1,x}^{\alpha,\beta}[qfg(x)] + \mathcal{H}_{1,x}^{\alpha,\beta}[q(x)]\mathcal{H}_{1,x}^{\alpha,\beta}[pfg(x)]\Big] \geq \\ &\mathcal{H}_{1,x}^{\alpha,\beta}[r(x)]\Big[\mathcal{H}_{1,x}^{\alpha,\beta}[pf(x)]\mathcal{H}_{1,x}^{\alpha,\beta}[qg(x)] + \mathcal{H}_{1,x}^{\alpha,\beta}[qf(x)]\mathcal{H}_{1,x}^{\alpha,\beta}[pg(x)]\Big]. \end{aligned} \tag{26}$$

Again, putting $u = r$ and $v = q$, into Lemma 2, we obtain

$$\begin{aligned} &\mathcal{H}_{1,x}^{\alpha,\beta}[r(x)]\mathcal{H}_{1,x}^{\alpha,\beta}[qfg(x)] + \mathcal{H}_{1,x}^{\alpha,\beta}[q(x)]\mathcal{H}_{1,x}^{\alpha,\beta}[rfg(x)] \geq \\ &\mathcal{H}_{1,x}^{\alpha,\beta}[rf(x)]\mathcal{H}_{1,x}^{\alpha,\beta}[qg(x)] + \mathcal{H}_{1,x}^{\alpha,\beta}[qf(x)]\mathcal{H}_{1,x}^{\alpha,\beta}[rg(x)], \end{aligned} \tag{27}$$

Multiplying both sides of Equation (27) by $\mathcal{H}_{1,x}^{\alpha,\beta}[p(x)]$, we have

$$\begin{aligned} &\mathcal{H}_{1,x}^{\alpha,\beta}[p(x)]\Big[\mathcal{H}_{1,x}^{\alpha,\beta}[r(x)]\mathcal{H}_{1,x}^{\alpha,\beta}[qfg(x)] + \mathcal{H}_{1,x}^{\alpha,\beta}[q(x)]\mathcal{H}_{1,x}^{\alpha,\beta}[rfg(x)]\Big] \geq \\ &\mathcal{H}_{1,x}^{\alpha,\beta}[p(x)]\Big[\mathcal{H}_{1,x}^{\alpha,\beta}[rf(x)]\mathcal{H}_{1,x}^{\alpha,\beta}[qg(x)] + \mathcal{H}_{1,x}^{\alpha,\beta}[qf(x)]\mathcal{H}_{1,x}^{\alpha,\beta}[rg(x)]\Big]. \end{aligned} \tag{28}$$

With the same arguments as in Equations (26) and (28), we can write

$$\begin{aligned} &\mathcal{H}_{1,x}^{\alpha,\beta}[q(x)]\Big[\mathcal{H}_{1,x}^{\alpha,\beta}[r(x)]\mathcal{H}_{1,x}^{\alpha,\beta}[pfg(x)] + \mathcal{H}_{1,x}^{\alpha,\beta}[p(x)]\mathcal{H}_{1,x}^{\alpha,\beta}[rfg](x)\Big] \geq \\ &\mathcal{H}_{1,x}^{\alpha,\beta}[q(x)]\Big[\mathcal{H}_{1,x}^{\alpha,\beta}[rf(x)]\mathcal{H}_{1,x}^{\alpha,\beta}[pg(x)] + \mathcal{H}_{1,x}^{\alpha,\beta}[pf(x)]\mathcal{H}_{1,x}^{\alpha,\beta}[rg(x)]\Big]. \end{aligned} \tag{29}$$

Adding Inequalities (26), (28) and (29), we obtain Inequality (24). □

Lemma 3. *Let f and g be two integrable and synchronous functions on $[1,\infty)$, and $u, v : [1,\infty) \to [0,\infty)$. Then, for all $x > 1$, $\beta, \varphi \in (0,1]$, and $\alpha, \phi > 0$, we have*

$$\begin{aligned} &\mathcal{H}_{1,x}^{\alpha,\beta}[u(x)]\mathcal{H}_{1,x}^{\phi,\varphi}[vfg(x)] + \mathcal{H}_{1,x}^{\phi,\varphi}[v(x)]\mathcal{H}_{1,x}^{\alpha,\beta}[ufg(x)] \geq \\ &\mathcal{H}_{1,x}^{\alpha,\beta}[uf(x)]\mathcal{H}_{1,x}^{\phi,\varphi}[vg(x)] + \mathcal{H}_{1,x}^{\phi,\varphi}[vf(x)]\mathcal{H}_{1,x}^{\alpha,\beta}[ug(x)]. \end{aligned} \tag{30}$$

Proof. Multiplying both sides of Equation (22) by $\frac{1}{\varphi^{\phi}\Gamma(\phi)\sigma}e^{\left[\frac{\varphi-1}{\varphi}(\ln x - \ln\sigma)\right]}(\ln x - \ln\sigma)^{\phi-1}$, $\sigma \in (1,x), x > 1, \phi, \varphi > 0$, which remains positive (in view of the argument mentioned above in the proof of Lemma 2). Then, integrating the resulting identity with respect to σ from 1 to x, we have

$$\begin{aligned} &\mathcal{H}_{1,x}^{\alpha,\beta}[ufg(x)]\frac{1}{\varphi^{\phi}\Gamma(\phi)}\int_1^x e^{\left[\frac{\varphi-1}{\varphi}(\ln x-\ln\sigma)\right]}(\ln x - \ln\sigma)^{\phi-1}v(\sigma)\frac{d\sigma}{\sigma} \\ &+ \mathcal{H}_{1,x}^{\alpha,\beta}[u(x)]\frac{1}{\varphi^{\phi}\Gamma(\phi)}\int_1^x e^{\left[\frac{\varphi-1}{\varphi}(\ln x-\ln\sigma)\right]}(\ln x - \ln\sigma)^{\phi-1}v(\sigma)f(\sigma)g(\sigma)\frac{d\sigma}{\sigma} \\ &\geq \mathcal{H}_{1,x}^{\alpha,\beta}[uf(x)]\frac{1}{\varphi^{\phi}\Gamma(\phi)}\int_1^x e^{\left[\frac{\varphi-1}{\varphi}(\ln x-\ln\sigma)\right]}(\ln x - \ln\sigma)^{\phi-1}v(\sigma)g(\sigma)\frac{d\sigma}{\sigma} \\ &+ \mathcal{H}_{1,x}^{\alpha,\beta}[ug(x)]\frac{1}{\varphi^{\phi}\Gamma(\phi)}\int_1^x e^{\left[\frac{\varphi-1}{\varphi}(\ln x-\ln\sigma)\right]}(\ln x - \ln\sigma)^{\phi-1}v(\sigma)f(\sigma)\frac{d\sigma}{\sigma}. \end{aligned} \tag{31}$$

This completes the proof of Inequality (30). □

Theorem 2. *Let f and g be two integrable and synchronous functions on $[1,\infty)$, and $r,p,q : [1,\infty) \to [0,\infty)$. Then, for all $x > 1$, $\beta,\varphi \in (0,1]$, and $\alpha,\phi > 0$, we have*

$$\begin{aligned}
&\mathcal{H}_{1,x}^{\alpha,\beta}[r(x)]\Big[\mathcal{H}_{1,x}^{\alpha,\beta}[q(x)]\,\mathcal{H}_{1,x}^{\phi,\varphi}[pfg(x)] + 2\,\mathcal{H}_{1,x}^{\alpha,\beta}[p(x)]\,\mathcal{H}_{1,x}^{\phi,\varphi}[qfg(x)] \\
&+ \mathcal{H}_{1,x}^{\phi,\varphi}[q(x)]\,\mathcal{H}_{1,x}^{\alpha,\beta}[pfg(x)]\Big] \\
&+ \Big[\mathcal{H}_{1,x}^{\alpha,\beta}[p(x)]\,\mathcal{H}_{1,x}^{\phi,\varphi}[q(x)] + \mathcal{H}_{1,x}^{\phi,\varphi}[p(x)]\,\mathcal{H}_{1,x}^{\alpha,\beta}[q(x)]\Big]\,\mathcal{H}_{1,x}^{\alpha,\beta}[rfg(x)] \geq \\
&\mathcal{H}_{1,x}^{\alpha,\beta}[r(x)]\Big[\mathcal{H}_{1,x}^{\alpha,\beta}[pf(x)]\,\mathcal{H}_{1,x}^{\phi,\varphi}[qg(x)] + \mathcal{H}_{1,x}^{\phi,\varphi}[qf(x)]\,\mathcal{H}_{1,x}^{\alpha,\beta}[pg(x)]\Big] + \\
&\mathcal{H}_{1,x}^{\alpha,\beta}[p(x)]\Big[\mathcal{H}_{1,x}^{\alpha,\beta}[rf(x)]\,\mathcal{H}_{1,x}^{\phi,\varphi}[qg(x)] + \mathcal{H}_{1,x}^{\phi,\varphi}[qf(x)]\,\mathcal{H}_{1,x}^{\alpha,\beta}[rg(x)]\Big] + \\
&\mathcal{H}_{1,x}^{\alpha,\beta}[q(x)]\Big[\mathcal{H}_{1,x}^{\alpha,\beta}[rf(x)]\,\mathcal{H}_{1,x}^{\phi,\varphi}[pg(x)] + \mathcal{H}_{1,x}^{\phi,\varphi}[pf(x)]\,\mathcal{H}_{1,x}^{\alpha,\beta}[rg(x)]\Big].
\end{aligned} \tag{32}$$

Proof. To prove this theorem, we put $u = p$ and $v = q$ into Lemma 3, and we obtain

$$\begin{aligned}
&\mathcal{H}_{1,x}^{\alpha,\beta}[p(x)]\,\mathcal{H}_{1,x}^{\phi,\varphi}[qfg(x)] + \mathcal{H}_{1,x}^{\phi,\varphi}[q(x)]\,\mathcal{H}_{1,x}^{\alpha,\beta}[pfg(x)] \geq \\
&\mathcal{H}_{1,x}^{\alpha,\beta}[pf(x)]\,\mathcal{H}_{1,x}^{\phi,\varphi}[qg(x)] + \mathcal{H}_{1,x}^{\phi,\varphi}[qf(x)]\,\mathcal{H}_{1,x}^{\alpha,\beta}[pg(x)].
\end{aligned} \tag{33}$$

Now, multiplying both sides of Equation (33) by $\mathcal{H}_{1,x}^{\alpha,\beta}[r(x)]$, we have

$$\begin{aligned}
&\mathcal{H}_{1,x}^{\alpha,\beta}[r(x)]\Big[\mathcal{H}_{1,x}^{\alpha,\beta}[p(x)]\,\mathcal{H}_{1,x}^{\phi,\varphi}[qfg(x)] + \mathcal{H}_{1,x}^{\phi,\varphi}[q(x)]\,\mathcal{H}_{1,x}^{\alpha,\beta}[pfg(x)]\Big] \geq \\
&\mathcal{H}_{1,x}^{\alpha,\beta}[r(x)]\Big[\mathcal{H}_{1,x}^{\alpha,\beta}[pf(x)]\,\mathcal{H}_{1,x}^{\phi,\varphi}[qg(x)] + \mathcal{H}_{1,x}^{\phi,\varphi}[qf(x)]\,\mathcal{H}_{1,x}^{\alpha,\beta}[pg(x)]\Big],
\end{aligned} \tag{34}$$

Now, putting $u = r$ and $v = q$ into Lemma 3, we obtain

$$\begin{aligned}
&\mathcal{H}_{1,x}^{\alpha,\beta}[r(x)]\,\mathcal{H}_{1,x}^{\phi,\varphi}[qfg(x)] + \mathcal{H}_{1,x}^{\phi,\varphi}[q(x)]\,\mathcal{H}_{1,x}^{\alpha,\beta}[rfg(x)] \geq \\
&\mathcal{H}_{1,x}^{\alpha,\beta}[rf(x)]\,\mathcal{H}_{1,x}^{\phi,\varphi}[qg(x)] + \mathcal{H}_{1,x}^{\phi,\varphi}[qf(x)]\,\mathcal{H}_{1,x}^{\alpha,\beta}[rg(x)].
\end{aligned} \tag{35}$$

Multiplying both sides of Equation (35) by $\mathcal{H}_{1,x}^{\alpha,\beta}[p(x)]$, we have

$$\begin{aligned}
&\mathcal{H}_{1,x}^{\alpha,\beta}[p(x)]\Big[\mathcal{H}_{1,x}^{\alpha,\beta}[r(x)]\,\mathcal{H}_{1,x}^{\phi,\varphi}[qfg(x)] + \mathcal{H}_{1,x}^{\phi,\varphi}[q(x)]\,\mathcal{H}_{1,x}^{\alpha,\beta}[rfg(x)]\Big] \geq \\
&\mathcal{H}_{1,x}^{\alpha,\beta}[p(x)]\Big[\mathcal{H}_{1,x}^{\alpha,\beta}[rf(x)]\,\mathcal{H}_{1,x}^{\phi,\varphi}[qg(x)] + \mathcal{H}_{1,x}^{\phi,\varphi}[qf(x)]\,\mathcal{H}_{1,x}^{\alpha,\beta}[rg(x)]\Big].
\end{aligned} \tag{36}$$

Arguing as for Equations (34) and (36), we obtain

$$\begin{aligned}
&\mathcal{H}_{1,x}^{\alpha,\beta}[q(x)]\Big[\mathcal{H}_{1,x}^{\alpha,\beta}[r(x)]\,\mathcal{H}_{1,x}^{\phi,\varphi}[pfg(x)] + \mathcal{H}_{1,x}^{\phi,\varphi}[p(x)]\,\mathcal{H}_{1,x}^{\alpha,\beta}[rfg(x)]\Big] \geq \\
&\mathcal{H}_{1,x}^{\alpha,\beta}[q(x)]\Big[\mathcal{H}_{1,x}^{\alpha,\beta}[rf(x)]\,\mathcal{H}_{1,x}^{\phi,\varphi}[pg(x)] + \mathcal{H}_{1,x}^{\phi,\varphi}[pf(x)]\,\mathcal{H}_{1,x}^{\alpha,\beta}[rg(x)]\Big].
\end{aligned} \tag{37}$$

Adding Inequalities (34), (36) and (37), we obtain Inequality (32). □

Remark 5. *We assume f, g, r, p and q satisfy the following conditions:*

1. *the functions f and g are asynchronous on $[1,\infty)$;*
2. *the functions r, p, q are negative on $[1,\infty)$;*
3. *two of the functions r, p, q are positive and the third is negative on $[1,\infty)$.*

Then, Inequalities (24) and (32) are reversed.

4. Some Other Fractional Integral Inequalities

Now, we give some other fractional integral inequalities using generalized proportional Hadamard fractional integral operators.

Theorem 3. *Suppose that f, g, and h are positive and continuous functions on $[1,\infty)$, such that*

$$(g(\tau)-g(\sigma))\left(\frac{f(\sigma)}{h(\sigma)}-\frac{f(\tau)}{h(\tau)}\right)\geq 0, \quad \tau,\sigma\in(1,x)\ \ x>1. \tag{38}$$

Then, for all $x>1$, $\alpha>0$ and $\beta\in(0,1]$, we have

$$\frac{\mathcal{H}_{1,x}^{\alpha,\beta}[f(x)]}{\mathcal{H}_{1,x}^{\alpha,\beta}[h(x)]}\geq\frac{\mathcal{H}_{1,x}^{\alpha,\beta}[gf(x)]}{\mathcal{H}_{1,x}^{\alpha,\beta}[gh(x)]}. \tag{39}$$

Proof. Since f,g, and h are three positive and continuous functions on $[1,\infty)$, by Equation (38) we obtain

$$g(\tau)\frac{f(\sigma)}{h(\sigma)}+g(\sigma)\frac{f(\tau)}{h(\tau)}-g(\sigma)\frac{f(\sigma)}{h(\sigma)}-g(\tau)\frac{f(\tau)}{h(\tau)}\geq 0, \quad \tau,\sigma\in(0,x)x>0. \tag{40}$$

Multiplying both sides of Equation (40) by $h(\sigma)h(\tau)$, we have

$$g(\tau)f(\sigma)h(\tau)-g(\tau)f(\tau)h(\sigma)-g(\sigma)f(\sigma)h(\tau)+g(\sigma)f(\tau)h(\sigma)\geq 0. \tag{41}$$

Now, multiplying Equation (41) by $\psi(x,\tau)$ defined by Equation (20), then integrating the resulting identity with respect to τ from 1 to x, we obtain

$$\begin{aligned}
&\frac{f(\sigma)}{\beta^{\alpha}\Gamma(\alpha)}\int_1^x e^{\left[\frac{\beta-1}{\beta}(\ln x-\ln\tau)\right]}(\ln x-\ln\tau)^{\alpha-1}g(\tau)h(\tau)\frac{d\tau}{\tau}\\
&-\frac{h(\sigma)}{\beta^{\alpha}\Gamma(\alpha)}\int_1^x e^{\left[\frac{\beta-1}{\beta}(\ln x-\ln\tau)\right]}(\ln x-\ln\tau)^{\alpha-1}g(\tau)f(\tau)\frac{d\tau}{\tau}\\
&-\frac{f(\sigma)g(\sigma)}{\beta^{\alpha}\Gamma(\alpha)}\int_1^x e^{\left[\frac{\beta-1}{\beta}(\ln x-\ln\tau)\right]}(\ln x-\ln\tau)^{\alpha-1}h(\tau)\frac{d\tau}{\tau}\\
&+\frac{h(\sigma)g(\sigma)}{\beta^{\alpha}\Gamma(\alpha)}\int_1^x e^{\left[\frac{\beta-1}{\beta}(\ln x-\ln\tau)\right]}(\ln x-\ln\tau)^{\alpha-1}f(\tau)\frac{d\tau}{\tau}\geq 0.
\end{aligned} \tag{42}$$

It follows from Equation (42) that

$$\begin{aligned}
&f(\sigma)\mathcal{H}_{1,x}^{\alpha,\beta}[gh(x)]+g(\sigma)h(\sigma)\mathcal{H}_{1,x}^{\alpha,\beta}[f(x)]\\
&-g(\sigma)f(\sigma)\mathcal{H}_{1,x}^{\alpha,\beta}[h(x)]-h(\sigma)\mathcal{H}_{1,x}^{\alpha,\beta}[gf(x)]\geq 0.
\end{aligned} \tag{43}$$

Again, let us multiply Equation (43) by $\psi(x,\sigma)$ as defined by Equation (20), which remains positive because for all $\sigma\in(1,x)$, $(x>1),\alpha,\beta>0$. Then, integrating the resulting identity with respect to σ from 1 to x, we obtain

$$\begin{aligned}
&\mathcal{H}_{1,x}^{\alpha,\beta}[f(x)]\mathcal{H}_{1,x}^{\alpha,\beta}[gh(x)]-\mathcal{H}_{1,x}^{\alpha,\beta}[h(x)]\mathcal{H}_{1,x}^{\alpha,\beta}[gf(x)]\\
&-\mathcal{H}_{1,x}^{\alpha,\beta}[gf(x)]\mathcal{H}_{1,x}^{\alpha,\beta}[h(x)]+\mathcal{H}_{1,x}^{\alpha,\beta}[gh(x)]\mathcal{H}_{1,x}^{\alpha,\beta}[f(x)]\geq 0,
\end{aligned} \tag{44}$$

which implies that

$$\mathcal{H}_{1,x}^{\alpha,\beta}[f(x)]\mathcal{H}_{1,x}^{\alpha,\beta}[gh(x)]\geq\mathcal{H}_{1,x}^{\alpha,\beta}[h(x)]\mathcal{H}_{1,x}^{\alpha,\beta}[gf(x)]. \tag{45}$$

This completes the proof of the theorem. □

Theorem 4. *Suppose that f, g, and h are positive and continuous functions on $[1,\infty)$, such that*

$$(g(\tau)-g(\sigma))\left(\frac{f(\sigma)}{h(\sigma)}-\frac{f(\tau)}{h(\tau)}\right)\geq 0,\quad \tau,\sigma\in(1,x)\ \ x>1, \tag{46}$$

Then, for all $x>1$, $\beta,\varphi\in(0,1]$, and $\alpha,\phi>0$, we have

$$\frac{\mathcal{H}_{1,x}^{\alpha,\beta}[f(x)]\mathcal{H}_{1,x}^{\phi,\varphi}[gf(x)]+\mathcal{H}_{1,x}^{\phi,\varphi}[f(x)]\mathcal{H}_{1,x}^{\alpha,\beta}[gh(x)]}{\mathcal{H}_{1,x}^{\alpha,\beta}[h(x)]\mathcal{H}_{1,x}^{\phi,\varphi}[gf(x)]+\mathcal{H}_{1,x}^{\phi,\varphi}[h(x)]\mathcal{H}_{1,x}^{\alpha,\beta}[gf(x)]}\geq 1. \tag{47}$$

Proof. Multiplying Equation (43) by $\frac{1}{\varphi^{\phi}\Gamma(\phi)\sigma}e^{\left[\frac{\varphi-1}{\varphi}(\ln x-\ln\sigma)\right]}(\ln x-\ln\sigma)^{\phi-1},\sigma\in(1,x)$, $x>1,\phi,\varphi>0$, which is always positive, then integrating the resulting identity with respect to σ from 1 to x, we have

$$\begin{aligned}&\mathcal{H}_{1,x}^{\alpha,\beta}[hg(x)]\frac{1}{\varphi^{\phi}\Gamma(\phi)}\int_1^x e^{\left[\frac{\varphi-1}{\varphi}(\ln x-\ln\sigma)\right]}(\ln x-\ln\sigma)^{\phi-1}f(\sigma)\frac{d\sigma}{\sigma}\\&-\mathcal{H}_{1,x}^{\alpha,\beta}[gf(x)]\frac{1}{\varphi^{\phi}\Gamma(\phi)}\int_1^x e^{\left[\frac{\varphi-1}{\varphi}(\ln x-\ln\sigma)\right]}(\ln x-\ln\sigma)^{\phi-1}h(\sigma)\frac{d\sigma}{\sigma}\\&-\mathcal{H}_{1,x}^{\alpha,\beta}[h(x)]\frac{1}{\varphi^{\phi}\Gamma(\phi)}\int_1^x e^{\left[\frac{\varphi-1}{\varphi}(\ln x-\ln\sigma)\right]}(\ln x-\ln\sigma)^{\phi-1}gf(\sigma)\frac{d\sigma}{\sigma}\\&+\mathcal{H}_{1,x}^{\alpha,\beta}[f(x)]\frac{1}{\varphi^{\phi}\Gamma(\phi)}\int_1^x e^{\left[\frac{\varphi-1}{\varphi}(\ln x-\ln\sigma)\right]}(\ln x-\ln\sigma)^{\phi-1}fh(\sigma)\frac{d\sigma}{\sigma}\geq 0.\end{aligned} \tag{48}$$

This gives us the following relation:

$$\begin{aligned}&\mathcal{H}_{1,x}^{\phi,\varphi}[f(x)]\mathcal{H}_{1,x}^{\alpha,\beta}[gh(x)]-\mathcal{H}_{1,x}^{\phi,\varphi}[h(x)]\mathcal{H}_{1,x}^{\alpha,\beta}[gf(x)]\\&-\mathcal{H}_{1,x}^{\phi,\varphi}[gf(x)]\mathcal{H}_{1,x}^{\alpha,\beta}[h(x)]+\mathcal{H}_{1,x}^{\phi,\varphi}[gh(x)]\mathcal{H}_{1,x}^{\alpha,\beta}[f(x)]\geq 0.\end{aligned} \tag{49}$$

From Equation (49), we obtain

$$\begin{aligned}&\mathcal{H}_{1,x}^{\phi,\varphi}[f(x)]\mathcal{H}_{1,x}^{\alpha,\beta}[gh(x)]+\mathcal{H}_{1,x}^{\phi,\varphi}[gh(x)]\mathcal{H}_{1,x}^{\alpha,\beta}[f(x)]\\&\geq\mathcal{H}_{1,x}^{\phi,\varphi}[h(x)]\mathcal{H}_{1,x}^{\alpha,\beta}[gf(x)]+\mathcal{H}_{1,x}^{\phi,\varphi}[gf(x)]\mathcal{H}_{1,x}^{\alpha,\beta}[h(x)].\end{aligned} \tag{50}$$

This yields Inequality (47). This completes the proof of the theorem. □

Remark 6. *If we take $\alpha=\phi$ and $\beta=\varphi$ in Theorem 4, then we obtain Theorem 3.*

Theorem 5. *Suppose that f and h are two positive continuous functions such that $f\leq h$ on $[1,\infty)$. If $\frac{f}{h}$ is decreasing and f is increasing on $[1,\infty)$, then, for any $p\geq 1$, $x>1$, $\alpha>0$ and $\beta\in(0,1]$, we have*

$$\frac{\mathcal{H}_{1,x}^{\alpha,\beta}[f(x)]}{\mathcal{H}_{1,x}^{\alpha,\beta}[h(x)]}\geq\frac{\mathcal{H}_{1,x}^{\alpha,\beta}[f^p(x)]}{\mathcal{H}_{1,x}^{\alpha,\beta}[h^p(x)]}. \tag{51}$$

Proof. Now, by taking $g=f^{p-1}$ in Theorem 3, we obtain

$$\frac{\mathcal{H}_{1,x}^{\alpha,\beta}[f(x)]}{\mathcal{H}_{1,x}^{\alpha,\beta}[h(x)]}\geq\frac{\mathcal{H}_{1,x}^{\alpha,\beta}[ff^{p-1}(x)]}{\mathcal{H}_{1,x}^{\alpha,\beta}[hf^{p-1}(x)]}. \tag{52}$$

Since $f\leq h$ on $[1,\infty)$, we have

$$hf^{p-1}\leq h^p. \tag{53}$$

Multiplying Equation (53) by $\psi(x,\tau)$ defined by Equation (20), then integrating the resulting identity with respect to τ from 1 to x, we obtain

$$\frac{1}{\beta^{\alpha}\Gamma(\alpha)}\int_1^x e^{\left[\frac{\beta-1}{\beta}(\ln x-\ln\tau)\right]}(\ln x-\ln\tau)^{\alpha-1}f^{p-1}h(\tau)\frac{d\tau}{\tau} \tag{54}$$

$$\leq \frac{1}{\beta^{\alpha}\Gamma(\alpha)}\int_1^x e^{\left[\frac{\beta-1}{\beta}(\ln x-\ln\tau)\right]}(\ln x-\ln\tau)^{\alpha-1}h^{p}(\tau)\frac{d\tau}{\tau}, \tag{55}$$

which implies that

$$\mathcal{H}_{1,x}^{\alpha,\beta}[hf^{p-1}(x)] \leq \mathcal{H}_{1,x}^{\alpha,\beta}[h^p(x)]. \tag{56}$$

Thus, we have

$$\frac{\mathcal{H}_{1,x}^{\alpha,\beta}[ff^{p-1}(x)]}{\mathcal{H}_{1,x}^{\alpha,\beta}[hf^{p-1}(x)]} \geq \frac{\mathcal{H}_{1,x}^{\alpha,\beta}[f^p(x)]}{\mathcal{H}_{1,x}^{\alpha,\beta}[h^p(x)]}. \tag{57}$$

From Equations (52) and (57), we obtain Equation (51). □

Theorem 6. *Suppose that f and h are two positive continuous functions such that $f \leq h$ on $[1,\infty)$. If $\frac{f}{h}$ is decreasing and f is increasing on $[1,\infty)$, then, for any $p \geq 1$, $x > 1$, $\beta,\varphi \in (0,1]$, $\alpha,\phi > 0$, we have*

$$\frac{\mathcal{H}_{1,x}^{\alpha,\beta}[f(x)]\mathcal{H}_{1,x}^{\phi,\varphi}[h^p(x)] + \mathcal{H}_{1,x}^{\phi,\varphi}[f(x)]\mathcal{H}_{1,x}^{\alpha,\beta}[h^p(x)]}{\mathcal{H}_{1,x}^{\alpha,\beta}[h(x)]\mathcal{H}_{1,x}^{\phi,\varphi}[f^p(x)] + \mathcal{H}_{1,x}^{\phi,\varphi}[h(x)]\mathcal{H}_{1,x}^{\alpha,\beta}[f^p(x)]} \geq 1. \tag{58}$$

Proof. Taking $g = f^{p-1}$ in Theorem 4, we obtain

$$\frac{\mathcal{H}_{1,x}^{\alpha,\beta}[f(x)]\mathcal{H}_{1,x}^{\phi,\varphi}[hf^{p-1}(x)] + \mathcal{H}_{1,x}^{\phi,\varphi}[f(x)]\mathcal{H}_{1,x}^{\alpha,\beta}[hf^{p-1}(x)]}{\mathcal{H}_{1,x}^{\alpha,\beta}[h(x)]\mathcal{H}_{1,x}^{\phi,\varphi}[f^p(x)] + \mathcal{H}_{1,x}^{\phi,\varphi}[h(x)]\mathcal{H}_{1,x}^{\alpha,\beta}[f^p(x)]} \geq 1, \tag{59}$$

then, by hypothesis, $f \leq h$ on $[1,\infty)$, which implies that

$$hf^{p-1} \leq h^p. \tag{60}$$

Now, multiplying both sides of Equation (60) by $\frac{1}{\varphi^{\phi}\Gamma(\phi)\sigma}e^{\left[\frac{\varphi-1}{\varphi}(\ln x-\ln\sigma)\right]}(\ln x-\ln\sigma)^{\phi-1}$, $\sigma \in (1,x), x > 1, \phi,\varphi > 0$, which remains positive. Then, integrating the resulting identity with respect to σ from 1 to x, we have

$$\begin{aligned}&\frac{1}{\varphi^{\phi}\Gamma(\phi)}\int_1^x e^{\left[\frac{\varphi-1}{\varphi}(\ln x-\ln\sigma)\right]}(\ln x-\ln\sigma)^{\phi-1}hf^{p-1}(\sigma)\frac{d\sigma}{\sigma}\\ &\leq \frac{1}{\varphi^{\phi}\Gamma(\phi)}\int_1^x e^{\left[\frac{\varphi-1}{\varphi}(\ln x-\ln\sigma)\right]}(\ln x-\ln\sigma)^{\phi-1}h^p(\sigma)\frac{d\sigma}{\sigma}.\end{aligned} \tag{61}$$

Integrating both sides of Equation (61) with respect to σ over 1 to x, we have

$$\mathcal{H}_{1,x}^{\phi,\varphi}[hf^{p-1}(x)] \leq \mathcal{H}_{1,x}^{\phi,\varphi}[h^p(x)]. \tag{62}$$

Multiplying both sides of Equation (62) by $\mathcal{H}_{1,x}^{\alpha,\beta}[f(x)]$, we obtain

$$\mathcal{H}_{1,x}^{\alpha,\beta}[f(x)]\mathcal{H}_{1,x}^{\phi,\varphi}[hf^{p-1}(x)] \leq \mathcal{H}_{1,x}^{\alpha,\beta}[f(x)]\mathcal{H}_{1,x}^{\phi,\varphi}[h^p(x)]. \tag{63}$$

Hence, from Equations (56) and (63), we obtain

$$\begin{aligned}&\mathcal{H}_{1,x}^{\alpha,\beta}[f(x)]\mathcal{H}_{1,x}^{\phi,\varphi}[hf^{p-1}(x)] + \mathcal{H}_{1,x}^{\phi,\varphi}[f(x)]\mathcal{H}_{1,x}^{\alpha,\beta}[hf^{p-1}(x)]\\ &\leq \mathcal{H}_{1,x}^{\alpha,\beta}[f(x)]\mathcal{H}_{1,x}^{\phi,\varphi}[h^p(x)] + \mathcal{H}_{1,x}^{\phi,\varphi}[f(x)]\mathcal{H}_{1,x}^{\alpha,\beta}[h^p(x)].\end{aligned} \tag{64}$$

From Equations (59) and (64), we obtain Equation (58). This ends the proof of the theorem. □

5. Concluding Remarks

In [42], the authors proposed the concept of generalized proportional fractional integral operators with exponential kernels. Following this, Rahman et al. [44] worked on these operators and established some fractional inequalities for convex functions by considering Hadamard proportional fractional integrals. In this study, we obtained some fractional integral inequalities for the extended Chebyshev function by considering the generalized proportional Hadamard fractional integral operator. The inequalities investigated in this paper represent novel contributions in the fields of fractional calculus and generalized proportional Hadamard fractional integral operators. They are also expected to lead to some applications for determining the uniqueness of fractional differential equation solutions. We also believe that the findings of this study will help to solve additional practical problems in mathematical physics, statistics, and approximation theory.

Author Contributions: Conceptualization, V.L.C., A.B.N., S.K.P., C.C. and A.D.K.; methodology, V.L.C., A.B.N., S.K.P., C.C. and A.D.K.; validation, V.L.C., A.B.N., S.K.P., C.C. and A.D.K.; formal analysis, V.L.C., A.B.N., S.K.P., C.C. and A.D.K.; investigation, V.L.C., A.B.N., S.K.P., C.C. and A.D.K.; writing—original draft preparation, V.L.C., A.B.N., S.K.P., C.C. and A.D.K.; writing—review and editing, V.L.C., A.B.N., S.K.P., C.C. and A.D.K. All authors have read and agreed to the published version of the manuscript.

Funding: This research received no external funding

Acknowledgments: We warmly thank the three reviewers and the associate editor for the thorough and constructive comments on the paper.

Conflicts of Interest: The authors declare no conflict of interest.

References

1. Anastassiou, G.A. *Fractional Differentiation Inequalities*; Springer: New York, NY, USA, 2009.
2. Dahmani, Z. On Minkowski and Hermite-Hadamard integral inequalities via fractional integration. *Ann. Funct. Anal.* **2010**, *1*, 51–58. [CrossRef]
3. Dahmani, Z. New inequalities in fractional integrals. *Int. J. Nonlinear Sci.* **2010**, *9*, 493–497.
4. Dahmani, Z. The Riemann-Liouville operator to generate some new inequalities. *Int. J. Nonlinear Sci.* **2011**, *12*, 452–455.
5. Dahmani, Z. Some results associated with fractional integrals involving the extended Chebyshev. *Acta Univ. Apulensis Math. Inform.* **2011**, *27*, 217–224.
6. Dahmani, Z.; Tabharit, L.; Taf, S. New generalisations of Gruss inequality using Riemann-Liouville fractional integrals. *Bull. Math. Anal. Appl.* **2010**, *2*, 93–99.
7. Denton, Z.; Vatsala, A.S. Fractional integral inequalities and application. *Comput. Math. Appl.* **2010**, *59*, 1087–1094. [CrossRef]
8. Chinchane, V.L.; Pachpatte, D.B. A note on some integral inequalities via Hadamard integral. *J. Fract. Calc. Appl.* **2013**, *4*, 1–5.
9. Chinchane, V.L. Certain inequalities using Saigo fractional integral operator. *Facta Univ. Ser. Math. Inform.* **2014**, *29*, 343–350.
10. Kiryakova, V. On two Saigo's fractional integral operator in the class of univalent functions. *Fract. Calc. Appl. Anal.* **2006**, *9*, 159–176.
11. Purohit, S.D.; Raina, R.K. Chebyshev type inequalities for the Saigo fractional integral and their q-analogues. *J. Math. Inequal.* **2013**, *7*, 239–249. [CrossRef]
12. Saigo, M. A remark on integral operators involving the Gauss hypergeometric functions. *Math. Rep. Kyushu Univ.* **1978**, *11*, 135–143.
13. Che, Y.; Abo Keir, M. Study on the training model of football movement trajectory drop point based on fractional differential equation. *Appl. Math. Nonlinear Sci.* **2021** . [CrossRef]
14. Nisar, K.S.; Rahman, G.; Khan, A. Some new inequalities for generalized fractional conformable integral operators. *Adv. Differ. Equ.* **2019**, *2019*, 427. [CrossRef]
15. Nisar, K.S.; Logeswari, K.; Vijayaraj, V.; Baskonus, H.M.; Ravichandran, C. Fractional Order Modeling the Gemini Virus in Capsicum annuum with Optimal Control. *Fractal Fract.* **2022**, *6*, 61.. [CrossRef]
16. Katugampola, U.N. A new approach to generalized fractional derivatives. *Bull. Math. Anal. Appl.* **2014**, *6*, 1–15.
17. Nale, A.B.; Panchal, S.K.; Chinchane, V.L. Certain fractional integral inequalities using generalized Katugampola fractional integral operator. *Malaya J. Math.* **2020**, *3*, 791–797.
18. Set, E.; Choi, J.; Mumcu, L. Chebyshev type inequalities involving generalized Katugampola fractional integral operators. *Tamkang J. Math.* **2019**, *4*, 381–390. [CrossRef]

19. Chinchane, V.L. New approach to Minkowski fractional inequalities using generalized K-fractional integral operator. *J. Indian. Math. Soc.* **2018**, *85*, 32–41. [CrossRef]
20. Chinchane, V.L. On Chebyshev type inequalities using generalized K-fractional integral operator. *Progr. Fract. Differ. Appl.* **2017**, *3*, 1–8. [CrossRef]
21. Prabhakaran, A.R.; Rao, K.S. Saigo operator of fractional integration of Hypergeometric functions. *Int. J. Pure Appl. Math.* **2012**, *81*, 755–763.
22. Virchenko, N.; Lisetska, O. On some fractional integral operators involving generalized Gauss hypergeometric functions. *Appl. Appl. Math.* **2010**, *5*, 1418–1427.
23. Aghili, A. Complete solution for the time fractional diffusion problem with mixed boundary conditions by operational method. *Appl. Math. Nonlinear Sci.* **2021**, *6*, 9–20. [CrossRef]
24. Akdemir, A.; Deniz, E.; Yüksel, E. On Some Integral Inequalities via Conformable Fractional Integrals. *Appl. Math. Nonlinear Sci.* **2021**, *6*, 489–498.. [CrossRef]
25. Gao, W.; Veeresha, P.;Cattani, C.; Baishya, C.; Baskonus, H.M. Modified Predictor—Corrector method for the numerical solution of a fractional-order SIR model with 2019 nCoV. *Fractal Fract.* **2022**, *6*, 92. [CrossRef]
26. Gao, J.; Alotaibi, F.; Ismail, R. The model of sugar metabolism and exercise energy expenditure based on fractional linear regression equation. *Appl. Math. Nonlinear Sci.* **2021** , 1–9. [CrossRef]
27. Rezazadeh, H.; Odabasi, M.; Tariq, K.U.; Abazari, R.; Baskonus, H.M. On the conformable nonlinear Schrödinger equation with second order spatiotemporal and group velocity dispersion coefficients *Chin. J. Phys.* **2021**, *72*, 403–414. [CrossRef]
28. Veeresha, P.; Malagi, N.S.; Prakasha, D.G.; Baskonus, H.M. An efficient technique to analyze the fractional model of vector-borne diseases. *Phys. Scr.* **2022**, *97*, 054004. [CrossRef]
29. Veeresha, P.; Ilhan, E.; Prakasha, D.G.; Baskonus, H.M.; Wei, G. A new numerical investigation of fractional order susceptible-infected-recovered epidemic model of childhood disease. *Alex. Eng. J.* **2022**, *61*, 1747–1756. [CrossRef]
30. Wen, L.; Liu, H.; Chen, J.; Fakieh, B.; Shorman, S. Fractional linear regression equation in agricultural disaster assessment model based on geographic information system analysis technology. *Appl. Math. Nonlinear Sci.* **2021** , 1–10.. [CrossRef]
31. Wei, G.; Baskonus, H.M. Deeper investigation of modified epidemiological computer virus model containing the Caputo operator. *Chaos Solitons Fractals* **2022**, *158*, 112050. [CrossRef]
32. Zhirong, G.; Alghazzawi, D. Optimal solution of fractional differential equations in solving the relief of college students' mental obstacles. *Appl. Math. Nonlinear Sci.* **2021** , 1–8. [CrossRef]
33. Xiao, Y.; Liu, J.; Alkhathlan, A. Informatisation of educational reform based on fractional differential equations. *Appl. Math. Nonlinear Sci.* **2021** , 1–11.. [CrossRef]
34. Anber, A.; Dahmani, Z.; Bendoukha, B. New integral inequalities of Feng Qi type via Riemann-Liouville fractional integration. *Facta Univ. (Nis) Ser. Math. Inform.* **2012**, *27*, 13–22.
35. Panchal, S.K.; Chinchane, V.L.; Nale, A.B. On weighted fractional inequalities using generalized Katugampola fractional integral operator. *Fract. Differ. Calc.* **2020**, *2*, 255–266. [CrossRef]
36. Andrić, M.; Farid, G.; Pečarić, J.; Siddique, U. Generalized Minkowski-type fractional inequalities involving extended Mittag-Leffler function. *J. Indian Math. Soc.* **2020**, *3–4*, 137–147. [CrossRef]
37. Rahman, G.; Abdejawad, T.; Khan, A.; Nisar, K.S. Some fractional proportional integral inequalities. *J. Inequal. Appl.* **2019**, *2019*, 244. [CrossRef]
38. Rahman, G.; Khan, A.; Abdejawad, T.; Nisar, K.S. The Minkowski inequalities via generalized proportional fractional integral operators. *Adv. Differ. Equ.* **2019**, *2019*, 287. [CrossRef]
39. Rahman, G.; Nisar, K.S.; Abdejawad, T.; Ullah, S. Certain fractional proportional integral inequalities via convex functions. *Mathematics* **2020**, *8*, 222. [CrossRef]
40. Atangana, A.; Baleanu, D. New fractional derivatives with nonlocal and non-singular kernel: Theory and application to heat transfer model. *Therm. Sci.* **2016**, *20*, 763–769. [CrossRef]
41. Jarad, F.; Ugurlu, E.; Abdeljawad, T.; Baleanu, D. On a new class of fractional operators. *Adv. Differ. Equ.* **2017**, *247*, 1–16. [CrossRef]
42. Jarad, F.; Abdejawad, T.; Alzabut, J. Generalized fractional derivatives generated by a class of local proportional derivatives. *Eur. Phys. J. Spec. Top.* **2017**, *226*, 3457–3471. [CrossRef]
43. Rahman, G.; Nisar, K.S.; Abdejawad, T. Certain Hadamard proportional fractional integral inequalities. *Mathematics* **2020**, *8*, 504. [CrossRef]
44. Rahman, G.; Abdejawad, T.; Jarad, F.; Khan, A.; Nisar, K.S. Certain inequalities via generalized proportional Hadamard fractional integral operators. *Adv. Differ. Equ.* **2019**, *2019*, 454. [CrossRef]
45. Nisar, K.S.; Rahman, G.; Mehrez, K. Chebyshev type inequalities via generalized fractional conformable integrals. *J. Inequal. Appl.* **2019**, *245*, 1–9. [CrossRef]
46. Nisar, K.S.; Tassadiq, A.; Rahman, G.; Khan, A. Some inequalities via fractional conformable integral operators. *J. Inequal. Appl.* **2019**, *2019*, 271. [CrossRef]
47. Tassaddiq, A.; Rahman, G.; Nisar, K.S.; Samraiz, M. Certain fractional conformable inequalities for the weighted and the extended Chebyshev functionals. *Adv. Differ. Equ.* **2020**, *96*, 1–9. [CrossRef]
48. Caputo, M.; Fabrizio, M. A new Definition of Fractional Derivative without Singular Kernel. *Progr. Fract. Differ. Appl.* **2015**, *1*, 73–85.

49. Losada, J.; Nieto, J.J. Properties of a New Fractional Derivative without Singular Kernel. *Progr. Fract. Differ. Appl.* **2015**, *1*, 87–92.
50. Nale, A.B.; Panchal, S.K.; Chinchane, V.L. Some Minkowski-type inequalities using generalized proportional Hadamard fractional integral operators. *Filomat* **2021**, *35*, 2973–2984. [CrossRef]
51. Kukushkin, M.V. Abstract fractional Calculus for m-accretive operators. *Int. J. Appl. Math.* **2021**, *34*, 1–41. [CrossRef]
52. Yosida, K. *Functional Analysis*; Springer: Berlin/Heidelberg, Germany; New York, NY, USA, 1980.
53. Chebyshev, P.L. Sur les expressions approximatives des intégrales définies par les autes entre les mêmes limites. *Proc. Math. Soc. Charkov* **1882**, *2*, 93–98.
54. Anastassiou, G.A.; Hooshmandasl, M.R.; Ghasemi A.; Moftakharzadeh, F. Montgomery identities for fractional integrals and related fractional inequalities. *J. Inequal. Pure Appl. Math.* **2009**, *10*, 97.
55. Belarbi, S.; Dahmani, Z. On some new fractional integral inequality. *J. Inequal. Pure Appl. Math.* **2009**, *10*, 86.
56. Dragomir, S.S. Some integral inequalities of Grüss type. *Indian J. Pure Appl. Math.* **2002**, *31*, 397–415.
57. Samko, S.G.; Kilbas, A.A.; Marichev, O.I. *Fractional Integral and Derivative Theory and Application*; Gordon and Breach: Yverdon, Switzerland, 1993.

axioms

MDPI

Article

Deformed Mathematical Objects Stemming from the q-Logarithm Function

Ernesto P. Borges [1,*] **and Bruno G. da Costa** [2]

1 Instituto de Física, Universidade Federal da Bahia, Salvador 40170-115, BA, Brazil
2 Instituto Federal de Educação, Ciência e Tecnologia do Sertão Pernambucano, Petrolina 56314-520, PE, Brazil; bruno.costa@ifsertao-pe.edu.br
* Correspondence: ernesto@ufba.br

Abstract: Generalized numbers, arithmetic operators, and derivative operators, grouped in four classes based on symmetry features, are introduced. Their building element is the pair of q-logarithm/q-exponential inverse functions. Some of the objects were previously described in the literature, while others are newly defined. Commutativity, associativity, and distributivity, and also a pair of linear/nonlinear derivatives, are observed within each class. Two entropic functionals emerge from the formalism, and one of them is the nonadditive Tsallis entropy.

Keywords: deformed numbers; deformed algebras; deformed calculus; nonadditive entropy

Citation: Borges, E.P.; da Costa, B.G. Deformed Mathematical Objects Stemming from the q-Logarithm Function. *Axioms* **2022**, *11*, 138. https://doi.org/10.3390/axioms11030138

Academic Editor: Hans J. Haubold

Received: 26 January 2022
Accepted: 6 March 2022
Published: 16 March 2022

Publisher's Note: MDPI stays neutral with regard to jurisdictional claims in published maps and institutional affiliations.

Copyright: © 2022 by the authors. Licensee MDPI, Basel, Switzerland. This article is an open access article distributed under the terms and conditions of the Creative Commons Attribution (CC BY) license (https://creativecommons.org/licenses/by/4.0/).

1. Introduction

Extensivity of an entropy is expressed as S being proportional to the number N of elements of the system. The hypervolume Ω of the phase space of a system composed by independent subsystems increases with the product of the hypervolumes μ_i of the corresponding subspaces of its elements ($\mu_i > 1$). For identical and independent subsystems, the phase space exponentially increases with the number of elements, $\Omega = \mu_1^N$, and thus the Boltzmann entropy is proportional to N: $S = k \ln \Omega = Nk \ln \mu_1$, i.e., it is extensive. Correlations between subsystems make the hypervolume of the phase space smaller than that of the product of the hypervolumes of its subsystems, and particular kinds of strong correlations make the phase space asymptotically increase as a power law, at a much slower rate than the exponential law; in these cases the Boltzmann entropy is no longer extensive. For such special cases, —and there are plenty of observational, experimental, and numerical examples— the nonadditive entropy S_q [1] becomes proportional to N, recovering extensivity, which is a central property for connecting with thermodynamics (see details and further implications of extensivity in Ref. [2]). The mathematical property that plays this role is a generalized multiplication operator defined in Ref. [3]. The present paper identifies four classes of generalized algebras associated with the nonextensive formalism in a broader point of view. One of them contains the above-mentioned generalized multiplication. These developments hopefully help to understand the underlying mathematical structures that support the nonextensive statistical mechanics.

The Tsallis nonadditive entropy S_q has induced investigations on deformed mathematical structures aiming to represent relations of the nonextensive framework through expressions formally similar to the standard Boltzmann-Gibbs (BG) statistical mechanics. The definition of the generalized logarithm function (the q-logarithm) [4]

$$\ln_q x \equiv \frac{x^{1-q} - 1}{1 - q}, \quad (x > 0), \tag{1}$$

allowed to rewrite $S_q \equiv k\,(q-1)^{-1}\left(1-\sum_i^W p_i^q\right)$ (in its discrete version) as

$$\begin{aligned} S_q &= -k\sum_i^W p_i^q \ln_q p_i, \\ &= k\sum_i^W p_i \ln_q (1/p_i) \end{aligned} \tag{2}$$

(sum over W microstates, each one labeled i, with their corresponding probabilities p_i, k is a positive constant, $q \in \mathbb{R}$ is the generalizing entropic index). Ordinary formalism is recovered as $q \to 1$ ($\ln_1 x = \ln x$; $S_1 = S_{\rm BG} = k\sum_i^W p_i \ln 1/p_i$), equiprobability yields $S_q[p_i = 1/W] = k \ln_q W$. The q-logarithm presents the limiting cases

$$\lim_{x\to 0^+} \ln_q x = \begin{cases} \frac{-1}{1-q}, & q < 1, \\ -\infty, & q \geq 1, \end{cases} \tag{3}$$

$$\lim_{x\to \infty} \ln_q x = \begin{cases} \infty, & q \leq 1, \\ \frac{1}{q-1}, & q > 1. \end{cases} \tag{4}$$

Its inverse, the q-exponential, is

$$\exp_q(x) = \begin{cases} [1+(1-q)x]^{\frac{1}{1-q}}\,\theta\!\left(x+\frac{1}{1-q}\right), & q < 1, \\ e^x, & q = 1, \\ \dfrac{1}{[1-(q-1)x]^{\frac{1}{q-1}}\,\theta\!\left(\frac{1}{q-1}-x\right)}, & q > 1, \end{cases} \tag{5}$$

($\theta(x)$ is the Heaviside step function) that is more compactly written as $\exp_q(x) = [1+(1-q)x]_+^{1/(1-q)}$, with the symbol $[\cdot]_+ \equiv \max\{0,\cdot\}$, — the subscript symbol $_+$ encompasses the Heaviside function. The Heaviside step function $\theta(x)$ defines the cutoff condition: the q-exponential is set to zero for $q < 1$ and $x < -1/(1-q)$, and diverges for $q > 1$ and $x > 1/(q-1)$. In the following we use either notations $\exp_q(x)$ or e_q^x, equivalently. Some properties of q-logarithm and q-exponential functions may be found in [2,5–7].

The q-logarithm of a product splits into a nonadditive form for $q \neq 1$:

$$\ln_q(xy) = \ln_q x + \ln_q y + (1-q)\ln_q x \ln_q y. \tag{6}$$

This property triggered the definition of new generalized arithmetic operators: (*i*) what if the right hand side (r.h.s.) of this expression is viewed as the definition of a generalized addition of q-logarithms? Answer: Equation (4) of [3], Equation (7) of [8], Equation (25) of the present work. (*ii*) What should be the argument of the q-logarithm of the left hand side (l.h.s.) of (6) if its r.h.s. were an ordinary addition, instead of the generalized addition just defined? Answer: Equation (7) of [3], Equation (8) of [8], Equation (48) of the present work. Since then, these operators have usually been referred to as q-addition and q-multiplication, or, more colloquially, q-sum and q-product. This q-multiplication is the one that makes S_q extensive, as mentioned previously, and it is not distributive with respect to the q-addition, and Nivanen et al. [8] identified additional deformed operators, recovering the distributivity [their Equations (24)–(28)]. In an extension of that work by the same authors with collaborators [9], the q-multiplication and the q-addition were identified as belonging to two different classes, and further operators were defined.

Examples of mathematical developments along these lines include: spiral generalizations of the trigonometric and hyperbolic functions through the extension of Euler's formula to the complex domain [10], generalization of derivative operators [3,11], generalizations of Fourier transforms, representations of the Dirac delta function [12,13], two parameter extensions for the logarithm and exponential and their related algebras [14,15] etc. Other deformed mathematical structures were introduced, particularly the Kaniadakis formalism [16–19], from which some of the developments within the nonextensive context have been inspired. Generalization of algebras has been recently proposed [20], conforming to the group entropy theory [21].

Examples of physical systems described by nonextensive statistical mechanics include: anomalous diffusion of cold atoms in dissipative optical lattices [22], anomalous diffusion in granular matter [23], experimental high energy physics [24], and observational high energy physics in cosmic rays [25]. An up-to-date bibliography may be found at the site [26].

The present paper revisits generalized algebras and calculus motivated by the nonextensive formalism in a broader point of view. It identifies the basic arithmetic operators for four complementary classes, and defines a pair of linear/nonlinear derivative for each one. A connection with entropic functionals is established. The starting point is the definition of the generalized numbers.

The paper is organized as follows. Section 2 introduces four deformed numbers, by combining the pair of the inverse logarithm/exponential functions and their generalized forms. Section 3 explores each class of deformed arithmetics, derived from the generalized numbers. Section 4 is dedicated to the deformed calculus emerged from the infinitesimal deformed differences. Two possibilities are focused: a linear and a nonlinear deformed derivative. A connection between these structures with entropic functionals is addressed in Section 5. Particularly, the nonadditive entropy S_q is alternatively obtained through a procedure that uses one of the generalized powers defined in Section 3. Section 6 draws our final remarks and points towards new perspectives. Throughout the text, many expressions use symbols designed for compactness. Some of them appear in their explicit forms in the Appendix A.

2. Deformed q-Numbers

One fundamental mathematical object deserves a special attention within the present context, namely, the very concept of number. This was implicitly advanced within the nonextensive formalism in Ref. [10], through the variable $\zeta_q = \ln e_q^z$ ($z \in \mathbb{C}$) used in the generalization of Euler's formula, that may be read as a complex generalized number [see Equation (22) of [10], Equation (10a) of the present work]. Deformed numerical sets (q-natural $\mathbb{N}_q$, q-integer $\mathbb{Z}_q$, q-rational $\mathbb{Q}_q$, q-real $\mathbb{R}_q$ numbers) were considered following Peano-like axioms and generalized arithmetic operators were consistently defined [27]. Those generalized numbers are a transformation of the so-called *Q-analog of n* — Q standing for quantum, within the context of quantum calculus (we write it with upper case Q to avoid confusion with the present lower case index q) [28]:

$$[n]_Q = \frac{Q^n - 1}{Q - 1}, \tag{7}$$

from which we borrow the idea of a q-number. This connection had been previously realized, see [29]. Deformation of reals had also been reported in Ref. [30].

Given a continuous, analytical, monotonous, invertible function $f(x)$ generalized through a real parameter q that recovers the ordinary case as a limiting procedure (in this context, $q \to 1$), we introduce the generalized numbers through four combinations, such as the ordinary case identically recovered:

$$[x]_q = f(f_q^{-1}(x)), \tag{8a}$$
$$_q[x] = f_q(f^{-1}(x)), \tag{8b}$$
$$\{x\}_q = f^{-1}(f_q(x)), \tag{8c}$$
$$_q\{x\} = f_q^{-1}(f(x)). \tag{8d}$$

The adopted notation obeys the following criteria: the square brackets are used when f_q^{-1} (or f^{-1}) is the argument of f (or f_q) and the curly brackets are used when f_q (or f) is the argument of f^{-1} (or f_q^{-1}); the function labeled as f is arbitrary. The deformation parameter q is used as a subscripted *postfix* if the *inner* function is deformed, referred to as i-number,

Equations (8a) and (8c), and as a subscripted *prefix* if the *outer* function is deformed, referred to as o-number, Equations (8b) and (8d) (in analogy with the notation employed for the generalized hypergeometric series — in that case, prefix for the numerator, postfix for the denominator).

The pair of i/o numbers are inverse of each other, and thus

$$_q\left[\,[x]_q\,\right] = \left[\,_q[x]\,\right]_q = {_q}\left\{\,\{x\}_q\,\right\} = \left\{\,_q\{x\}\,\right\}_q = x. \tag{9}$$

To be more specific to the case we are focusing upon, we define $f(x) = \ln x$, and, consequently, $f^{-1}(x) = \mathrm{e}^x$. It follows the le-numbers (l stands for logarithm and e stands for exponential, 'le' expresses the order in which the functions are taken)

$$[x]_q = \ln \mathrm{e}_q^x \quad \text{(ile-number)}, \tag{10a}$$

$$_q[x] = \ln_q \mathrm{e}^x \quad \text{(ole-number)}, \tag{10b}$$

and the el-numbers

$$\{x\}_q = \mathrm{e}^{\ln_q x} \quad \text{(iel-number)}, \tag{11a}$$

$$_q\{x\} = \mathrm{e}_q^{\ln x} \quad \text{(oel-number)}. \tag{11b}$$

Equation (11) is constrained to $x \in \mathbb{R}_+$. This limitation can be overcome, allowing $x \in \mathbb{R}$, in analogy to what was done in Ref. [31], by ad hoc redefining the el-numbers as

$$\{x\}_q = \operatorname{sign}(x)\, \mathrm{e}^{\ln_q |x|} \quad \text{(iel-number)}, \tag{12a}$$

$$_q\{x\} = \operatorname{sign}(x)\, \mathrm{e}_q^{\ln |x|} \quad \text{(oel-number)}, \tag{12b}$$

with $\operatorname{sign}(x) = x/|x|$ and $\operatorname{sign}(0) \equiv 0$. The present work uses the el-numbers as defined by Equation (12), but expressions are easily rewritten in its simpler form (11) by taking into consideration the restricted domain.

The le-numbers have one fixed point ($[x] = x$) at $[0]_q = 0$, and $_q[0] = 0$ (ile and ole, respectively) for all values of $q \neq 1$. The iel-numbers have two fixed points ($\{x\} = x$) for $q < 1$, at $\{\pm 1\}_q = \pm 1$, — zero is not a fixed point for iel-numbers, since $\nexists\, \{0\}_q$ ($\lim_{x\to 0^-} \{x\}_q = -\mathrm{e}^{-1/(1-q)}$, $\lim_{x\to 0^+} \{x\}_q = \mathrm{e}^{-1/(1-q)}$) —, and three fixed points for $q \geq 1$, at $\{0\}_q = 0$ and $\{\pm 1\}_q = \pm 1$. The oel-numbers have three fixed points, at $_q\{0\} = 0$, and $_q\{\pm 1\} = \pm 1$. Due to the cutoff condition of the q-exponential, $_{q<1}\left\{|x| < \mathrm{e}^{1/(q-1)}\right\} = 0$, and due to the absolute values, the el-numbers are odd, for both i and o deformed numbers ($\{-x\} = -\{x\}$). le-numbers and el-numbers are monotonous crescent with the ordinary numbers, i.e., if $x > y$, $[x] > [y]$ and $\{x\} > \{y\}$ for both i and o deformed numbers. An exception may apply for the oel-numbers: it may happen $x > y$ but $_q\{x\} = {_q}\{y\} = 0$ for $q < 1$ within the cutoff region, $|x| \leq \exp\left(-1/(1-q)\right)$ and $|y| \leq \exp\left(-1/(1-q)\right)$. The inverse relations between ile/ole and iel/oel numbers expressed by Equation (9) are valid outside the cutoff regions. Figure 1 illustrates the four q-numbers. These deformed numbers also satisfy the identities

$$\left[\ln x\right]_q = \ln\left(_q\{x\}\right), \tag{13a}$$

$$_q\left[\ln x\right] = \ln\left(\{x\}_q\right) = \ln_q x, \tag{13b}$$

$$\left[\ln_q x\right]_q = \ln_q\left(_q\{x\}\right) = \ln x, \tag{13c}$$

$$_q\left[\ln_q x\right] = \ln_q\left(\{x\}_q\right), \tag{13d}$$

$$\{\exp x\}_q = \exp\left({}_q[x]\right), \tag{14a}$$

$$_q\{\exp x\} = \exp\left([x]_q\right) = \exp_q x, \tag{14b}$$

$$\{\exp_q x\}_q = \exp_q\left({}_q[x]\right) = \exp x, \tag{14c}$$

$$_q\{\exp_q x\} = \exp_q\left([x]_q\right). \tag{14d}$$

Whenever convenient and not ambiguous, for the sake of compactness of notation, we henceforth may occasionally use the symbols $\langle x\rangle_q$ to denote the i-numbers (either $[x]_q$ or $\{x\}_q$), and $_q\langle x\rangle$ to denote the o-numbers (either $_q[x]$ or $_q\{x\}$), and the most general case $\langle x\rangle$, without subscripts, to denote any of the four generalized numbers (not to be confound with mean value or the bra-ket symbols). The expressions 'generalized number' and 'generalized variable' are used interchangeably, just as the convenience of the context, without restricting ourselves to the rigorous mathematical distinction these concepts may have.

The following sections explore the connections of these deformed numbers with their corresponding arithmetics and calculus.

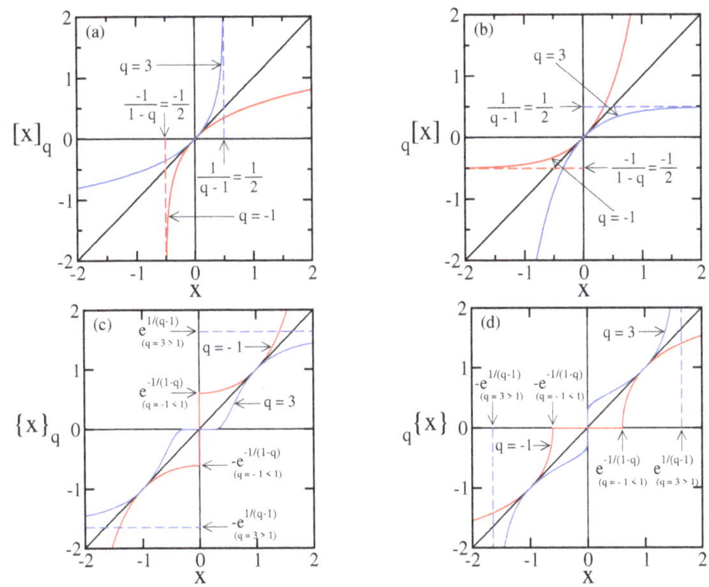

Figure 1. q-numbers , illustrated with $q = -1$ (red), 1 (black), 3 (blue). (**a**) ile-number; $[x \le -1/(1-q)]_{q<1} \to -\infty$, illustrated by the vertical red asymptote for $q = -1$; $[x \ge 1/(q-1)]_{q>1} \to \infty$, illustrated by the vertical blue asymptote for $q = 3$. (**b**) ole-number; $\lim_{x\to-\infty}{}_{q<1}[x] = -1/(1-q)$, illustrated by the horizontal red asymptote for $q = -1$: $\lim_{x\to\infty}{}_{q>1}[x] = 1/(q-1)$, illustrated by the horizontal blue asymptote for $q = 3$. (**c**) iel-number; $\lim_{x\to 0^\pm}\{x\}_{q<1} = \pm e^{-1/(1-q)}$; illustrated for $q = -1$; $\lim_{x\to\pm\infty}\{x\}_{q>1} = \pm e^{1/(q-1)}$, illustrated by the horizontal blue asymptotes for $q = 3$; (**d**) oel-number; $_{q<1}\{|x| \le e^{-1/(1-q)}\} = 0$, illustrated for $q = -1$; $_{q>1}\{|x| \ge e^{1/(q-1)}\} \to \mathrm{sign}(x)\infty$, illustrated by the vertical blue asymptotes for $q = 3$.

3. Deformed q-Arithmetics

Starting from the generalized numbers (10) and (12) we identify four generalized classes of arithmetics. In this paper, the designation q-addition, q-multiplication, etc., are ambiguous, and thus we introduce a different notation: the ile-, ole-, iel-, and oel- arithmetic operators. Particularly, and partially anticipating the results of the next subsections, the deformed addition and subtraction of Ref. [3] belong to the ole-algebra (here symbolized by ${}_{[q]}\oplus$ and ${}_{[q]}\ominus$), considered in Section 3.2, and the deformed multiplication and division of Ref. [3] belong to the oel-algebra (here symbolized by ${}_{\{q\}}\otimes$ and ${}_{\{q\}}\oslash$), considered in Section 3.3. By q-arithmetics we generically denote the set of the four arithmetics described in this paper. They can also be referred to as q-algebras, understood as algebras over the real numbers, or some subset of the reals.

An i-arithmetic operator is defined as the i-number of the ordinary arithmetic operator of the corresponding o-numbers, and, complementary, an o-arithmetic operator is defined as the o-number of the ordinary arithmetic operator of the corresponding i-numbers. The generating rules follow the lines of the κ-arithmetic operators of Kaniadakis [16–18], more generally expressed by Equation (1) of [32] (also in [20]), and are

$$\text{i-arithmetics:} \qquad x \bigcirc_{\langle q \rangle} y = \Big\langle {}_q\langle x \rangle \circ {}_q\langle y \rangle \Big\rangle_q, \tag{15a}$$

$$\text{o-arithmetics:} \qquad x \;{}_{\langle q \rangle}\!\bigcirc y = {}_q\Big\langle \langle x \rangle_q \circ \langle y \rangle_q \Big\rangle. \tag{15b}$$

The symbol $\circ$, a small circle without subscripts, represents any general usual arithmetic operator, $\circ \in \{+, -, \times, /\}$; its generalized version is represented by a larger circle $\bigcirc$, with bracket subscripts: prefixed/postfixed, square/curly, in consonance with the case. To avoid ambiguity in notation, the generalized *operators* are represented within a circle with their subscripts within brackets. The generalized *numbers* are represented within brackets, with their subscripts without brackets.

Some general relations are valid for all cases (the symbol N without subscript generically represents the neutral element of the addition for any of the four arithmetics $N \in \{N_{[+]}, {}_{[+]}N, N_{\{+\}}, {}_{\{+\}}N\}$; similarly to I, the neutral element of the multiplication; A, the absorbing element of the multiplication): the neutral element of the deformed addition N, such as $x \oplus N = x$, is the corresponding deformed zero ($N_{[+]} = [0]_q$, ${}_{[+]}N = {}_q[0]$, $N_{\{+\}} = \{0\}_q$, ${}_{\{+\}}N = {}_q\{0\}$); the deformed additive opposite of x, written as $\ominus x \equiv 0 \ominus x$, such that $x \oplus (\ominus x) = N$, and $x \ominus y = x \oplus (\ominus y)$. Similarly, the neutral element of the deformed multiplication I, $x \otimes I = x$, is the corresponding deformed unity ($I_{[\times]} = [1]_q$, ${}_{[\times]}I = {}_q[1]$, $I_{\{\times\}} = \{1\}_q$, ${}_{\{\times\}}I = {}_q\{1\}$). The deformed multiplicative inverse of x, written as $I \oslash x$, is such that $x \otimes (I \oslash x) = I$, and $x \oslash y = x \otimes (I \oslash y)$. The absorbing element A of the deformed multiplication, such that $x \otimes A = A$, coincides with the neutral element N of the corresponding deformed addition ($A_{[\times]} = N_{[+]}$, ${}_{[\times]}A = {}_{[+]}N$, $A_{\{\times\}} = N_{\{+\}}$, ${}_{\{\times\}}A = {}_{\{+\}}N$). The deformed addition and multiplication are commutative ($x \oplus y = y \oplus x$, $x \otimes y = y \otimes x$), associative $[x \oplus (y \oplus z) = (x \oplus y) \oplus z$, $x \otimes (y \otimes z) = (x \otimes y) \otimes z]$, and the deformed multiplication is (left and right) distributive with respect to the deformed addition $\big[x \otimes (y \oplus z) = (x \otimes y) \oplus (x \otimes z)$, $(y \oplus z) \otimes x = (y \otimes x) \oplus (z \otimes x)\big]$ [32]. Some constraints may apply to these relations according to the case, to be detailed in the next subsections.

3.1. ile-Arithmetics

The ile-algebraic operators follow from the generating rule expressed by (15a). The ile-addition is

$$\begin{aligned} x \oplus_{[q]} y &= \big[\, {}_q[x] + {}_q[y] \,\big]_q, \\ &= \ln \exp_q\big(\ln_q \mathrm{e}^x + \ln_q \mathrm{e}^y\big). \end{aligned} \tag{16}$$

The neutral element of the ile-addition is $N_{[+]} = [0]_q = 0$, and consequently the opposite ile-additive of y is

$$\ominus_{[q]}\, y = \frac{1}{1-q} \ln\Big(2 - \mathrm{e}^{(1-q)y}\Big). \tag{17}$$

The ile-difference (15a) with $\ominus_{\langle q\rangle} = \ominus_{[q]}$,

$$\begin{aligned} x \ominus_{[q]} y &= \left[\, {}_q[x] - {}_q[y] \,\right]_q, \\ &= \ln \exp_q\left(\ln_q e^x - \ln_q e^y \right), \end{aligned} \tag{18}$$

consistently satisfies $x \ominus_{[q]} y = x \oplus_{[q]} (\ominus_{[q]} y)$ for all q.

The ile-multiplication is

$$\begin{aligned} x \otimes_{[q]} y &= \left[\, {}_q[x]\ {}_q[y] \,\right]_q, \\ &= \ln \exp_q\left(\ln_q e^x\ \ln_q e^y \right), \end{aligned} \tag{19}$$

with its neutral ile-multiplicative element $I_{[\times]} = [1]_q = (1-q)^{-1}\ln(2-q)$ for $q < 2$ ($[1]_{q\neq 1} \neq 1$, $\nexists I_{[\times]}$ for $q \geq 2$), and its ile-absorbing element $A_{[\times]} = [0]_q = 0$, for all q. The ile-division is

$$\begin{aligned} x \oslash_{[q]} y &= \left[\frac{{}_q[x]}{{}_q[y]} \right]_q, \\ &= \ln \exp_q\left(\frac{\ln_q e^x}{\ln_q e^y} \right), \end{aligned} \tag{20}$$

and $\nexists\ I_{[\times]} \oslash_{[q]} 0$.

The ile-power of x is defined as the ile-multiplication of n identical factors x,

$$x \circledS_{[q]} n = \prod_{[q]}^{n} x = \left[\left({}_q[x] \right)^n \right]_q. \tag{21}$$

Its analytical extension from $n \in \mathbb{N}$ to $y \in \mathbb{R}$ is written as

$$x \circledS_{[q]} y = \ln \exp_q\left((\ln_q e^x)^y \right), \quad (x > 0), \tag{22}$$

with the particular cases: $x \circledS_{[q]} 0 = [1]_q$ $(x \neq 0)$, $x \circledS_{[q]} 1 = x$ $(x \neq 0)$, $1 \circledS_{[q]} y \neq 1$ (for $q \neq 1$), $\lim_{x\to 0^+}(x \circledS_{[q]} y) = 0$ $(y > 0)$, $\lim_{x\to 0^+}(x \circledS_{[q]} y) \to \infty$ $(y < 0)$, and the trivial case $x \circledS_{[1]} y = x^y$. The ile-power is right-distributive with respect to the ile-multiplication: $(x \otimes_{[q]} y) \circledS_{[q]} z = (x \circledS_{[q]} z) \otimes_{[q]} (y \circledS_{[q]} z)$.

The repeated generalized addition defines a different generalized multiplication that can be named as dot-multiplication, identified by the symbol $\odot$, to distinguish it from the previous generalized multiplication (or times-multiplication), symbolized by $\otimes$ [Equation (19) for the ile class]. The repeated ile-addition is given by

$$\begin{aligned} n \odot_{[q]} y &= \sum_{[q]}^{n} y, \\ &= \left[\sum^{n} {}_q[y] \right]_q, \\ &= \ln \exp_q\left(n \ln_q e^y \right), \end{aligned} \tag{23}$$

where we have used the generalized summation symbol for the ile class, $\sum_{[q]}$, compatible with the notation adopted in this work. Analytical extension from $n \in \mathbb{N}$ to $x \in \mathbb{R}$ yields the non commutative generalized ile-dot-multiplication:

$$x \odot_{[q]} y = \frac{1}{1-q} \ln\left(x\, e^{(1-q)y} - (x-1) \right)_+. \tag{24}$$

The dot-multiplication with the unity has two behaviors, due to its non-commutativity. The trivial case ($1 \odot y = y$) holds for the four classes (for the ile-dot-multiplication of this subsection, as well as for the ole-, iel-, and oel- of the subsections to come). The other case, $x \odot 1$, connects the dot-multiplication with the deformed numbers. The ile-dot-multiplication with unity results $x \odot_{[q]} 1 = \left[x \, {}_q[1] \right]_q$, with ${}_q[1] = \ln_q e = (e^{1-q} - 1)/(1 - q) \neq 1$ for $q \neq 1$. Repeated ile-dot-multiplication defines ile-dot-power, not explicitly shown here.

The generating rule (15b) defines the ole-algebraic operators. The ole-addition (or ole-sum) is

$$\begin{aligned} x \,{}_{[q]}\!\oplus y &= {}_q\left[\, [x]_q \,+\, [y]_q \,\right], \\ &= \ln_q \exp\left(\ln e_q^x + \ln e_q^y \right), \\ &= x + y + (1-q)xy. \end{aligned} \tag{25}$$

Its neutral ole-additive element is ${}_{[+]}N = {}_q[0] = 0$ and the opposite ole-additive element ${}_{[q]}\!\ominus y$ such as $y \,{}_{[q]}\!\oplus (0 \,{}_{[q]}\!\ominus y) = 0$ is

$$ {}_{[q]}\!\ominus y = \frac{-y}{1+(1-q)y}, \quad (y \neq 1/(q-1)), \tag{26}$$

and, consequently, the ole-subtraction is

$$\begin{aligned} x \,{}_{[q]}\!\ominus y &= {}_q\left[\, [x]_q \,-\, [y]_q \,\right], \\ &= \ln_q \exp\left(\ln e_q^x - \ln e_q^y \right), \\ &= \frac{x-y}{1+(1-q)y} \end{aligned} \tag{27}$$

provided $y \neq 1/(q-1)$. These are the generalized addition and subtraction of Ref. [3], referred to as q-sum and q-difference, respectively (see also Section 3.3.3 of Ref. [2]).

From (15b), the ole-product

$$\begin{aligned} x \,{}_{[q]}\!\otimes y &= {}_q\left[\, [x]_q \, [y]_q \,\right], \\ &= \ln_q \exp\left(\ln e_q^x \, \ln e_q^y \right), \end{aligned} \tag{28}$$

and its neutral ole-multiplicative element ${}_{[\times]}I = {}_q[1] = \dfrac{e^{1-q}-1}{1-q} \neq 1$ for $q \neq 1$, together with the ole-division,

$$\begin{aligned} x \,{}_{[q]}\!\oslash y &= {}_q\left[\frac{[x]_q}{[y]_q} \right], \\ &= \ln_q \exp\left(\frac{\ln e_q^x}{\ln e_q^y} \right), \end{aligned} \tag{29}$$

are coherent with the ole-multiplicative inverse element ${}_{[\times]}I \,{}_{[q]}\!\oslash y = \ln_q \exp\left((\ln e_q^y)^{-1} \right)$. The ole-absorbing element is ${}_{[\times]}A = {}_q[0] = 0$. The generalized diamond multiplication defined by Equation (24) of Ref. [27] is related to the ole-multiplication as $x \,{}_{[q]}\!\otimes y = (x \diamond_q y) \,{}_{[q]}\!\otimes 1$, and this expression connects the distributivity property of the diamond multiplication with respect to the ole-addition (Equation (28) of Ref. [27]) and the distributivity of the ole-multiplication with respect to this generalized addition.

The ole-power (the repeated ole-multiplication),

$$ x \,{}_{[q]}\!\circledast n = {}_{\{q\}}\prod^{n} x = {}_q\left[\left([x]_q \right)^n \right], \tag{30}$$

after analytic continuation, becomes

$$ x \,{}_{[q]}\!\circledast y = \ln_q \exp\left(\ln e_q^x \right)^y, \quad (x > 0), \tag{31}$$

with $x \;{}_{[q]}\!\oslash 0 = {}_q[1]$ $(x \neq 0)$, $x \;{}_{[q]}\!\oslash 1 = x$ $(x \neq 0)$, $1 \;{}_{[q]}\!\oslash y \neq 1$ (for $q \neq 1$), $\lim_{x \to 0^+}(x \;{}_{[q]}\!\oslash y) = 0$ $(y > 0)$, $\lim_{x \to 0^+}(x \;{}_{[q]}\!\oslash y) \to \infty$ ($y < 0$ and $q < 1$), $\lim_{x \to 0^+}(x \;{}_{[q]}\!\oslash y) = 1/(q-1)$ ($y < 0$ and $q > 1$), $x \;{}_{[1]}\!\oslash y = x^y$. The ole-power is right-distributive with respect to the ole-multiplication: $(x \;{}_{[q]}\!\otimes y) \;{}_{[q]}\!\oslash z = (x \;{}_{[q]}\!\oslash z) \;{}_{[q]}\!\otimes (y \;{}_{[q]}\!\oslash z)$.

The repeated ole-addition has been defined in Ref. [3], and reads

$$\begin{aligned} n \;{}_{[q]}\!\odot y &= {}_{[q]}\!\sum^{n} y, \\ &= {}_q\!\left[\sum^{n} [y]_q\right], \\ &= \frac{(1+(1-q)y)_+^n - 1}{1-q}. \end{aligned} \tag{32}$$

This is identical to Equation (8) of Ref. [9]. Analytical extension into the real domain yields the non commutative ole-dot-multiplication:

$$x \;{}_{[q]}\!\odot y = \frac{(1+(1-q)y)_+^x - 1}{1-q}. \tag{33}$$

The ole-dot-multiplication with the unity is expressed by $x \;{}_{[q]}\!\odot 1 = {}_q\!\left[x\ [1]_q\right]$, with $[1]_q = \ln \exp_q(1) = (1-q)^{-1}\ln(2-q) \neq 1$ for $q \neq 1$ and $q < 2$. This relation connects the ole-dot-multiplication and the le deformed numbers with the Q-analog of n (7): $n \;{}_{[q]}\!\odot 1 = (Q^n - 1)/(Q-1)$, with $Q = 2-q$. The ole-dot power naturally follows from the repeated ole-dot-multiplication, not shown here.

3.2. *iel-Arithmetics*

According to the generating rule for i-algebras (15a), the iel-addition is

$$\begin{aligned} x \oplus_{\{q\}} y &= \left\{ {}_q\{x\} + {}_q\{y\} \right\}_q, \\ &= \mathrm{sign}(x+y)\, \exp\!\left(\ln_q\left|\mathrm{sign}(x)\, e_q^{\ln|x|} + \mathrm{sign}(y)\, e_q^{\ln|y|}\right|\right). \end{aligned} \tag{34}$$

The cutoff of the q-exponential (5) imposes restrictions on the domain of (34). Its neutral iel-additive element $N_{\{+\}} = \{0\}_q$, is

$$\begin{aligned} &N_{\{+\}} \to 0, \quad q \geq 1, \\ &N_{\{+\}} \leq e^{\frac{-1}{1-q}}, \quad q < 1. \end{aligned} \tag{35}$$

For $q < 1$, there are infinite neutral iel-additive elements, including the zero. The iel-difference reads

$$\begin{aligned} x \ominus_{\{q\}} y &= \left\{ {}_q\{x\} - {}_q\{y\} \right\}_q, \\ &= \mathrm{sign}(x-y)\, \exp\!\left(\ln_q\left|\mathrm{sign}(x)\, e_q^{\ln|x|} - \mathrm{sign}(y)\, e_q^{\ln|y|}\right|\right). \end{aligned} \tag{36}$$

The opposite iel-additive element is

$$\ominus_{\{q\}} y = \begin{cases} \left.\begin{array}{ll} -y, & \text{if } |y| > \exp\left(\frac{-1}{1-q}\right) \\ -\mathrm{sign}(y)\exp\left(\frac{-1}{1-q}\right), & \text{if } |y| \leq \exp\left(\frac{-1}{1-q}\right) \end{array}\right\} & q < 1, \\ -y, & q = 1, \\ \left.\begin{array}{ll} -y, & \text{if } |y| < \exp\left(\frac{1}{q-1}\right) \\ -\mathrm{sign}(y)\exp\left(\frac{1}{q-1}\right), & \text{if } |y| \geq \exp\left(\frac{1}{q-1}\right) \end{array}\right\} & q > 1. \end{cases} \tag{37}$$

The iel-multiplication and the iel-division are

$$\begin{aligned} x \otimes_{\{q\}} y &= \left\{ {}_q\{x\}\ {}_q\{y\} \right\}_q, \\ &= \mathrm{sign}(xy)\ \exp\!\left(\ln_q\!\left(e_q^{\ln|x|}\ e_q^{\ln|y|}\right)\right), \end{aligned} \tag{38}$$

$$\begin{aligned} x \oslash_{\{q\}} y &= \left\{ \frac{{}_q\{x\}}{{}_q\{y\}} \right\}_q, \\ &= \mathrm{sign}(x/y)\ \exp\!\left(\ln_q\!\left(\frac{e_q^{\ln|x|}}{e_q^{\ln|y|}}\right)\right). \end{aligned} \tag{39}$$

The neutral element of the iel-multiplication is $I_{\{\times\}} = \{1\}_q = 1$.

The iel-absorbing element coincides with the neutral iel-additive element, $A_{\{\times\}} = N_{\{+\}}$ (35).

The repeated iel-multiplication (iel-power) is given by

$$x \circledcirc_{\{q\}} n = \prod_{\{q\}}^{n} x = \left\{ \left({}_q\{x\}\right)^n \right\}_q, \tag{40}$$

which is rewritten as (after analytical extension from $n \in \mathbb{N}$ to $y \in \mathbb{R}$)

$$x \circledcirc_{\{q\}} y = \exp\left(\ln_q\left(e_q^{\ln|x|}\right)^y\right), \quad (x > 0), \tag{41}$$

with the particular cases $x \circledcirc_{\{q\}} 0 = 1\ (x \neq 0)$, $x \circledcirc_{\{q\}} 1 = x\ (x \neq 0)$, $1 \circledcirc_{\{q\}} y = 1\ (y \neq 0)$, $\lim_{x\to 0^+}(x \circledcirc_{\{q\}} y) = \exp\left(-1/(1-q)\right)$ $(y > 0$ and $q < 1)$, $\lim_{x\to 0^+}(x \circledcirc_{\{q\}} y) \to \infty$ $(y < 0$ and $q < 1)$, $\lim_{x\to 0^+}(x \circledcirc_{\{q\}} y) = 0$ $(y > 0$ and $q > 1)$, $\lim_{x\to 0^+}(x \circledcirc_{\{q\}} y) = \exp\left(-1/(1-q)\right)$ $(y < 0$ and $q > 1)$, $x \circledcirc_{\{1\}} y = x^y$. The iel-power is right-distributive with respect to the iel-multiplication: $(x \otimes_{\{q\}} y) \circledcirc_{\{q\}} z = (x \circledcirc_{\{q\}} z) \otimes_{\{q\}} (y \circledcirc_{\{q\}} z)$.

The repeated iel-addition defines the iel-dot-multiplication:

$$\begin{aligned} n \odot_{\{q\}} y &= \sum_{\{q\}}^{n} y, \\ &= \left\{ \sum^{n} {}_q\{y\} \right\}_q, \\ &= \mathrm{sign}(y)\ \exp(\ln_q n)\ |y|^{n^{1-q}}. \end{aligned} \tag{42}$$

Analytical extension from $n \in \mathbb{N}$ to $x \in \mathbb{R}_+$ can be represented by

$$x \odot_{\{q\}} y = \mathrm{sign}(y)\ \exp(\ln_q x)\ |y|^{x^{1-q}}, \quad (x > 0). \tag{43}$$

The iel-number is connected to the iel-dot-multiplication by $x \odot_{\{q\}} 1 = \left\{ x\ {}_q\{1\} \right\}_q = \{x\}_q$, since ${}_q\{1\} = 1$.

3.3. oel-Arithmetics

The oel-arithmetic operators derives from (15b):

$$
\begin{aligned}
x \,_{\{q\}}\!\oplus y &= {}_q\{\, \{x\}_q + \{y\}_q \,\}, \\
&= \operatorname{sign}(x+y)\, \exp_q\!\left(\ln\left|\operatorname{sign}(x)\, e^{\ln_q |x|} + \operatorname{sign}(y)\, e^{\ln_q |y|}\right|\right), && (44) \\
x \,_{\{q\}}\!\ominus y &= {}_q\{\, \{x\}_q - \{y\}_q \,\}, \\
&= \operatorname{sign}(x-y)\, \exp_q\!\left(\ln\left|\operatorname{sign}(x)\, e^{\ln_q |x|} - \operatorname{sign}(y)\, e^{\ln_q |y|}\right|\right), && (45) \\
x \,_{\{q\}}\!\otimes y &= {}_q\{\, \{x\}_q \{y\}_q \},\\
&= \operatorname{sign}(xy)\, \exp_q\!\left(\ln\left|e^{\ln_q |x|}\, e^{\ln_q |y|}\right|\right), && (46) \\
x \,_{\{q\}}\!\oslash y &= {}_q\left\{ \frac{\{x\}_q}{\{y\}_q} \right\}, \\
&= \operatorname{sign}(x/y)\, \exp_q\!\left(\ln\left|\frac{e^{\ln_q |x|}}{e^{\ln_q |y|}}\right|\right). && (47)
\end{aligned}
$$

Equations (46) and (47) can be rearranged as

$$
x \,_{\{q\}}\!\otimes y = \operatorname{sign}(xy) \left(|x|^{1-q} + |y|^{1-q} - 1\right)_+^{\frac{1}{1-q}} \qquad (48)
$$

and

$$
x \,_{\{q\}}\!\oslash y = \operatorname{sign}(x/y) \left(|x|^{1-q} - |y|^{1-q} + 1\right)_+^{\frac{1}{1-q}}. \qquad (49)
$$

The oel-product and the oel-ratio were defined in Ref. [3], referred to as q-product and q-ratio, respectively (see also Section 3.3.2 of Ref. [2]). The cutoff that appears in (48) defines regions in which the oel-arithmetical operators are ill-defined. Figures 2 and 3 show the regions for which the cutoff applies for the oel-addition and oel-multiplication, respectively. The first column of each (Figures a and c) shows instances for $q < 1$, and the second column (Figures b and d), for $q > 1$. The first line (Figures a and b) exhibits the cutoff regions with a shaded pattern for one typical value of the parameter q. The second line (Figures c and d) display superimposed curves of the borders of the cutoff regions for various values of q, without shading them, otherwise they would be confusing; they follow the same pattern of the corresponding Figures a and b, respectively. The cutoff regions are closed for $q < 1$ (illustrated with $q = -1$ by Figures 2a and 3a), and they are open and not connected, lying on the outer side delimited by the bounding curves, for $q > 1$ (illustrated with $q = 3$ by Figures 2b and 3b). The second line of the figures help us to understand the effect of the deforming parameter q on the cutoff regions. As q approaches unity from below (Figures 2c and 3c), the cutoff regions become smaller and eventually vanish. For the oel-addition, Figure 2c, the borders of the cutoff region approach the second bisector ($y = -x$), and, for the oel-multiplication, Figure 3c, they approach the origin $(0, 0)$. As q approaches unity from above (Figures 2d and 3d), the cutoff regions move away from the origin. At $q = 1$, no pair of numbers (x, y) fall within the cutoff regions, and the ordinary arithmetic operators are defined everywhere.

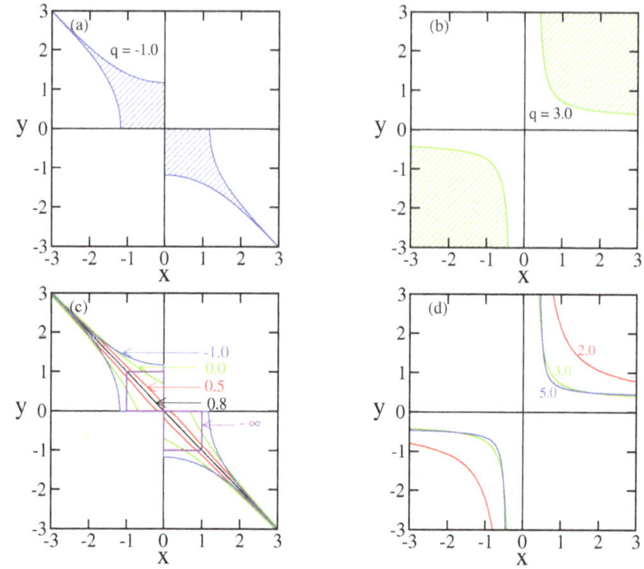

Figure 2. Cutoff regions for the oel-addition (44). Left column: $q < 1$, right column: $q > 1$. Top line: the shaded regions correspond to the cutoff regions of the oel-addition. (**a**) $q = -1$. (**b**) $q = 3$. Bottom line: the curves represent the cutoff borders. Regions are not shaded to avoid excessively heavy representation. Their pattern is similar to (**a**,**b**): for $q < 1$, the cutoff regions lie inside the corresponding closed curves, and for $q > 1$, the cutoff regions lie outside the corresponding curves. (**c**) Different values of $q < 1$ (indicated). The cutoff region shrinks and eventually collapses at $y = -x$ as $q \to 1^-$. (**d**) Different values of $q > 1$ (indicated). As $q \to 1^+$, the non connected regions depart from the origin, and there are no cutoff regions.

The distributivity of the oel-multiplication with respect to the oel-addition is valid whenever the cutoff conditions of the l.h.s. and the r.h.s. of $x \,{}_{\{q\}}\!\otimes (y \,{}_{\{q\}}\!\oplus z) = (x \,{}_{\{q\}}\!\otimes y) \,{}_{\{q\}}\!\oplus (x \,{}_{\{q\}}\!\otimes z)$ are not met. As q approaches unity, even from below or from above, the distributivity of the oel-multiplication with respect to the oel-addition is valid for all real values (x, y, z).

The neutral oel-additive element is ${}_{\{+\}}N = {}_{q}\{0\} = 0$ for $q \geq 1$, and $\nexists\, {}_{\{+\}}N \mid {}_{\{+\}}N \,{}_{\{q\}}\!\oplus x = x$ for $q < 1$. As a consequence, there is no opposite oel-additive element for $q < 1$. For $q \geq 1$, ${}_{\{q\}}\!\ominus y = -y$. The absorbing element ${}_{\{\times\}}A = {}_{q}\{0\} = 0$ for $q \geq 1$, and $\nexists\, {}_{\{\times\}}A \mid {}_{\{\times\}}A \,{}_{\{q\}}\!\otimes x = 0$ for $q < 1$ and $|x| > 1$. If $q < 1$, and $|x| < 1$ the cutoff of (48) (see (5)) implies that zero is an absorbing element, and, in this case, differently from the other three generalized algebras, ${}_{\{+\}}N \neq {}_{\{\times\}}A$. The neutral multiplicative element of the oel-multiplication is ${}_{\{\times\}}I = {}_{q}\{1\} = 1$, for all values of q. The inverse oel-multiplicative element is

$$1 \,{}_{\{q\}}\!\oslash y = \begin{cases} \operatorname{sign}(y)\left(2 - |y|^{1-q}\right)^{\frac{1}{1-q}}, & \text{if } |y| < 2^{\frac{1}{1-q}}, \\ 0, & \text{otherwise.} \end{cases} \tag{50}$$

This implies the unorthodox property $\lim_{y\to 0^+} (1 \,{}_{\{q\}}\!\oslash y) \to 2^{1/(1-q)}$, for $q < 1$.

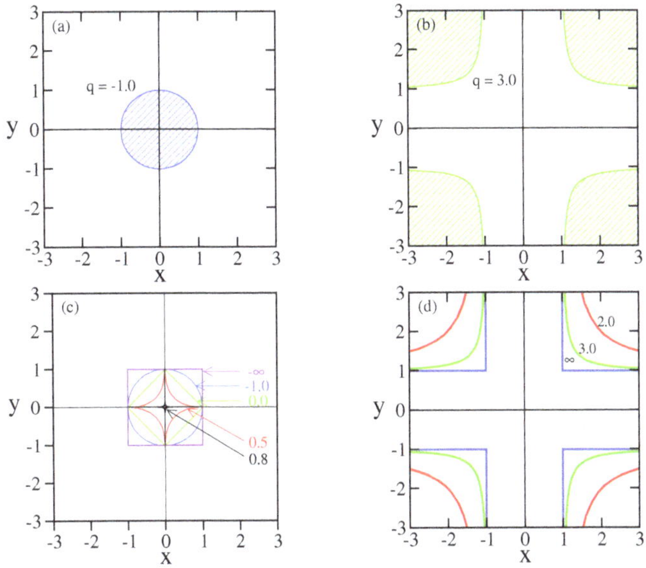

Figure 3. Cutoff regions for the oel-multiplication (46). Left column: $q < 1$, right column: $q > 1$. Top line: the shaded regions correspond to the cutoff regions of the oel-multiplication. (**a**) $q = -1$. (**b**) $q = 3$. Bottom line: the curves represent the cutoff borders, $|y| = (1 - |x|^{1-q})^{1/(1-q)}$. Regions are not shaded to avoid excessively heavy representation. Their pattern is similar to the adopted in (**a**) or (**b**): for $q < 1$, the cutoff regions lie inside the corresponding closed curves, and for $q > 1$, the cutoff regions lie outside the corresponding curves. (**c**) Different values of $q < 1$ (indicated). The cutoff region shrinks and eventually collapses at $(0,0)$ as $q \to 1^-$, when the curves coincide with the axes. (**d**) Different values of $q > 1$ (indicated). As $q \to 1^+$, the non connected regions depart from the origin, and there are no cutoff regions.

The oel-power, previously defined in Ref. [3] (with different symbols), is written as

$$x \;_{\{q\}}\!\oslash\, n = {}_{\{q\}}\!\prod^{n} x = {}_{q}\{(\{x\}_q)^n\}, \quad (x > 0). \tag{51}$$

This operator also appears as Equation (8) of Ref. [9]. We make an analytical extension from $n \in \mathbb{N}$ to $y \in \mathbb{R}$, and the oel-power can also be written as

$$x \;_{\{q\}}\!\oslash\, y = \exp_q(y \ln_q x), \quad (x > 0). \tag{52}$$

Particular cases are $x \;_{\{q\}}\!\oslash\, 0 = 1$ $(x \neq 0)$, $x \;_{\{q\}}\!\oslash\, 1 = x$ $(x \neq 0)$, $1 \;_{\{q\}}\!\oslash\, y = 1$ $(y \neq 0)$, $\lim_{x\to 0^+}(x \;_{\{q\}}\!\oslash\, y) = 0$ $(y \geq 1, q < 1)$, $\lim_{x\to 0^+}(x \;_{\{q\}}\!\oslash\, y) = \exp_q(-y/(1-q))$ $(y < 1, q < 1)$, $\lim_{x\to 0^+}(x \;_{\{q\}}\!\oslash\, y) = 0$ $(y > 0, q > 1)$, $\lim_{x\to 0^+}(x \;_{\{q\}}\!\oslash\, y) \to \infty$ $(y < 0, q > 1)$, and, as always, $x \;_{\{1\}}\!\oslash\, y = x^y$. The oel-power is right-distributive with respect to the oel-multiplication: $(x \;_{\{q\}}\!\otimes\, y) \;_{\{q\}}\!\oslash\, z = (x \;_{\{q\}}\!\oslash\, z) \;_{\{q\}}\!\otimes\, (y \;_{\{q\}}\!\oslash\, z)$.

The repeated oel-addition is

$$ {}_{\{q\}}\!\sum^{n} y = {}_{q}\left\{ \sum^{n} \{y\}_q \right\}. \tag{53}$$

Its analytical extension from $n \in \mathbb{N}$ to $x \in \mathbb{R}_+$ defines the non commutative oel-dot-multiplication:

$$x \;_{\{q\}}\!\odot\; y = \mathrm{sign}(y)\left((1-q)\ln x + |y|^{1-q}\right)_+^{\frac{1}{1-q}}, \quad (x > 0). \tag{54}$$

The oel-number is connected to the oel-dot-multiplication by $x \;_{\{q\}}\!\odot 1 = {}_q\{x\,\{1\}_q\} = {}_q\{x\}$, since $\{1\}_q = 1$.

4. Deformed q-Calculus

Following the lines of Ref. [3] (see also Sections II.C and II.D of [33]), we connect the deformed algebra with deformed calculus, and define the deformed differentials of ordinary numbers:

$$\mathrm{d}_{[q]}x = \lim_{x'\to x}\left(x' \ominus_{[q]} x\right), \tag{55a}$$

$$_{[q]}\mathrm{d}x = \lim_{x'\to x}\left(x' \;_{[q]}\!\ominus x\right), \tag{55b}$$

$$\mathrm{d}_{\{q\}}x = \lim_{x'\to x}\left(x' \ominus_{\{q\}} x\right), \tag{55c}$$

$$_{\{q\}}\mathrm{d}x = \lim_{x'\to x}\left(x' \;_{\{q\}}\!\ominus x\right). \tag{55d}$$

The definitions of the corresponding deformed differences, Equations (18), (27), (36), and (45), lead to

$$\mathrm{d}_{[q]}x = \mathrm{d}\left({}_q[x]\right), \tag{56a}$$

$$_{[q]}\mathrm{d}x = \mathrm{d}\left([x]_q\right), \tag{56b}$$

$$\mathrm{d}_{\{q\}}x = \mathrm{d}\left({}_q\{x\}\right), \tag{56c}$$

$$_{\{q\}}\mathrm{d}x = \mathrm{d}\left(\{x\}_q\right), \tag{56d}$$

i.e., the deformed differential of an ordinary variable (l.h.s. of (56)) is equal to the ordinary differential of the corresponding complementary deformed variable (r.h.s. of (56)): the i-differential of a variable is equal to the ordinary differential of an o-variable, (56a) and (56c), and the o-differential of a variable is equal to the ordinary differential of an i-variable, (56b) and (56d). All the deformed differentials given by (56) can be arranged as the product of the ordinary differential $\mathrm{d}x$ by a deforming function $h_\delta(x)$, with $\delta \in \{\mathrm{ile}, \mathrm{ole}, \mathrm{iel}, \mathrm{oel}\}$ representing the deformation ($\mathrm{d}_{[q]}x = h_{\mathrm{ile}}(x)\,\mathrm{d}x$, $_{[q]}\mathrm{d}x = h_{\mathrm{ole}}(x)\,\mathrm{d}x$, $\mathrm{d}_{\{q\}}x = h_{\mathrm{iel}}(x)\,\mathrm{d}x$, $_{\{q\}}\mathrm{d}x = h_{\mathrm{oel}}(x)\,\mathrm{d}x$). Their explicit forms are

$$h_{\mathrm{ile}}(x) = \mathrm{e}^{(1-q)x}, \tag{57a}$$

$$h_{\mathrm{ole}} = \frac{1}{1+(1-q)x}, \quad \left(x \neq \frac{-1}{1-q}\right), \tag{57b}$$

$$h_{\mathrm{iel}}(x) = \frac{1}{x}\left(1+(1-q)\ln x\right)^{\frac{q}{1-q}}, \quad (x > 0), \tag{57c}$$

$$h_{\mathrm{oel}}(x) = \frac{1}{x^q}\exp\left(\frac{x^{1-q}-1}{1-q}\right), \quad (x > 0). \tag{57d}$$

A pair of generalized derivatives of a function $f(x)$, holding a duality nature between them, stem from each of the deformed differentials, according to which variable the deformed differential applies on: whether on the independent variable x, — and thus a linear deformed derivative —, generically represented by $\mathrm{D}_\delta f(x)$, or on the dependent variable f, — and thus a nonlinear deformed derivative — generically represented by $\widetilde{\mathrm{D}}_\delta f(x)$, resulting in eight different cases:

1. ile-Derivatives
Linear ile-derivative:

$$\mathrm{D}_{\mathrm{ile}} f(x) \equiv \frac{\mathrm{d} f(x)}{\mathrm{d}_{[q]}\, x} = \frac{1}{h_{\mathrm{ile}}(x)} \frac{\mathrm{d} f(x)}{\mathrm{d}x}, \tag{58a}$$

Nonlinear ile-derivative:

$$\widetilde{\mathrm{D}}_{\mathrm{ile}} f(x) \equiv \frac{\mathrm{d}_{[q]}\, f(x)}{\mathrm{d}x} = h_{\mathrm{ile}}\big(f(x)\big) \frac{\mathrm{d} f(x)}{\mathrm{d}x}. \tag{58b}$$

2. ole-Derivatives
Linear ole-derivative:

$$\mathrm{D}_{\mathrm{ole}} f(x) \equiv \frac{\mathrm{d} f(x)}{{}_{[q]}\mathrm{d}\, x} = \frac{1}{h_{\mathrm{ole}}(x)} \frac{\mathrm{d} f(x)}{\mathrm{d}x}, \tag{59a}$$

Nonlinear ole-derivative:

$$\widetilde{\mathrm{D}}_{\mathrm{ole}} f(x) \equiv \frac{{}_{[q]}\mathrm{d}\, f(x)}{\mathrm{d}x} = h_{\mathrm{ole}}\big(f(x)\big) \frac{\mathrm{d} f(x)}{\mathrm{d}x}. \tag{59b}$$

3. iel-Derivatives
Linear iel-derivative:

$$\mathrm{D}_{\mathrm{iel}} f(x) \equiv \frac{\mathrm{d} f(x)}{\mathrm{d}_{\{q\}}\, x} = \frac{1}{h_{\mathrm{iel}}(x)} \frac{\mathrm{d} f(x)}{\mathrm{d}x}, \tag{60a}$$

Nonlinear iel-derivative:

$$\widetilde{\mathrm{D}}_{\mathrm{iel}} f(x) \equiv \frac{\mathrm{d}_{\{q\}}\, f(x)}{\mathrm{d}x} = h_{\mathrm{iel}}\big(f(x)\big) \frac{\mathrm{d} f(x)}{\mathrm{d}x}. \tag{60b}$$

4. oel-Derivatives
Linear oel-derivative: linear oel-derivative:

$$\mathrm{D}_{\mathrm{oel}} f(x) \equiv \frac{\mathrm{d} f(x)}{{}_{\{q\}}\mathrm{d}\, x} = \frac{1}{h_{\mathrm{oel}}(x)} \frac{\mathrm{d} f(x)}{\mathrm{d}x}, \tag{61a}$$

Nonlinear oel-derivative:

$$\widetilde{\mathrm{D}}_{\mathrm{oel}} f(x) \equiv \frac{{}_{\{q\}}\mathrm{d}\, f(x)}{\mathrm{d}x} = h_{\mathrm{oel}}\big(f(x)\big) \frac{\mathrm{d} f(x)}{\mathrm{d}x}. \tag{61b}$$

The duality between the linear and the nonlinear generalized derivatives is expressed by $\mathrm{D}_{\delta} f(x) = \widetilde{\mathrm{D}}_{\delta} f^{-1}(x)$. The el-derivatives are defined for $x > 0$. The ole-derivatives has been defined in Ref. [3], then referred to as q-derivative (the linear deformed derivative) and its dual q-derivative (the nonlinear deformed derivative). Particularly, the linear ole-derivative (59a) was used to generalize Fisher's information measure and the Cramer-Rao inequality [34]. The eigenfunction of the linear i/o-deformed derivative is the ordinary exponential of the o/i-deformed variable, which directly follows from (56). They are (written with the symbols $\langle\cdot\rangle$ representing either $[\cdot]$ or $\{\cdot\}$)

$$\frac{\mathrm{d}\, \exp\!\big({}_{q}\langle x\rangle\big)}{\mathrm{d}_{\langle q\rangle}\, x} = \frac{\mathrm{d}\, \exp\!\big({}_{q}\langle x\rangle\big)}{\mathrm{d}\,\big({}_{q}\langle x\rangle\big)} = \exp\!\big(\langle x\rangle_{q}\big) \tag{62a}$$

and

$$\frac{\mathrm{d}\,\exp(\langle x\rangle_q)}{{}_{\langle q\rangle}\mathrm{d}\,x} = \frac{\mathrm{d}\,\exp(\langle x\rangle_q)}{\mathrm{d}\,(\langle x\rangle_q)} = \exp(\langle x\rangle_q). \tag{62b}$$

Particularly, the q-exponential (5) is the eigenfunction of the linear ole-derivative, $\mathrm{D_{ole}} e_q^x = e_q^x$ (a particular case of (62b) with $e_q^x = e^{[x]_q}$, see (14b)). Alternatively, its ordinary derivative is $\mathrm{d}\,e_q^x\,/\,\mathrm{d}\,x = (e_q^x)^q$. The nonlinear deformed derivative of which the q-exponential is eigenfunction was defined in Ref. [11]:

$$\widetilde{\mathfrak{D}}_q f(u) = [f(u)]^{1-q}\,\frac{df(u)}{du}, \tag{63}$$

where we have used the symbol, $\widetilde{\mathfrak{D}}_q$ to distinguish it from the present deformed derivatives.

The integral of the inverse of a variable, $\int_1^x t^{-1}\mathrm{d}t$, is typically associated to, and frequently taken as the definition of, the logarithm function. The general nonlinear cases are

$$\frac{\mathrm{d}_{\langle q\rangle}\,\langle \ln x\rangle_q}{\mathrm{d}\,x} = \frac{{}_{\langle q\rangle}\mathrm{d}\;{}_q\langle \ln x\rangle}{\mathrm{d}\,x} = \frac{1}{x}. \tag{64}$$

The particular case of this equation for the nonlinear ole-derivative is (see (13b)): $\widetilde{\mathrm{D}}_{\mathrm{ole}} \ln_q x = 1/x$. Alternatively, the ordinary derivative of the q-logarithm is $\mathrm{d}\ln_q x\,/\,\mathrm{d}\,x = 1/x^q$. This expression yields an integral representation of the q-logarithm function,

$$\int_1^x t^{-q}\mathrm{d}t = \ln_q x. \tag{65}$$

The dual linear deformed derivative of (63), defined by Equation (25) of Ref. [33],

$$\mathfrak{D}_q f(x) = \frac{1}{x^{1-q}}\,\frac{\mathrm{d}f(x)}{\mathrm{d}x}, \tag{66}$$

operates on the q-logarithm similarly to the nonlinear ole-derivative: $\mathfrak{D}_q \ln_q x = 1/x$.

Generalized derivatives of a power (for the linear cases), or generalized powers (for the nonlinear cases), of q-numbers, are

$$\begin{aligned} \mathrm{D_i}\left({}_q\langle x\rangle^n\right) &= n\;{}_q\langle x\rangle^{n-1}, \\ \mathrm{D_o}\left(\langle x\rangle_q^n\right) &= n\;\langle x\rangle_q^{n-1}, \end{aligned} \tag{67}$$

$$\begin{aligned} \widetilde{\mathrm{D}}_{\mathrm{i}}\left(\langle x\rangle_q \otimes_{\langle q\rangle} n\right) &= \widetilde{\mathrm{D}}_{\mathrm{i}}\left(\langle x^n\rangle_q\right) &= n\,x^{n-1}, \\ \widetilde{\mathrm{D}}_{\mathrm{o}}\left({}_q\langle x\rangle\;{}_{\langle q\rangle}\!\otimes n\right) &= \widetilde{\mathrm{D}}_{\mathrm{o}}\left({}_q\langle x^n\rangle\right) &= n\,x^{n-1}. \end{aligned} \tag{68}$$

Second and higher deformed linear derivatives follow the usual rule, $\mathrm{D}_\delta^2 f(x) = \mathrm{D}_\delta\left[\mathrm{D}_\delta f(x)\right]$ and so on, but for the deformed nonlinear cases, second order derivatives (and similarly for higher order derivatives) are defined as

$$\widetilde{\mathrm{D}}_\delta^2 f(x) = h_\delta\left(f(x)\right)\frac{\mathrm{d}}{\mathrm{d}x}\left[h_\delta(f(x))\,\frac{\mathrm{d}f(x)}{\mathrm{d}x}\right]. \tag{69}$$

The product rule for the deformed linear derivatives is identical to the usual one, $\mathrm{D}_\delta(f(x)\,g(x)) = \mathrm{D}_\delta(f(x))\,g(x) + f(x)\,\mathrm{D}_\delta(g(x))$. The product rule for the deformed nonlinear derivatives is

$$\frac{1}{h_\delta(f(x)g(x))}\,\widetilde{\mathrm{D}}_\delta\Big(f(x)\,g(x)\Big) = \left(\frac{1}{h_\delta(f(x))}\,\widetilde{\mathrm{D}}_\delta\, f(x)\right)g(x) + f(x)\left(\frac{1}{h_\delta(g(x))}\,\widetilde{\mathrm{D}}_\delta\, g(x)\right). \tag{70}$$

The deformed antiderivatives, or indefinite deformed integrals, associated to the linear deformed derivatives are defined by

$$\int_{(\delta)}^{x} f(x')\,\mathrm{d}x' \equiv \int^{x} f(x')\,\mathrm{d}_{\delta}x', \tag{71}$$

$$= \int^{x} f(x')\,h_{\delta}(x')\,\mathrm{d}x' \tag{72}$$

(the symbol (δ) within parenthesis refers to the deformation, and not a limit of integration), so

$$\mathrm{D}_{\delta} \int_{(\delta)}^{x} f(x')\,\mathrm{d}x' = f(x) \tag{73}$$

and

$$\int_{(\delta)}^{x} \mathrm{D}_{\delta} f(x')\,\mathrm{d}x' = f(x) + C. \tag{74}$$

One possibility for defining the deformed antiderivatives associated to the nonlinear deformed derivatives, particularly following the definition used in [3] for the $\delta =$ ole case, is

$$\widetilde{\int_{(\delta)}^{x}} f(x')\,\mathrm{d}x' \equiv \int^{x} \frac{1}{h_{\delta}(f(x'))} f(x')\,\mathrm{d}x', \tag{75}$$

A significant weakness with this option is that the following important properties are not satisfied:

$$\widetilde{\mathrm{D}}_{\delta} \widetilde{\int_{(\delta)}^{x}} f(x')\,\mathrm{d}x' \neq f(x) \tag{76}$$

and

$$\widetilde{\int_{(\delta)}^{x}} \widetilde{\mathrm{D}}_{\delta} f(x')\,\mathrm{d}x' \neq f(x) + C. \tag{77}$$

5. Entropy Generator

The connection between entropies and derivatives was pointed out by Abe [35]. He observed that the Boltzmann-Gibbs entropy can be rewritten as (with $k = 1$)

$$S_1 = -\frac{\mathrm{d}}{\mathrm{d}\alpha} g(\alpha)\Big|_{\alpha=1} \tag{78}$$

with

$$g(\alpha) = \sum_{i}^{W} p^{\alpha}. \tag{79}$$

He realized that S_q entropy can be similarly recast through the Jackson's derivative of a function $f(x)$ [36]

$$\mathrm{D}_q^{(\mathrm{J})} f(x) \equiv \frac{f(qx) - f(x)}{qx - x} \tag{80}$$

(the same deformed derivative of quantum calculus [28]; Newtonian derivative is recovered as the limiting case $q \to 1$), so

$$S_q = -\mathrm{D}_q^{(\mathrm{J})} g(\alpha)\Big|_{\alpha=1}. \tag{81}$$

This property had been interpreted as expressing the association between Boltzmann-Gibbs entropy (S_1) to infinitesimal translations, and Tsallis entropy to finite dilations [2]. Abe applied this procedure a step further, and used a different derivative operator on $g(\alpha)$, generating a new symmetric entropic functional S_q^S with $q \leftrightarrow q^{-1}$ invariance. Following the same line, a two-parameter derivative operator was used to define a two-parameter $S_{q,q'}$

entropy, that recovers the previous $S_{q'}^{S}$, S_q and S_1 with convenient choices of the indices q and q' [37].

All the eight deformed derivatives (58)–(61) applied on (79) result in S_1 entropy with a multiplying function of the parameter q: $-\mathrm{D}_\delta g(\alpha)|_{\alpha=1} = h_\delta^\eta(1)\, S_1$, where D_δ represents any of the deformed (linear or nonlinear) derivatives (at this point we do not use the tilde for the nonlinear deformed derivatives), $h_\delta(1)$ is a particular value of the corresponding Equation (57), $\eta = -1$ for the linear deformed derivatives, and $\eta = +1$ for the nonlinear deformed derivatives. This is a consequence of the generalized derivatives being based on infinitesimal deformed translations, and the infinitesimal nature of the translation determines the entropy (except for a multiplicative constant), despite of the deformations.

A non-trivial result is obtained by inverting the procedure. Instead of applying one of the generalized derivatives on the generating function (79), we apply the ordinary Newtonian derivative on a generalized generating function:

$$S_q^\delta = -\frac{\mathrm{d}}{\mathrm{d}\alpha} g_\delta(\alpha;q)\bigg|_{\alpha=1}. \tag{82}$$

The generalized generating functions are obtained through the four generalized powers, (22), (31), (41), (52): $g_{\mathrm{ile}}(\alpha;q) = \sum_i^W (p_i \otimes_{[q]} \alpha)$, $g_{\mathrm{ole}}(\alpha;q) = \sum_i^W (p_i\, {}_{[q]}\!\otimes \alpha)$, $g_{\mathrm{iel}}(\alpha;q) = \sum_i^W (p_i \otimes_{\{q\}} \alpha)$, $g_{\mathrm{oel}}(\alpha;q) = \sum_i^W (p_i\, {}_{\{q\}}\!\otimes \alpha)$. The resulting functionals are

$$S_q^{\mathrm{ile}} = \sum_i {}_q[-p_i] \ln\left({}_q[p_i]\right), \tag{83a}$$

$$S_q^{\mathrm{ole}} = -\sum_i [p_i]_q \ln\left([p_i]_q\right) - (1-q)\sum_i p_i\, [p_i]_q \ln\left([p_i]_q\right), \tag{83b}$$

$$S_q^{\mathrm{iel}} = -\sum_i p_i \ln\left({}_q\{p_i\}\right) - (1-q)\sum_i p_i \ln p_i \ln\left({}_q\{p_i\}\right), \tag{83c}$$

$$S_q^{\mathrm{oel}} = -\sum_i p_i^q \ln\left(\{p_i\}_q\right). \tag{83d}$$

The use of the generalized derivatives essentially result in the same, $-\mathrm{D}_\delta g_\delta(\alpha;q)|_{\alpha=1} = h_\delta^\eta(1)\, S_q^\delta$, except for a multiplicative constant for the le cases, since $h_{\mathrm{iel}}(1) = h_{\mathrm{oel}}(1) = 1$. The certainty distribution originates non zero values for the le functionals: $S_q^{\mathrm{ile}}[p_i = 1; p_j = 0, \forall j \neq i] \neq 0$ for $q > 1$, and, $S_q^{\mathrm{ole}}[p_i = 1; p_j = 0, \forall j \neq i] \neq 0$ for $q < 1$, since ${}_q[1] \neq 1$ and $[1]_q \neq 1$. Additionally, the le functionals present negative values: S_q^{ile} presents negative values for $q < 1$, S_q^{ole} presents negative values for $q > 1$. Besides, there are ranges of values of q for which neither S_q^{ile} nor S_q^{ole} present a definite concavity (two instances: $q = 2.4$, for ile; $q = 2.3$, for ole). These are severe drawbacks and consequently (83a) and (83b) can not be considered as legitimate entropic forms.

The iel-functional S_q^{iel} fails on the expansibility property for $q < 1$ (adding events of zero probability), since ${}_{q<1}\{0\}$ is not defined. For $q > 1$, it is expansible, non negative and the certainty distribution ($p_i = 1; p_j = 0, \forall j \neq i$) implies $S_q^{\mathrm{iel}} = 0$, so, (83c) is admissible as an entropic form for $q > 1$.

The oel-functional (83d) is the nonadditive entropy S_q (see Equation (13b)), vastly considered in the literature. This result permits to amend a previous statement: S_q entropy, that is associated to finite dilations, can also be associated to infinitesimal translations, but in a deformed space expressed by the oel-power. Figure 4a illustrates the concavity for the two admissible entropic functionals, Equation (83c) with $q > 1$ and Equation (83d), for a two-state system. Figure 4b illustrates el-entropies as monotonically increasing functions of the number of states W for the equiprobable distribution, $p_i = 1/W, \forall i$, with the abscissa in logarithm scale, for which the usual case appears as a straight line.

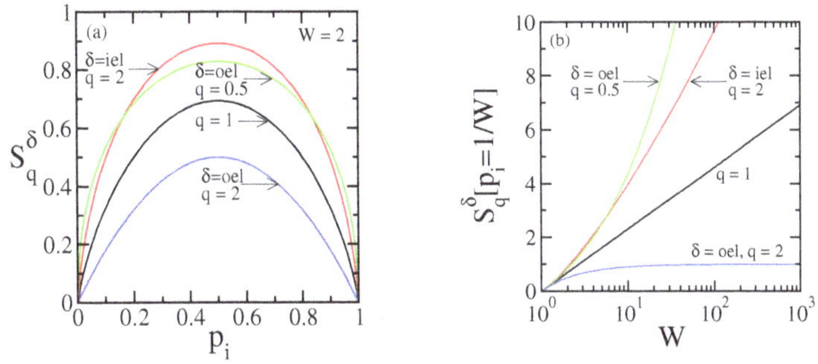

Figure 4. (**a**) el-Entropies for a two-state system. S_q^{iel} (83c) for $q = 2$ (red); $S_q^{\text{oel}} = S_q$ (83d), for $q = 0.5$ (green), $q = 2$ (blue); S_1 (black). S_q entropy is convex for $q < 0$, see [1]. (**b**) el-Entropies for equiprobable states as a function of W. Abscissa in log scale, for which the Boltzmann case is a straight line (black). S_q^{iel} for $q = 2$ (red), S_q^{oel} for $q = 0.5$ (green), $q = 2$ (blue).

6. Final Remarks

A forerunner of the transformations given by Equation (10) is the relation between Rényi entropy, $S_q^{\text{R}} = (1-q)^{-1} \ln\left(\sum_i^W p_i^q\right)$, and Tsallis entropy (2) (see Equation (8) of Ref. [1]), $S_q^{\text{R}} = [S_q]_q$, and, equivalently, $S_q = {}_q[S_q^{\text{R}}]$. Another instance of the transformation represented by the ile-number (10a) appeared in Equation (22) of Ref. [10] and allowed the generalization of trigonometric functions. The ole-number ${}_q[x]$ appeared as Equation (5) of Ref. [38], as the scaling factor of the generalized Kolmogorov-Nagumo average for expressing the Rényi entropy. A former example of connecting deformed numbers with deformed differential operators have appeared in Ref. [39], with the transformation (10a) and the deformed differential (59a), establishing an equivalence between a position-dependent mass system in a usual space and a constant mass within a deformed space. These works have been recently extended to the deformed version of the Fokker-Planck equation for inhomogeneous medium with position-dependent mass [33]. In addition, the use of the iel-number, Equation (11a), to the generalization of the Riemann's zeta function has been recently advanced in [40].

Expressions with operations belonging to one class of q-algebra may result in operations belonging to a different class. Some instances: the following are generalizations of the logarithm of a product as a sum of logarithms $(\ln(xy) = \ln x + \ln y)$:

$$\ln_q(xy) = \ln_q x \;{}_{[q]}\!\oplus\; \ln_q y, \tag{84a}$$

$$\ln_q(x \;{}_{\{q\}}\!\otimes\; y) = \ln_q x + \ln_q y, \tag{84b}$$

$$\ln(x \otimes_{\{q\}} y) = \ln x \;{}_{[q]}\!\oplus\; \ln y, \tag{84c}$$

$$\ln(x \;{}_{\{q\}}\!\otimes\; y) = \ln x \oplus_{[q]} \ln y. \tag{84d}$$

Generalizations of the logarithm of a power, $\ln x^y = y \ln x$, are

$$\ln_q(x \;{}_{\{q\}}\!\owedge\; y) = y \ln_q x. \tag{85a}$$

$$\ln(x \owedge_{\{q\}} y) = y \;{}_{[q]}\!\odot\; \ln x, \tag{85b}$$

$$\ln(x \;{}_{\{q\}}\!\owedge\; y) = y \odot_{[q]} \ln x. \tag{85c}$$

The counterpart of these expressions are generalizations of the exponential of a sum as a product of exponentials, $e^{x+y} = e^x e^y$:

$$e_q^{x \;{}_{[q]}\!\oplus\; y} = e_q^x \, e_q^y, \tag{86a}$$

$$e_q^{x+y} = e_q^x \;{}_{\{q\}}\!\otimes\; e_q^y, \tag{86b}$$

$$\mathrm{e}^{x\ {}_{[q]}\!\oplus\ y} = \mathrm{e}^x \otimes_{\{q\}} \mathrm{e}^y, \tag{86c}$$

$$\mathrm{e}^{x\ \oplus_{[q]}\ y} = \mathrm{e}^x\ {}_{\{q\}}\!\otimes \mathrm{e}^y, \tag{86d}$$

and the power of an exponential as the exponential of a product $\left((\mathrm{e}^x)^{\,y} = \mathrm{e}^{yx}\right)$:

$$\mathrm{e}_q^x\ {}_{\{q\}}\!\owedge\ y = \mathrm{e}_q^{yx}, \tag{87a}$$

$$\mathrm{e}^x \owedge_{\{q\}} y = \mathrm{e}^{y\ {}_{[q]}\!\odot\ x}, \tag{87b}$$

$$\mathrm{e}^x\ {}_{\{q\}}\!\owedge\ y = \mathrm{e}^{y\ \odot_{[q]}\ x}. \tag{87c}$$

Relations (84) are also valid for the logarithm, or for the q-logarithm, of a ratio, simply replacing ordinary or general products by ordinary or general ratios, and ordinary or general sums by ordinary or general differences. Similarly, relations (86) are also valid for the exponential, or for the q-exponential, of a difference, by replacing the operators accordingly. Sums of q-logarithm functions, Equation (84b), appeared in the literature prior to the definition of the q-product [3] —here called the oel-product, Equation (48)— within the context of the generalization of Boltzmann's molecular chaos hypothesis and the H Theorem, see Equation (16) of Ref. [41] and Equation (22) of Ref. [42]. The oel-product has shown to be a key ingredient to the generalization of the Fourier transform and the central limit theorem [43,44]. It is allowed to think that the present algebras may be relevant within these contexts.

Equation (85a) is the one referred to in the Introduction, that makes S_q extensive: consider a composed system for which its subsystems have $W_i > 1$ available states. If they are independent, the number of available states of the composed system is $W = \prod_i^N W_i$, and, besides, if they are identical, $W = W_1^N$. Correlations between the subsystems lead to a smaller number of available states for the composed system, and particular strong correlations represented by $W = W_1\ {}_{\{q\}}\!\owedge\ N$, with $q < 1$, makes $S_q = k \ln_q W = Nk \ln_q W_1$. This is a non trivial case of extensivity.

Different possibilities for generating rules of arithmetic operations, instead of (15), are ${}_q[x]\ {}_{[q]}\!\bigcirc\ {}_q[y] = {}_q[x \circ y]$, $[x]_q \bigcirc_{[q]} [y]_q = [x \circ y]_q$; These patterns are used in References [27,40].

Weberszpil, Lazo, and Helayël-Neto [45] have shown that the linear ole-derivative (59a) is the first order expansion of the Hausdorff derivative. Whether the other generalized derivatives are also connected to fractal derivatives and fractal metrics remains to be investigated.

Two of the functionals obtained with the recipe of applying the ordinary derivatives to a generalized version of the generating function, (82), result in admissible entropic forms corresponding to the el-class: S_q^{iel} (83c), and S_q^{oel} (83d). The other functionals (83a) and (83b) are not admissible to be considered as entropies, but this does not mean that the le-algebras or le-calculus they are based on are not feasible for other applications.

Extension to the complex domain of the deformed numbers still remains to be explored. Two-parameter generalization are not addressed here, we just advance a few lines. Two-parameter generalizations of numbers in accordance with the present developments are given by

$$\begin{aligned} {}_q[x]_{q'} &\equiv [x]_{q,q'} = \ln_q \exp_{q'}(x), && (88a)\\ {}_q\{x\}_{q'} &\equiv \{x\}_{q,q'} = \exp_q(\ln_{q'} x). && (88b) \end{aligned}$$

The use of the relatively uncommon subscripted prefix to represent the two parameter deformed number may be avoided, since there is no ambiguity with the symbol $\langle x\rangle_{q,q'}$. Two-parameter arithmetic operators follow straightforwardly:

$$x \bigcirc_{\langle q,q'\rangle} y = \left\langle\, \langle x\rangle_{q',q} \circ \langle y\rangle_{q',q} \,\right\rangle_{q,q'}, \tag{89}$$

for which, of course, all the previous developments are particular cases. The two-parameter algebra of Ref. [15] is obtained through a different generating rule than (89): it derives from the two-parameter generalized logarithm and exponential functions [14] (Equations (16) and (17) of [15]).

It also comes naturally the two-parameter derivative $D_{q,q'} f(x)$, with deformation on both the independent and dependent variables. A broader generalization of the derivatives can be defined by using not only deformations on the variation of the independent and dependent variable, but also on the ratio among them, with three parameters, in a rather intricate way, say: q for the deformed differential of the independent variable, q' for the deformed differential of the dependent variable, and q'' for the deformed ratio between them. A particular case with $q = q' = q''$ was shown in Ref. [46], and, more recently, in Ref. [47].

Finally, all the present scenario stands on the pair of q-logarithm/q-exponential functions, inverse of each other. The whole picture may be differently deformed by using different continuous, monotonous, and invertible pair of functions, in agreement with Equation (8).

Author Contributions: Conceptualization, E.P.B. and B.G.d.C.; methodology, E.P.B. and B.G.d.C.; software, E.P.B.; writing—original draft preparation, E.P.B.; writing—review and editing, E.P.B. and B.G.d.C. All authors have read and agreed to the published version of the manuscript.

Funding: This research received no external funding.

Data Availability Statement: Not applicable.

Acknowledgments: This work was partially supported by National Institute of Science and Technology for Complex Systems (INCT-SC). E.P.B. thanks C. Tsallis and Si Hyung Joo, and both authors thank I.S. Gomez for stimulating discussions.

Conflicts of Interest: The authors declare no conflict of interest.

Abbreviations

The following prefix abbreviations are used in this manuscript:

ile	inner logarithm exponential, see Equation (10a)
ole	outer logarithm exponential, see Equation (10b)
iel	inner exponential logarithm, see Equation (11a)
oel	outer exponential logarithm, see Equation (11b)

Appendix A. A Note on Notations—Explicit Expressions

The peculiar notation adopted in the present work is conceived for compactness, once the explicit forms of some equations may be large or cumbersome. The notation for the generalized numbers is inspired in the *Q-analog of n* [28], a generalized number represented within square brackets (7). The four classes of generalized numbers are grouped into two categories, one, the 'le' category, uses the generalized exponential (or its ordinary version) as argument of the ordinary logarithm (or its generalized version), and the other, the 'el' category, the other way around. We have used square brackets for the former, and curly brackets for the latter. Some ambiguity is unfortunately unavoidable, as square and curly brackets are also used with their usual meanings, and the reader must resolve it by the context. We refer to them as 'le' or 'el' concerning the order in which the logarithm/exponential functions appear. Despite of the unusualness, or even possibly strangeness, of the notation, we consider that it may help identify the classes more promptly than something like 'type 1', 'type 2', etc. Differently from the generalized numbers, we use the subscripts enclosed by their corresponding brackets, when dealing with generalized arithmetic operators, so the reader can easily identify the object being generalized, if it is a number or an operator. We have chosen prefix and postfix subscripts, to avoid using superscripts. These pair of subscripts may play a simplifying role if used appropriately, as illustrated by Equation (9). In the following we present explicit forms of some expressions, for the benefit of the interested reader. The notation $[\cdot]_+ \equiv \max\{0,\cdot\}$ is used here.

ile-number (Equation (10a))

$$[x]_q = \frac{1}{1-q} \ln\left(1+(1-q)x\right)_+ . \tag{A1}$$

ole-number (Equation (10b))

$${}_q[x] = \frac{e^{(1-q)x}-1}{1-q}. \tag{A2}$$

iel-number (Equation (12a))

$$\{x\}_q = \operatorname{sign}(x)\, \exp\left(\frac{|x|^{1-q}-1}{1-q}\right). \tag{A3}$$

oel-number (Equation (12b))

$${}_q\{x\} = \operatorname{sign}(x)\left(1+(1-q)\ln|x|\right)_+^{1/(1-q)} . \tag{A4}$$

ile-addition, ile-subtraction (Equations (16) and (18))

$$\begin{aligned} x \oplus_{[q]} y &= \frac{1}{1-q}\ln\left[1+(1-q)\left(\frac{e^{(1-q)x}-1}{1-q} \pm \frac{e^{(1-q)y}-1}{1-q}\right)\right]_+, & \text{(A5)} \\ &= \frac{1}{1-q}\ln\left[e^{(1-q)x} \pm e^{(1-q)y} \mp 1\right]_+. & \text{(A6)} \end{aligned}$$

ile-multiplication (Equation (19))

$$x \otimes_{[q]} y = \frac{1}{1-q}\ln\left[1+\frac{(e^{(1-q)x}-1)\,(e^{(1-q)y}-1)}{1-q}\right]_+ . \tag{A7}$$

ile-division (Equation (20))

$$x \oslash_{[q]} y = \frac{1}{1-q}\ln\left[1+(1-q)\,\frac{e^{(1-q)x}-1}{e^{(1-q)y}-1}\right]_+ . \tag{A8}$$

ile-power (Equation (22))

$$x \circledwedge_{[q]} y = \frac{1}{1-q}\ln\left[1+(1-q)\left(\frac{e^{(1-q)x}-1}{1-q}\right)^y\right]_+, \quad (x>0). \tag{A9}$$

ole-addition (see Equation (25))

ole-subtraction (see Equation (27))

ole-multiplication (Equation (28))

$$\begin{aligned} x\ {}_{[q]}\!\otimes y &= \frac{\exp\left[\frac{\ln[1+(1-q)x]_+ \ln[1+(1-q)y]_+}{1-q}\right]-1}{1-q}, & \text{(A10)} \\ &= \frac{[1+(1-q)x]_+^{\frac{1}{1-q}\ln[1+(1-q)y]_+}-1}{1-q}, & \text{(A11)} \\ &= \frac{[1+(1-q)y]_+^{\frac{1}{1-q}\ln[1+(1-q)x]_+}-1}{1-q}. & \text{(A12)} \end{aligned}$$

ole-division (Equation (29))

$$x\ _{[q]}\oslash\ y=\frac{\exp\left[(1-q)^{\frac{\ln[1+(1-q)x]_+}{\ln[1+(1-q)y]_+}}\right]-1}{1-q}. \tag{A13}$$

ole-power (Equation (31))

$$x\ _{[q]}\otimes\ y=\frac{\exp\left[(1-q)^{1-y}\ln^{y}[1+(1-q)x]_+\right]-1}{1-q},\quad (x>0). \tag{A14}$$

iel-addition, iel-subtraction (Equations (34) and (36))

$$\begin{aligned} x \oplus_{\{q\}} y &= \mathrm{sign}(x\pm y) \\ &\times \exp\left(\frac{\left|\mathrm{sign}(x)\left[1+(1-q)\ln|x|\right]_+^{\frac{1}{1-q}} \pm \mathrm{sign}(y)\left[1+(1-q)\ln|y|\right]_+^{\frac{1}{1-q}}\right|^{1-q}-1}{1-q}\right). \end{aligned} \tag{A15}$$

iel-multiplication (Equation (38))

$$x\otimes_{\{q\}} y=\mathrm{sign}(xy)\ \exp\left(\frac{\left|\left[1+(1-q)\ln|x|\right]_+\left[1+(1-q)\ln|y|\right]_+\right|-1}{1-q}\right). \tag{A16}$$

iel-division (Equation (39))

$$x\oslash_{\{q\}} y=\mathrm{sign}(x/y)\ \exp\left[(1-q)^{-1}\left(\left|\frac{\left[1+(1-q)\ln|x|\right]_+}{\left[1+(1-q)\ln|y|\right]_+}\right|-1\right)\right]. \tag{A17}$$

iel-power (Equation (41))

$$x\otimes_{\{q\}} y=\mathrm{sign}(x)\ \exp\left(\frac{\left|\left[1+(1-q)\ln|x|\right]_+^{y}\right|-1}{1-q}\right),\quad (x>0). \tag{A18}$$

oel-addition, oel-subtraction (Equations (44) and (45))

$$\begin{aligned} x\ _{\{q\}}\oplus\ y &= \mathrm{sign}(x\pm y) \\ &\times \left[1+(1-q)\ln\left|\mathrm{sign}(x)\ \exp\left(\frac{|x|^{1-q}-1}{1-q}\right)\pm\mathrm{sign}(y)\ \exp\left(\frac{|y|^{1-q}-1}{1-q}\right)\right|\right]_+^{\frac{1}{1-q}}. \end{aligned} \tag{A19}$$

oel-multiplication (see Equation (48))

oel-division (see Equation (49))

oel-power (Equation (52))

$$x \;{}_{\{q\}}\!\circledwedge\, y = (\mathrm{sign}(x))^{y} \left[y\,|x|^{1-q} - (y-1) \right]_{+}^{\frac{1}{1-q}}, \quad (x > 0). \tag{A20}$$

S_q^{ile} functional (Equation (83a))

$$S_q^{\text{ile}} = \sum_i^W \frac{\mathrm{e}^{-(1-q)p_i} - 1}{1-q} \ln\left[\frac{\mathrm{e}^{(1-q)p_i} - 1}{1-q} \right]. \tag{A21}$$

S_q^{ole} functional (Equation (83b))

$$\begin{aligned} S_q^{\text{ole}} = &- \sum_i^W \frac{\ln\left[1 + (1-q)p_i\right]}{1-q} \ln\left[\frac{\ln\left[1 + (1-q)p_i\right]}{1-q} \right] \\ &- \sum_i^W p_i \ln\left[1 + (1-q)p_i\right] \ln\left[\frac{\ln\left[1 + (1-q)p_i\right]}{1-q} \right] \end{aligned} \tag{A22}$$

S_q^{iel} functional (Equation (83c))

$$S_q^{\text{iel}} = - \sum_i^W \frac{p_i}{1-q} \ln\left[1 + (1-q)\ln p_i\right] - \sum_i^W p_i \ln p_i \ln\left[1 + (1-q)\right] \ln p_i\right] \tag{A23}$$

S_q^{oel} functional (Equation (83d), see also Equation (2))

References

1. Tsallis, C. Possible generalization of Boltzmann-Gibbs statistics. *J. Stat. Phys.* **1988**, *52*, 479–487. [CrossRef]
2. Tsallis, C. *Introduction to Nonextensive Statistical Mechanics*; Springer: New York, NY, USA, 2009.
3. Borges, E.P. A possible deformed algebra and calculus inspired in nonextensive thermostatistics. *Phys. A* **2004**, *340*, 95–101; Corrigenda. *Phys. A* **2021**, *581*, 126206. [CrossRef]
4. Tsallis, C. What are the numbers that experiments provide? *Quimica Nova* **1994**, *17*, 468–471.
5. Yamano, T. Some properties of q-logarithm and q-exponential functions in Tsallis statistics. *Phys. A* **2002**, *305*, 486–496. [CrossRef]
6. Naudts, J. Deformed exponentials and logarithms in generalized thermostatistics. *Phys. A* **2002**, *316*, 323–334. [CrossRef]
7. Naudts, J. *Generalised Thermostatistics*; Springer: London, UK, 2011.
8. Nivanen, L.; Le Méhauté A.; Wang, Q.A. Generalized algebra within a nonextensive statistics. *Rep. Math. Phys.* **2003**, *52*, 437–444. [CrossRef]
9. Nivanen, L.; Wang, Q.A.; Le Méhauté, A.; El Kaabouchi, A.; Basillais, P.; Donati, J.D.; Lacroix, A.; Paulet, J.; Perriau, S.; Sime Chuisse, S.; et al. Hierarchical structure of operations defined in nonextensive algebra. *Rep. Math. Phys.* **2009**, *63*, 279–288. [CrossRef]
10. Borges, E.P. On a q-generalization of circular and hyperbolic functions. *J. Phys. A Math. Gen.* **1998**, *31*, 5281–5288. [CrossRef]
11. Nobre, F.D.; Rego-Monteiro, M.A.; Tsallis, C. Nonlinear relativistic and quantum equations with a common type of solution. *Phys. Rev. Lett.* **2011**, *106*, 140601. [CrossRef]
12. Jauregui, M.; Tsallis, C. New representations of π and Dirac delta using the nonextensive-statistical-mechanics q-exponential function. *J. Math. Phys.* **2010**, *51*, 063304. [CrossRef]
13. Sicuro, G.; Tsallis, C. q-Generalized representation of the d-dimensional Dirac delta and q-Fourier transform. *Phys. Lett. A* **2017**, *381*, 2583–2587. [CrossRef]
14. Schwämmle, V.; Tsallis, C. Two-parameter generalization of the logarithm and exponential functions and Boltzmann-Gibbs-Shannon entropy. *J. Math. Phys.* **2007**, *48*, 113301. [CrossRef]
15. Cardoso, P.G.S.; Borges, E.P.; Lobão, T.C.P.; Pinho, S.T.R. Nondistributive algebraic structures derived from nonextensive statistical mechanics. *J. Math. Phys.* **2008**, *49*, 093509. [CrossRef]
16. Kaniadakis, G. Non-linear kinetics underlying generalized statistics. *Phys. A* **2001**, *296*, 405–425. [CrossRef]
17. Kaniadakis, G. Statistical mechanics in the context of special relativity. *Phys. Rev. E* **2002**, *66*, 056125. [CrossRef]
18. Kaniadakis, G.; Scarfone, A.M. A new one-parameter deformation of the exponential function. *Phys. A* **2002**, *305*, 69–75. [CrossRef]
19. Kaniadakis, G.; Lissia, M.; Scarfone, A.M. Deformed logarithms and entropies. *Phys. A* **2004**, *340*, 41–49. [CrossRef]

20. Gomez, I.S.; Borges, E.P. Algebraic structures and position-dependent mass Schrödinger equation from group entropy theory. *Lett. Math. Phys.* **2021**, *111*, 43. [CrossRef]
21. Tempesta, P. Group entropies, correlation laws, and zeta functions. *Phys. Rev. E* **2011**, *84*, 021121. [CrossRef]
22. Lutz, E.; Renzonni, F. Beyond Boltzmann-Gibbs statistical mechanics in optical lattices. *Nat. Phys.* **2013**, *9*, 615–619. [CrossRef]
23. Combe, G.; Richefeu, V.; Stasiak, M.; Atman, A.P.F. Experimental validation of a nonextensive scaling law in confined granular media. *Phys. Rev. Lett.* **2015**, *115*, 238301. [CrossRef] [PubMed]
24. Wong, C.-Y.; Wilk, G. Tsallis fits to p_T spectra and multiple hart scattering in p p collisions at the LHC. *Phys. Rev. D* **2013**, *87*, 114007. [CrossRef]
25. Yalcin, G.C.; and Beck, C. Generalized statistical mechanics of cosmic rays: Application to positron-electron spectral indices. *Sci. Rep.* **2018**, *8*, 1764. [CrossRef] [PubMed]
26. Available online: http://tsallis.cat.cbpf.br/TEMUCO.pdf (accessed on 25 January 2022).
27. Lobão, T.C.P.; Cardoso, P.C.S.; Pinho, S.T.R.; Borges, E.P. Some properties of deformed q-numbers. *Braz. J. Phys.* **2009**, *39*, 402–407. [CrossRef]
28. Kac, V.; Cheung, P. *Quantum Calculus*; Springer: New York, NY, USA, 2002.
29. Tsallis, C. Nonextensive physics: A possible connection between generalized statistical mechanics and quantum groups. *Phys. Lett. A* **1994**, *195*, 329–334. [CrossRef]
30. Kalogeropoulos, N. Distributivity and deformation of the reals from Tsallis entropy. *Phys. A* **2012**, *391*, 1120–1127. [CrossRef]
31. Tsallis, C.; Queirós, S.M.D. Nonextensive statistical mechanics and central limit theorems I—Convolution of independent random variables and q-product. In *Complexity, Metastability and Nonextensivity*; Abe, S., Herrmann, H.J., Quarati, P., Rapisarda, A., Tsallis, C. , Eds.; American Institute of Physics: College Park, MA, USA, 2007; Volume 965, p. 8.
32. Jizba, P.; Korbel, J. When Shannon and Khinchin meet Shore and Johnson: Equivalence of information theory and statistical inference axiomatics. *Phys. Rev. E* **2020**, *101*, 042126. [CrossRef]
33. da Costa, B.G.C.; Gomez, I.S.; Borges, E.P. Deformed Fokker-Planck equation: Inhomogeneous medium with a position-dependent mass. *Phys. Rev. E* **2020**, *102*, 062105. [CrossRef]
34. Pennini, F.; Plastino, A.; Ferri, G.L. Fisher information, Borges operators, and q-calculus. *Phys. A* **2008**, *387*, 5778–5785. [CrossRef]
35. Abe, S. A note on the q-deformation-theoretic aspect of the generalized entropies in nonextensive physics. *Phys. Lett. A* **1997**, *224*, 326–330. [CrossRef]
36. Jackson, J.H. On q-definite integrals. *Quart. J. Pure Appl. Math.* **1910**, *41*, 193–203.
37. Borges, E.P.; Roditi, I. A family of nonextensive entropies. *Phys. Lett. A* **1998**, *246*, 399–402. [CrossRef]
38. Czachor, M.; Naudts, J. Thermostatistics based on Kolmogorov-Nagumo averages: unifying framework for extensive and nonextensive generalizations. *Phys. Lett. A* **2002**, *298*, 369–374. [CrossRef]
39. da Costa, B.G.; Borges, E.P. A position-dependent mass harmonic oscillator and deformed space. *J. Math. Phys.* **2018**, *59*, 042101. [CrossRef]
40. Borges, E.P.; Kodama, T.; Tsallis, C. Along the Lines of Nonadditive Entropies: q-prime numbers and q-Zeta Functions. *Entropy* **2022**, *24*, 60. [CrossRef] [PubMed]
41. Lima, J.A.S.; Silva R.; Plastino, A.R. Nonextensive Thermostatistics and the H Theorem. *Phys. Rev. Lett.* **2001**, *86*, 2938–2941. [CrossRef] [PubMed]
42. Lavagno, A. Relativistic nonextensive thermodynamics. *Phys. Lett. A* **2002**, *301*, 13–18. [CrossRef]
43. Umarov, S.; Tsallis, C.; Steinberg, S. On a q-central limit theorem consistent with nonextensive statistical mechanics. *Milan J. Math.* **2008**, *76*, 307–328. [CrossRef]
44. Umarov, S.; Tsallis, C.; Gell-Mann, M.; Steinberg, S. Generalization of symmetric α-stable Lévy distributions for $q > 1$. *J. Math. Phys.* **2010**, *51*, 033502. [CrossRef]
45. Weberszpil, J.; Lazo, M.J.; Helayël-Neto, J.A. On a connection between a class of q-deformed algebras and the Hausdorff derivative in a medium with fractal metric. *Phys. A* **2015**, *436*, 399–404. [CrossRef]
46. Kalogeropoulos, N. Algebra and calculus for Tsallis thermo-statistics. *Phys. A* **2005**, *356*, 408–418. [CrossRef]
47. Czachor, M. Unifying aspects of generalized calculus. *Entropy* **2020**, *22*, 1180. [CrossRef] [PubMed]

axioms

MDPI

Article

Hot Spots in the Weak Detonation Problem and Special Relativity

Satyanad Kichenassamy

Laboratoire de Mathématiques de Reims (CNRS, UMR9008), Université de Reims Champagne-Ardenne, Moulin de la Housse, B.P. 1039, CEDEX 2, F-51687 Reims, France; satyanad.kichenassamy@univ-reims.fr

Abstract: *Problem statement:* The initiation of a detonation in an explosive gaseous mixture in the high activation energy regime, in three space dimensions, typically leads to the formation of a singularity at one point, the "hot spot". It would be suitable to have a description of the physical quantities in a full neighborhood of the hot spot. *Results of this paper:* (1) To achieve this, it is necessary to replace the blow-up time, or time when the hot spot first occurs, by the blow-up surface in four dimensions, which is the set of all hot spots for a class of observers related to one another by a Lorentz transformation. (2) A local general solution of the nonlinear system of PDE modeling fluid flow and chemistry, with a given blow-up surface, is obtained by the method of Fuchsian reduction. *Advantages of this solution:* (i) Earlier approximate solutions are contained in it, but the domain of validity of the present solution is larger; (ii) it provides a signature for this type of ignition mechanism; (iii) quantities that remain bounded at the hot spot may be determined, so that, in principle, this model may be tested against measurements; (iv) solutions with any number of hot spots may be constructed. The impact on numerical computation is also discussed.

Keywords: weak detonation; high activation regime; nonlinear PDEs; Fuchsian reduction analysis; Lorentz transformation; blow-up; hot spot; chemically reactive flows

PACS: 11.30.Cp; 47.40.Rs; 82.33.Vx

MSC: 80A25; 35Q07; 35B44

Citation: Kichenassamy, S. Hot Spots in the Weak Detonation Problem and Special Relativity. *Axioms* **2021**, *10*, 311. https://doi.org/10.3390/axioms10040311

Academic Editor: Hans J. Haubold

Received: 4 October 2021
Accepted: 15 November 2021
Published: 19 November 2021

Publisher's Note: MDPI stays neutral with regard to jurisdictional claims in published maps and institutional affiliations.

Copyright: © 2021 by the author. Licensee MDPI, Basel, Switzerland. This article is an open access article distributed under the terms and conditions of the Creative Commons Attribution (CC BY) license (https://creativecommons.org/licenses/by/4.0/).

1. Introduction

Observations of ignitions in explosive gaseous systems indicate that the process is initiated in localized regions called reaction centers, or "hot spots". This phenomenon is ubiquitous, from the combustion engine to stellar explosions [1–4], including engines using carbon-free fuels such as ammonia or hydrogen [5–7]. A given observer typically sees one of these hot spots to first form at a definite point in space and time. We have pointed out earlier [8] (Section 10.4) that the set of hotspots in the weak detonation problem forms a blow-up pattern in the sense of Fuchsian reduction [9]. Here, we give a more detailed description of the solutions of the relevant system of nonlinear PDEs, indicating quantities that could be measurable. We also show that the notion of blow-up time, namely, the time when a singularity first occurs in a given inertial frame, is physically meaningless. As a consequence, the hot spot in the laboratory frame loses physical significance and must be replaced by the set of all hot spots seen by different observers. They form the blow-up surface in the sense of reduction theory [8,10], namely, the set of points at which a given solution presents a singularity. As we have shown in related contexts, reduction theory not only gives a precise description of singularity formation, it also enables its control by boundary action [11] and accounts for the possible concentration of energy or other quantities at the different blow-up points [12].

We first review the modeling assumptions (Section 2) and earlier results on the hot spot problem showing how our approach improves upon them (Section 3). Then, we perform a reduction analysis of the model, leading to a local representation of the general solution (Sections 4 and 5). The effect of Lorentz transformations on blow-up patterns is then described in a general set-up, common to all applications (Section 6). Section 7 is devoted to a discussion of the results and outlines perspectives for further work. Section 8 summarizes the conclusions. The derivation of the equations from first principles, and their non-dimensionalization, are given in Appendices A and B respectively.

The main new results are: (i) The hot spot first recorded by a given observer is not the cause of ignition. (ii) The present solution of the relevant equations has a wider domain of validity than earlier ones, that are recovered as special limiting cases. (iii) Our approach provides a set of measurable quantities that may be viewed as a signature for this type of ignition mechanism. (iv) Special relativity is relevant for purely geometric and kinematical reasons, even when the relativistic effects on the chemistry are negligible.

2. The Weak Detonation Problem: Modeling Assumptions

Let us first review the modeling assumptions that seem to represent the current consensus [13]. For background information, see also references [1,4,14,17–30,32,34].

Since ignition occurs on a very small scale, it is reasonable to assume that dissipative and convective effects may be neglected. In that case, near each hot spot, the behavior of the gas may be assumed to be close to a spatially homogeneous explosion. This leads to the following assumptions:

(A1) One considers small perturbations of a uniform state.
(A2) The chemistry is modeled by a one-step, strongly exothermic irreversible reaction with the Arrhenius reaction rate.
(A3) The activation energy is large.
(A4) The reaction progresses so fast that the diffusion and convection effects are negligible.
(A5) Reactants and products are perfect fluids with the same specific heats; they may be considered as forming a single perfect fluid.

Assumption (A1) is usually expressed by saying that the detonation is "quasisteady". Since fluid elements have no time to drift appreciably away from the hot spot, it is usually called a "weak detonation". This detonation is similar to the "weak detonation" represented by a nearly vertical line connecting two points on two Hugoniot curves in the p–v diagram. See [21] (p. 19) for details on terminology. Because of (A2), (A4) and (A5), one focuses on the reactive Euler equations with the Arrhenius reaction rate, recalled in Appendix A. Assumptions (A2)–(A4) suggest a choice of scales, leading to a non-dimensional form of the equations, namely, the system (A6) in Appendix B. It is obtained in two steps. One first introduces non-dimensional variables t^* and $\mathbf{x}^* = (x_1^*, x_2^*, x_3^*)$ and dependent variables $\mathbf{u}^*$, T^*, p^*, ρ^* and y^*, that represent velocity, temperature, pressure, density and reactant mass fraction, respectively. One also introduces the dimensionless inverse activation energy θ, the ratio of specific heats γ and the non-dimensional heat release parameter β. Second, one expands $\mathbf{u}^*$, T^*, p^*, ρ^* and y^* in powers of θ^{-1} in the limit when θ is large.

$$\mathbf{u}^* = \frac{\mathbf{u}_1}{\theta} + \mathcal{O}(\frac{1}{\theta^2}) \tag{1a}$$

$$T^* = 1 + \frac{T_1}{\theta} + \mathcal{O}(\frac{1}{\theta^2}) \tag{1b}$$

$$p^* = 1 + \gamma\frac{p_1}{\theta} + \mathcal{O}(\frac{1}{\theta^2}) \tag{1c}$$

$$\rho^* = 1 + \frac{\rho_1}{\theta} + \mathcal{O}(\frac{1}{\theta^2}) \tag{1d}$$

$$y^* = 1 + \frac{y_1}{\theta} + \mathcal{O}(\frac{1}{\theta^2}). \tag{1e}$$

The factor γ was introduced in the expansion of p^* to make formulae simpler. This expansion reflects assumptions (A1) and (A3); θ is large and the variables describe a nearly uniform state. Neglecting the terms in $\frac{1}{\theta^2}$, it follows that the specific internal energy ε and the total specific energy e are comparable, because the squared velocity is of order $\frac{1}{\theta^2}$ and

$$e = \varepsilon_0 + \frac{1}{\theta}(c_v T_1 + q y_1),$$

where ε_0 is the specific internal energy in the reference state.

Inserting (1) into (A6) and keeping contributions of order $\frac{1}{\theta}$, we obtain that the first correction to the uniform state is governed by the following nonlinear system, in which the dependent variables are $(\rho_1, \mathbf{u}_1, p_1, T_1, y_1)$, that represent the departure of the non-dimensionalized density, fluid velocity, pressure, temperature and reactant mass fraction, respectively, from their values in the reference, constant state.

$$\gamma p_1 = \rho_1 + T_1 \tag{2a}$$

$$\partial_{t^*}\rho_1 + \operatorname{div}^* \mathbf{u}_1 = 0 \tag{2b}$$

$$\partial_{t^*}\mathbf{u}_1 + \nabla^* p_1 = 0 \tag{2c}$$

$$\partial_{t^*} y_1 = -\frac{1}{\beta}\exp T_1 \tag{2d}$$

$$\partial_{t^*} T_1 = (\gamma - 1)\partial_{t^*}\rho_1 + \gamma \exp T_1. \tag{2e}$$

More precisely, (A6a)–(A6c) and (A6e) directly imply (2a)–(2d); on the other hand, Equation (A7) yields, at leading order, the simple relation

$$\partial_{t^*}[T_1 - (\gamma - 1)p_1 + \beta y_1] = 0. \tag{3}$$

Using (2d), Equation (3) is equivalent to (2e).

It suffices to determine T_1, $\mathbf{u}_1$ and ρ_1 from (2b)–(2d); one may then obtain p_1 from (2a) and y_1 from (3).

Limits on the validity of this expansion may be estimated as follows:

1. Reactant depletion ($y = 0$) corresponds to $y_1 \approx -\theta$.
2. The term T_1/θ becomes comparable to unity when T_1 is of the order of θ.
3. The replacement of (A7) by (3) requires that $\rho_1\theta^{-1}\partial_{t^*}[T^* + \beta y^*]$ is small.
4. The replacement of D/Dt^* by ∂_{t^*} requires dropping $\theta^{-1}u_1\partial_{x^*}X \ll \partial_{t^*}X$ for each of the expressions X to which this material derivative is applied.

The solutions of system (2) blow up in finite time and the final phase of rapid increase in temperature, just before singularity formation, is called *thermal runaway*. The hot spot for a given observer is thus very close to the point in spacetime where the first singularity of the system appears in his/her frame. However, it is merely part of a weak detonation locus, or blow-up set, described in this paper by the equation $t = \psi(x, y, z)$, where ψ depends on space variables. This set represents "a locus of nonuniform ignition times, resulting from the nonuniform initial state, in which each fluid particle released its chemical energy at a different time" [28] (p. 1243). The hot spot in a given system corresponds to a spacetime point $(t_0, \mathbf{x}_0)$ such that ψ becomes minimum at $\mathbf{x}_0$. We shall see that this point of spacetime is not a Lorentz invariant.

Two aspects of this problem are somewhat unusual and make many standard tools in the study of partial differential equations inappropriate. First, this initial phase of the process does not propagate as a wave, even though the equations are of hyperbolic type. Indeed, the hot spots observed by different observers are not causally related to one another. Ignition leads to the formation of a blow-up pattern in the sense of [9]. The second difficulty is that the temperature does not become infinite; the model ceases to be valid as soon as the variables ρ_1, etc., become of the order of θ, or when the reactant is depleted ($y = 0$). Therefore, information on the limit as T_1 goes to infinity is indeed irrelevant; we need expressions that make sense for a large, but finite T_1. Before we obtain such expressions, let us review earlier approaches.

3. Earlier Results

Numerical work is made difficult by the blow-up singularity [15,22]. Therefore, perturbative approaches are preferred. Three perturbative methods of solution have been applied to this problem. The first method [16] consists, when there is only one space variable, called x^*, in introducing a shifted dimensionless time variable $\tilde{t} = t_d^* - t^*$, with t_d^* constant and a self-similar variable $s = x^*/\tilde{t}$, such that the hot spot develops at time $\tilde{t} = 0$, when the space coordinate vanishes. Thus, the fact that ignition at different points occurs at different times is neglected in the vicinity of the first hot spot. This leads to expansions involving $\tilde{t}^n(\ln\tilde{t})^m$, where n and m are integers. The expansions break down when $s = \mathcal{O}(\tilde{t}^{-1/2})$; this self-similar scaling "does not quite span the entire hot spot" [16] (p. 439) and variables of the form $x^*/\tilde{t}^m$, with appropriate m, must be introduced. The second method [14] considers the limit in which $1 \gg (\gamma-1)/\gamma \gg \theta^{-1}$, in one space dimension; in this limit, (2e) decouples from the other equations. The hot spot is again investigated using a self-similar variable, leading to a further restriction on the validity of the expansion. The third approach [28] is to insert a small parameter μ in front of the space derivatives, making the spatial derivatives less important than the time derivatives and using μ as the expansion parameter. The result is a formal expansion involving $\tau = \psi(x^*,\mu) - t^*$, where $\psi(x^*,\mu)$, representing the ignition time at the location x^*, is itself expanded in powers of μ. The method may be extended to three-dimensional situations. In all cases, the spatial gradient of ψ must have length less than unity. In addition, one requires $\theta^{-1} \ll \mu = \theta^{1/3} \ll 1$, see [28] (p. 1258). The solution is not uniform as $\tau \to 0$.

The upshot of the above results is the following: The initiation of a weak detonation appears to be well represented by a solution of system (2), with a logarithmic singularity on a set of the form $t^* = \psi(\mathbf{x}^*)$. Ignition appears to start first at a spacetime point $(t_0^*, \mathbf{x}_0^*)$, such that $t_0^* = \psi(\mathbf{x}_0^*)$ and $\psi(\mathbf{x}^*)$ has a minimum for $\mathbf{x}^* = \mathbf{x}_0^*$, but only if one restricts one's attention to a neighborhood of $\mathbf{x}_0^*$ that shrinks as t^* tends to the blow-up time $t_0^* = \psi(\mathbf{x}_0^*)$. It would be desirable to obtain a solution valid in a full neighborhood of the hot spot. This is the result of the present paper.

4. Strategy and Results

We obtain expansions describing singular solutions of the basic system (2), uniformly in the vicinity of the hot spot, yielding a domain of validity larger than that obtained via self-similar variables. We construct (Theorem 2) a convergent expansion for three-dimensional solutions that contains powers and logarithms of $\tau := \psi(\mathbf{x}^*) - t^*$, multiplied by functions of the space variables, assuming $|\nabla\psi|$ is small, which is appropriate near the minimum of ψ. If the hot spot appears for $\tau = 0$, at $\mathbf{x}^* = 0$, the expansion is valid in a set defined by inequalities of the form $\tau < 0$ and $|\mathbf{x}^*| < \delta$ (we write $|\mathbf{x}|$ for the usual length of a 3-vector $\mathbf{x}$). Therefore, it is valid in a full neighborhood of the origin. It contains five freely specifiable functions of three variables that are called singularity data, including ψ. This number is the greatest possible, since there are five unknowns (density, temperature and the three components of velocity) that determine pressure and reactant mass fraction. The arbitrary functions may be interpreted in terms of the asymptotics of T_1, $\mathbf{u}_1$ and ρ_1, because, even though they may become very large at the hot spot, there are combinations of these variables that have well-defined limits as $t^* \to \psi(\mathbf{x}^*)$ and these limits have simple expressions in terms of the arbitrary functions—more precisely, the five functions (two scalar functions ψ, σ_0 and the three components of a 3-vector $\mathbf{w}_0$). They are related to the asymptotics of the non-dimensionalized variables via

$$\rho_1/T_1 \to \frac{|\nabla\psi|^2}{\gamma - |\nabla\psi|^2},$$

$$\mathbf{u}_1 - \frac{\nabla\psi}{\gamma - |\nabla\psi|^2}T_1 \to \mathbf{w}_0 - \frac{\nabla\psi}{\gamma - |\nabla\psi|^2}\ln\frac{1 - |\nabla\psi|^2}{\gamma - |\nabla\psi|^2},$$

and

$$\rho_1 - \mathbf{u}_1 \cdot \nabla\psi \to \sigma_0 - \mathbf{w}_0 \cdot \nabla\psi.$$

The method leading to these results consists in integrating the system starting from the singularity. Thus, the solution is determined from the arbitrary functions in its singular expansion, i.e., by its singularity data on the blow-up surface, rather than by Cauchy data on some hypersurface away from the singularity (for the relation between Cauchy data and singularity data in a typical case, see [31]). While, in the Cauchy problem, the series solution is determined by its first few terms and contains only integral powers of the time variable, in singular problems such as this, the expansion involves logarithms and the arbitrary coefficients are not the first few coefficients of the series. More complicated functions, such as fractional powers, are required in some cases, but not here. There are general rules to perform the reduction and to predict the form of the solution [8]. In fact, a major advantage of the reduction technique is that it enables one to predict the form of the expansion to any order, without having to compute it, since this task is often unwieldy.

Simple applications of these results include the following.

(1) *Recovering earlier results:* Hot spots correspond to the minima of the function ψ, and the large activation energy regime is appropriate in a small neighborhood of these minima. This makes it easy to compare our solution to the three asymptotic approaches in Section 3. The results of the first method may be recovered by expanding our solution after introducing self-similar variables. Expanding our solution in powers of γ leads to the second method. Inserting the parameter μ, both in the expansion and in ψ, leads to the third method. Therefore, each of these is recovered as a particular limit of our solution. As already noted, the introduction of self-similar variables leads to a restriction in the domain of validity of the solution.

(2) *Better approximation for large γ:* The second application of our expansions is the determination of the limits of validity of large activation energy asymptotics. System (2) was obtained from the reactive Euler equations by assuming the activation energy parameter θ to be large. This implies that one replaces the material derivative $\partial_t + \mathbf{u} \cdot \nabla$ by ∂_t. Indeed, $\mathbf{u} \approx \mathbf{u}_1/\theta$, so that the spatial derivative term is of higher order in an expansion in powers of θ. Our results show that the neglected terms are smaller than the ones that have been kept if

$$\text{the gradient of } \psi \text{ is small compared with } \gamma,$$

where the detonation locus is given by $t^* = \psi(\mathbf{x}^*)$ in non-dimensional variables and γ is the ratio of specific heats. This suggests that this approximation could be better in cases where γ is large.

(3) *Signature of detonation:* The expansions obtained in this paper also enable one to compute quantities that remain finite at blow-up. They can be used as a characteristic signature of this ignition mechanism. This may also be used to monitor the quality of numerical schemes. More generally, the expansions may be used as a substitute for the numerical solutions precisely where the solution is large and mesh refinement may become unwieldy. More applications and perspectives are described in Section 7.

5. Reduction Analysis

We construct solutions of system (2) that become infinite when t reaches a value $\psi(\mathbf{x})$ that depends on space. This reflects the expectation that ignition does not occur simultaneously everywhere. We first introduce new variables, identify the leading form of the expansion of the solution and prove that solutions with this behavior exist. Furthermore, we show that they are uniquely determined by some of the lower-order terms in the expansion and that these terms have a simple interpretation.

5.1. Introduction of the Detonation Locus

Let us first introduce new space and time variables:

$$\tau = \psi(\mathbf{x}^*) - t^*; \quad \xi = (\xi_1, \xi_2, \xi_3) = (x_1^*, x_2^*, x_3^*).$$

We assume that $|\nabla\psi| < 1$. The derivation operators transform as follows (we let $\nabla_\xi = (\partial/\partial\xi_1, \partial/\partial\xi_2, \partial/\partial\xi_3)$):

$$\begin{aligned} \nabla^* &= \nabla_\xi + (\nabla^*\psi)\partial_\tau \\ \partial_t^* &= -\partial_\tau \\ \operatorname{div}^* \mathbf{u}_1 &= \operatorname{div}_\xi \mathbf{u}_1 + (\nabla^*\psi)\cdot\partial_\tau \mathbf{u}_1. \end{aligned}$$

In particular, $\nabla^*\psi = \nabla_\xi\psi$. For any expression $F(\xi)$ that does not depend on τ, we write ∇F for $\nabla_\xi F$. The set of spacetime points where $\tau = 0$ represents the locus where the temperature becomes infinite. The actual detonation front is the locus where $T_1 = \eta\theta$, where η is a constant of order unity.

System (2), expressed in the new variables, can be written as

$$\gamma p_1 = \rho_1 + T_1 \tag{4a}$$
$$\partial_\tau[\rho_1 - \mathbf{u}_1\cdot\nabla\psi] = \operatorname{div}_\xi \mathbf{u}_1 \tag{4b}$$
$$\partial_\tau[\mathbf{u}_1 - p_1\nabla\psi] = \nabla_\xi p_1 \tag{4c}$$
$$\partial_\tau T_1 = (\gamma-1)\partial_\tau p_1 - \exp T_1 \tag{4d}$$
$$\partial_\tau y_1 = \frac{1}{\beta}\exp T_1. \tag{4e}$$

We now transform this system into an equivalent form that is easier to analyze.

Theorem 1. *System (4) is equivalent to the system*

$$\gamma p_1 = \rho_1 + T_1 \tag{5a}$$
$$\partial_\tau \rho_1 = \frac{1}{B}[A - (1-B)\exp T_1] \tag{5b}$$
$$\partial_\tau \mathbf{u}_1 = (A - \exp T_1)\frac{\nabla\psi}{B} + \nabla_\xi p_1 \tag{5c}$$
$$\partial_\tau T_1 = (\gamma-1)\frac{A}{B} - \alpha\exp T_1 \tag{5d}$$
$$\partial_\tau y_1 = \frac{1}{\beta}\exp T_1, \tag{5e}$$

where

$$\alpha = \frac{\gamma - |\nabla\psi|^2}{1-|\nabla\psi|^2} \tag{6a}$$
$$A = \operatorname{div}_\xi \mathbf{u}_1 + (\nabla\psi)\cdot\nabla_\xi p_1 \tag{6b}$$
$$B = 1 - |\nabla\psi|^2. \tag{6c}$$

Proof. Equations (5a) and (5e) are identical with (4a) and (4e). Using (6), Equation (4c) yields (5c). The other equations are transformed as follows. Replace (4b) by the linear combination (4b)+$(\nabla\psi)\cdot$(4c); this yields

$$\partial_\tau\left[\rho_1 - p_1|\nabla\psi|^2\right] = A,$$

where A is defined in (6). Using (4a), this relation is equivalent to

$$\partial_\tau\left[(\gamma - |\nabla\psi|^2)\rho_1 - |\nabla\psi|^2 T_1\right] = \gamma A. \tag{7}$$

Using (4a), Equation (4d) may be written as

$$\partial_\tau[T_1 - (\gamma-1)\rho_1] = -\gamma\exp T_1.$$

Replace Equations (7) and (4e) by their linear combinations with coefficients $(\gamma - 1, \gamma - |\nabla\psi|^2)$ and $(1, |\nabla\psi|^2)$, respectively; this yields (5b) and (5d). Retracing one's steps, one may conversely derive (4) from (5). This completes the proof. □

Remark 1. *System (6) does not give the derivative of p_1 directly. However, this derivative may be obtained by adding (5b) and (5c) and using (5a). We obtain*

$$\partial_\tau p_1 = (A - \exp T_1)/B. \tag{8}$$

5.2. Removing the Leading Singularity

In the right-hand side of Equation (5d) for T_1, the most important term should be the exponential, since the process is driven by the reaction. Let us use this observation to obtain some heuristic information on the appropriate behavior of the variables near the singularity, before we set out to construct solutions with this behavior. If the exponential is dominant in the right-hand side of (5d), then $\partial_\tau T_1 \approx -\alpha \exp T_1$ and we expect $T_1 \approx \ln(1/\alpha\tau)$. In that case, we have $\partial_\tau \rho_1 \approx (B-1)B^{-1}\exp T_1 \approx (B-1)/(\alpha B\tau)$, hence, we expect $\rho_1 \approx k\ln(1/\tau)$, with

$$k = \frac{1-B}{\alpha B} = \frac{|\nabla\psi|^2}{\gamma - |\nabla\psi|^2} = \frac{|\nabla\psi|^2}{\alpha B} = \frac{\gamma}{\alpha B} - 1. \tag{9}$$

Finally, $\partial_\tau \mathbf{u}_1 \approx -B^{-1}e^{T_1}\nabla\psi$; hence, the expected behavior $\mathbf{u}_1 \approx (\alpha B)^{-1}\nabla\psi\ln(1/\tau)$. Now, the term $\nabla_\xi p_1 = \gamma^{-1}(\rho_1 + T_1)$ contains terms involving $\ln\tau$; hence, $\mathbf{u}_1$ involves terms in $\tau\ln\tau$. These considerations suggest the introduction of renormalized variables $(\Phi, R, \mathbf{U})$ by letting

$$T_1 = \ln\frac{1}{\alpha\tau} + \varphi_1(\xi)\tau\ln\tau + \tau\Phi(\xi,\tau) \tag{10a}$$

$$\rho_1 = \frac{|\nabla\psi|^2}{\alpha B}\ln\frac{1}{\tau} + \sigma_0(\xi) + \sigma_1(\xi)\tau\ln\tau + \tau R(\xi,\tau) \tag{10b}$$

$$\mathbf{u}_1 = \frac{\nabla\psi}{\alpha B}\ln\frac{1}{\tau} + \mathbf{w}_0(\xi) + \mathbf{w}_1(\xi)\tau\ln\tau + \tau\mathbf{U}(\xi,\tau). \tag{10c}$$

These new dependent variables are renormalized in the sense that the "infinite part" of the solution was subtracted off *and* the remainder has been divided by an appropriate power of τ. This analysis is an application of the general procedure described in [8] (§ 1.4). Using (2a), we now have

$$p_1 = (\rho_1 + T_1)/\gamma = \frac{1}{\alpha B}\ln\frac{1}{\tau} + (\sigma_0 - \ln\alpha)/\gamma + (\sigma_1 + \varphi_1)/\gamma + \tau(\Phi + R)/\gamma. \tag{11}$$

The main result of this section states that, once ψ, σ_0 and $\mathbf{w}_0$ are given, with $|\nabla\psi| < 1$, the solution is completely and uniquely determined; σ_1, φ_1 and $\mathbf{w}_1$ may be found in closed form and the renormalized variables have expansions that may be computed inductively to any order, while the corresponding series converges if the data are analytic, or represent a very smooth function if the data are themselves very smooth.

Theorem 2. *System (5) admits, near the origin, a family of solutions given by power series in τ, $\tau\ln\tau$ and $\tau(\ln\tau)^2$, with coefficients depending on ξ, provided that $|\nabla\psi| \leq 1$ and*

$$\sigma_1 = -\frac{\tilde{A}}{\alpha B^2}[\gamma - 1 + (\gamma+1)B] \tag{12a}$$

$$\varphi_1 = -\frac{\gamma-1}{2B}\tilde{A} \tag{12b}$$

$$\mathbf{w}_1 = -\frac{\tilde{A}}{2\alpha B^2}\nabla\psi[\gamma - 1 + 2B] - \frac{1}{\gamma}\nabla k, \tag{12c}$$

where k is given by (9) *and*

$$\tilde{A} = \frac{(\gamma - |\nabla\psi|^2)\Delta\psi + 4\sum_{i,j}\psi_i\psi_j\psi_{ij}}{(\gamma - |\nabla\psi|^2)^2}. \tag{13}$$

The functions σ_0 *and* $\mathbf{w}_0$ *and the function* ψ *may be chosen arbitrarily; they determine all the other terms in the expansion.*

Proof. For any quantity X, Taylor's expansion gives an analytic function G_2 such that $e^X = 1 + X + X^2 G_1(X)$. Therefore,

$$\begin{aligned} e^{T_1} &= \frac{1}{\alpha\tau}\exp[\varphi_1\tau\ln\tau + \tau\Phi] \\ &= \frac{1}{\alpha\tau} + \frac{\varphi_1}{\alpha}\ln\tau + \frac{\Phi + G_2}{\alpha}, \end{aligned}$$

where

$$G_2 = G_2(\xi, \tau, \tau\ln\tau, \tau(\ln\tau)^2, \Phi) = \frac{\tau}{\alpha}(\varphi_1\ln\tau + \Phi)^2 G_1(\varphi_1\tau\ln\tau + \tau\Phi). \tag{14}$$

Additionally, (6), (10c) and (24) yield

$$A = \tilde{A}\ln\frac{1}{\tau} + A_0 + A_1\tau\ln\tau + \tau\mathscr{A}, \tag{15}$$

where

$$\begin{aligned} \tilde{A} &= \mathrm{div}_\xi\left(\frac{\nabla\psi}{\gamma - |\nabla\psi|^2}\right) + \nabla\psi\cdot\nabla_\xi\frac{1}{\gamma - |\nabla\psi|^2} \\ &= \frac{\Delta\psi}{\gamma - |\nabla\psi|^2} + 2\sum_j \psi_j\partial_j\frac{1}{\gamma - |\nabla\psi|^2} \\ &= (\gamma - |\nabla\psi|^2)^{-2}\left[(\gamma - |\nabla\psi|^2)\Delta\psi + 4\sum_{j,k}\psi_i\psi_k\psi_{jk}\right], \end{aligned}$$

and

$$\begin{aligned} A_0 &= \mathrm{div}_\xi\,\mathbf{w}_0 + \nabla\psi\cdot\nabla_\psi(\sigma_0 - \ln\alpha)/\gamma && (16a) \\ A_1 &= \mathrm{div}_\xi\,\mathbf{w}_1 + \nabla\psi\cdot\nabla_\xi(\sigma_1 + \varphi_1)/\gamma && (16b) \\ \mathscr{A} &= \mathrm{div}_\xi\,\mathbf{U} + \nabla\psi\cdot\nabla_\xi(\Phi + R)/\gamma. && (16c) \end{aligned}$$

Note that

$$\gamma\tilde{A} = (k+1)\Delta\psi + 2\nabla\psi\cdot\nabla k.$$

We may now determine φ_1, σ_1 and $\mathbf{w}_1$, and obtain the reduced equations for R, Φ and $\mathbf{U}$. We let

$$\mathscr{D} = \tau\partial_\tau.$$

We substitute (10) into each of the Equations (5b)–(5d). The following calculations have a common pattern. In each case, the choice of the leading terms in (10) ensures that the leading order terms in the resulting equation vanish. The vanishing of the next terms (in $\tau\ln\tau$) determines φ_1, σ_1 and $\mathbf{u}_1$. Finally, dividing by τ, one obtains a singular system for the renormalized unknowns, to which solutions are given by an existence result for singular initial-value problems. We now carry out this program.

From Equation (5b), we obtain

$$\mathscr{D}\rho_1 = A\tau/B - |\nabla\psi|^2\frac{\tau}{B}\exp(T_1),$$

hence, using (10b) and (9),

$$-k+\sigma_1\tau(1+\ln\tau)+\tau(\mathscr{D}+1)R=\frac{\tau}{B}A-k[1+\varphi_1\tau\ln\tau+\tau(\Phi+G_2)]. \tag{17}$$

The first term on the right cancels with one of the terms on the left. Equating the coefficients of $\tau\ln\tau$, we obtain

$$\begin{aligned}\sigma_1 &= -\frac{\tilde{A}}{B}-k\varphi_1\\ &= \frac{\tilde{A}}{2B}[k(\gamma-1)-2]\\ &= \frac{\tilde{A}}{2B}\frac{(\gamma-1)|\nabla\psi|^2-2(\gamma-|\nabla\psi|^2)}{\gamma-|\nabla\psi|^2}\\ &= \frac{\tilde{A}}{2B}\frac{(\gamma+1)|\nabla\psi|^2-2\gamma}{\alpha B}=-\frac{\tilde{A}}{\alpha B^2}[\gamma-1+(\gamma+1)B].\end{aligned}$$

This proves (12a).
Dividing (17) by τ and using (9), the reduced equation for R is obtained:

$$(\mathscr{D}+1)R+k\Phi=-\sigma_1+\frac{1}{B}[A_0+A_1\tau\ln\tau+\tau\mathscr{A}]-kG_2. \tag{18}$$

From (5d), we obtain

$$\mathscr{D}T_1=(\gamma-1)\frac{\tau A}{B}-\tau\alpha e^{T_1},$$

hence,

$$-1+\varphi_1\tau(1+\ln\tau)+\tau(\mathscr{D}+1)\Phi=\frac{\gamma-1}{B}(\tau A)-1-\varphi_1\tau\ln\tau-\tau(\Phi+G_2). \tag{19}$$

Equating the coefficients of $\tau\ln\tau$ and using (13), we obtain $\varphi_1=-\frac{\gamma-1}{B}\tilde{A}-\varphi_1$, or

$$\varphi_1=-\frac{\gamma-1}{2B}\tilde{A}.$$

This proves (12b).
Dividing (19) by τ, the reduced equation for Φ is obtained:

$$(\mathscr{D}+2)\Phi=-\varphi_1+\frac{\gamma-1}{B}[A_0+A_1\tau\ln\tau+\tau\mathscr{A}]-G_2. \tag{20}$$

From Equation (5c), we obtain

$$\mathscr{D}\mathbf{u}_1=\tau(A-e^{T_1})\frac{\nabla\psi}{B}+\tau\nabla_\xi p_1.$$

or

$$\begin{aligned}&-\frac{\nabla\psi}{\alpha B}+\tau\mathbf{w}_1(1+\ln\tau)+\tau(\mathscr{D}+1)\mathbf{U}\\ &=\frac{\nabla\psi}{\alpha B}[\tau\alpha A-1-\varphi_1\tau\ln\tau-\tau(\Phi+G_2)]\\ &\quad+\nabla_\xi\left[\frac{\tau\ln(1/\tau)}{\gamma-|\nabla\psi|^2}+(\sigma_0-\ln\alpha)\frac{\tau}{\gamma}+\frac{\varphi_1+\sigma_1}{\gamma}\tau^2\ln\tau+\frac{\tau^2}{\gamma}(\Phi+R)\right].\end{aligned} \tag{21}$$

Equating the coefficients of $\tau \ln \tau$, we obtain, since $(\alpha B)^{-1} = (k+1)/\gamma$,

$$\begin{aligned}
\mathbf{w}_1 &= -\frac{\nabla\psi}{B}\tilde{A} - \frac{\nabla\psi}{\alpha B}\varphi_1 - \nabla_\xi \frac{1}{\alpha B} \\
&= -\frac{\nabla\psi}{B}\tilde{A} + \frac{\nabla\psi}{\alpha B}(\gamma-1)\frac{\tilde{A}}{2B} - \frac{1}{\gamma}\nabla_\xi k \\
&= \frac{\tilde{A}}{2B}\nabla\psi\left[\frac{\gamma-1}{\gamma-|\nabla\psi|^2} - 2\right] - \frac{1}{\gamma}\nabla_\xi k \\
&= \frac{\tilde{A}}{2\alpha B^2}\nabla\psi\left[2|\nabla\psi|^2 - (\gamma+1)\right] - \frac{1}{\gamma}\nabla_\xi k.
\end{aligned}$$

This proves (12c).
Dividing (21) by τ, the reduced equation for $\mathbf{U}$ is obtained:

$$\begin{aligned}
&(\mathscr{D}+1)\mathbf{U} + \frac{\nabla\psi}{\alpha B}\Phi \\
= &-\mathbf{w}_1 + \frac{\nabla\psi}{B}[A_0 + A_1\tau\ln\tau + \tau\mathscr{A} - G_2/\alpha] \\
&+ \nabla_\xi\left[(\sigma_0 - \ln\alpha)/\gamma + \frac{\varphi_1+\sigma_1}{\gamma}\tau\ln\tau + \frac{\tau}{\gamma}(\Phi+R)\right].
\end{aligned} \tag{22}$$

It remains to prove that the renormalized unknowns admit the desired expansion. We start from the reduced system formed by (17), (20) and (22), namely

$$\begin{aligned}
(\mathscr{D}+1)R + k\Phi &= -\sigma_1 + \frac{1}{B}[A_0 + A_1\tau\ln\tau + \tau\mathscr{A}] - kG_2 \\
(\mathscr{D}+2)\Phi &= -\varphi_1 + \frac{\gamma-1}{B}[A_0 + A_1\tau\ln\tau + \tau\mathscr{A}] - G_2 \\
(\mathscr{D}+1)\mathbf{U} + \frac{\nabla\psi}{\alpha B}\Phi &= -\mathbf{w}_1 + \frac{\nabla\psi}{B}[A_0 + A_1\tau\ln\tau + \tau\mathscr{A} - G_2/\alpha] \\
&\quad + \nabla_\xi\left[(\sigma_0 - \ln\alpha)/\gamma + \frac{\varphi_1+\sigma_1}{\gamma}\tau\ln\tau + \frac{\tau}{\gamma}(\Phi+R)\right].
\end{aligned}$$

Letting $\mathscr{X} = (R, \Phi, \mathbf{U})^T$, this system has the general form

$$(\mathscr{D}+A)\mathscr{X} = \mathbf{F}(\tau, \tau\ln\tau, \tau(\ln\tau)^2, \mathscr{X}, \tau\nabla_\xi\mathscr{X}), \tag{23}$$

where

$$A = \begin{pmatrix} 1 & k & 0 & 0 & 0 \\ 0 & 2 & 0 & 0 & 0 \\ 0 & (\partial\psi/\partial\xi_1)/\alpha B & 1 & 0 & 0 \\ 0 & (\partial\psi/\partial\xi_2)/\alpha B & 0 & 1 & 0 \\ 0 & (\partial\psi/\partial\xi_3)/\alpha B & 0 & 0 & 1 \end{pmatrix}.$$

The essential features of this system are the presence of a factor of τ in front of every space derivative of a component of $\mathscr{X}$ and the fact that all the eigenvalues of A are positive. This system admits a formal series in powers of τ, $\tau(\ln\tau)$ and $\tau(\ln\tau)^2$ (using [8] (Theorem 2.4)), with coefficients depending only on ξ—they may be expressed explicitly in terms of the data characterizing the singularity, namely, ψ, σ_0 and $\mathbf{w}_0$. By [8] (Theorem 4.5), this series converges for a small τ; it is real for τ real and positive, because, by induction, all the terms of its expansion are. □

Remark 2. *We record the expression for p_1 that follows from (10) and (5a):*

$$p_1 = (\rho_1 + T_1)/\gamma = \frac{1}{\alpha B}\ln\frac{1}{\tau} + \frac{\sigma_0 - \ln\alpha}{\gamma} + \frac{\varphi_1+\sigma_1}{\gamma}\tau\ln\tau + \frac{\tau}{\gamma}(\Phi+R). \tag{24}$$

Remark 3. *The singularity data* $(\sigma_0, \mathbf{w}_0, \psi)$ *have a simple meaning at a point where* $\nabla\psi = 0$. *It is always possible to achieve this by performing a Lorentz transformation, since* $|\nabla\psi|$ *is small in the situation considered here. In that case, the leading terms in the expansions of* ρ_1 *and* $\mathbf{u}_1$ *vanish. Therefore the arbitrary functions* σ_0 *and* $\mathbf{w}_0$ *represent the density and Eulerian velocity at the first point of the hot spot. Further, we find that* φ_1 *and* $\mathbf{w}_1$ *vanish at this point and that* σ_1 *is proportional to* $\Delta\psi$.

6. Lorentz Transformation and Blow-Up Patterns

We now show that the hot spot, defined as the first spacetime point where a singularity forms, is not a Lorentz invariant; therefore, it is not an intrinsic feature of the ignition process. Since this is a purely kinematical phenomenon, of general applicability, we explain our result in general terms. To apply it to the ignition problem, it suffices to set $f = \psi$ in what follows.

Let us consider a physical phenomenon that exhibits a singularity that is observed to take place along a singular locus Σ described by an equation $t = f(x, y, z)$ with f smooth, in a given inertial system (S), by an observer who labels events in his/her local Minkowski space with coordinates (x, y, z, t). In this section, we establish that the first spacetime singularity for (S), corresponding to the minimum of f, is not the first spacetime singularity for another inertial observer. To see this, let us perform a Lorentz transformation. By translation of variables, we may assume that f admits a minimum for $x = y = z = 0$. Adding a constant to f, we may also assume that $f(0,0,0) = 0$. Since f is minimum at the origin, there, we have $\partial_x f = \partial_y f = \partial_z f = 0$. Since performing spatial rotations on coordinates does not change the value of the minimum of f, it suffices to consider a special Lorentz transformation in the x-direction:

$$x' = \gamma(x - vt); \quad t' = \gamma(t - vx/c^2); \quad y' = y,\ z' = z, \qquad \text{with } \gamma = (1 - v^2/c^2)^{-1/2}.$$

The inverse transformation is given by

$$x = \gamma(x' + vt'); \quad t = \gamma(t' + vx'/c^2); \quad y = y',\ z = z'.$$

Therefore, the equation of the singular set, namely, $t - f(x, y, z) = 0$, becomes

$$F(t', x', y', z', v) := \gamma(t' + vx'/c^2) - f(\gamma(x' + vt'), y', z') = 0. \tag{25}$$

Equation $F = 0$ is an implicit equation for the singularity locus as viewed in (S'). Since $\partial F/\partial t' = \gamma(1 - v f_x(\gamma(x' + vt'), y', z'))$ reduces to γ at the origin (because $f_x = 0$ there), the implicit function theorem enables one to locally solve equation $F = 0$ in the form $t' = g(x', y', z', v)$. Differentiating Equation (25) with respect to the primed variables, we obtain the following:

$$\partial_{x'} g = \frac{\partial_x f - v/c^2}{1 - v\partial_x f}, \partial_{y'} g = \frac{\gamma^{-1}\partial_y f}{1 - v\partial_x f}, \partial_{z'} g = \frac{\gamma^{-1}\partial_z f}{1 - v\partial_x f},$$

where $\partial_x f = \frac{\partial f}{\partial x}(x, y, z) = \frac{\partial f}{\partial x}(\gamma(x' + vt'), y', z')$, $\partial_{x'} g = \frac{\partial g}{\partial x'}(x', y', z', v)$, and similarly for $\partial_y f, \partial_z f, \partial_{y'} g, \partial_{z'} g$.

Now, the places where f exhibits an extremum ($\partial_x f = \partial_y f = \partial_z f = 0$, point D on Figure 1) do not coincide with those where g does ($\partial_{x'} g = \partial_{y'} g = \partial_{z'} g = 0$, point E on Figure 1). The location of the first singularity in the second system (S') is obtained by solving the equation $\partial_{x'} g = 0$ in (S'); the same spacetime point E would be obtained in (S) by solving $\partial_x f = v/c^2$ in (S). By contrast, in (S), the first singularity D satisfies $\partial_x f = 0$. The change in the spacetime point where the first singularity is observed may be seen geometrically in one space dimension (see Figure 1). Therefore, the first singularity in (S') does not correspond to the same spacetime point as the first singularity in (S). The first hot spot in a given inertial system is not the cause of ignition and has no intrinsic physical significance. By contrast, the blow-up set is a well-defined geometric object and its geometric characteristics are physically meaningful.

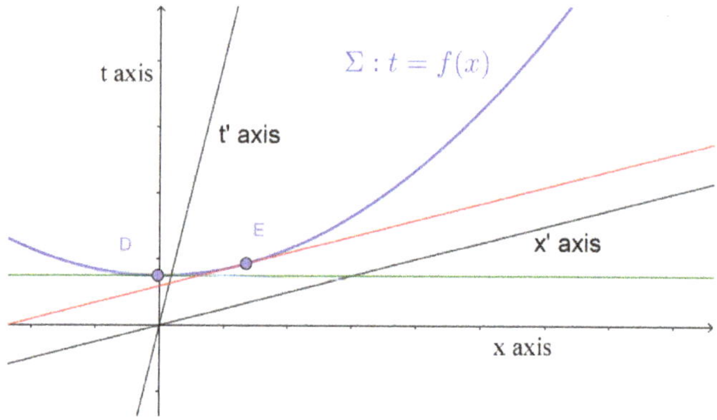

Figure 1. *Illustration of the transformation of the first hot spot under a Lorentz transformation.* The singular set Σ (in blue) has equation $t = f(x)$ in a two-dimensional Minkowski space, for an observer with rectangular coordinates (t, x). The time of the first singularity corresponds, for this observer, to the spacetime point D. For another inertial observer, his/her time and space axes are slanted lines, as indicated, and the first singularity is observed at a different spacetime point E, that may be constructed by finding the tangent to Σ (in red) parallel to the x'-axis.

7. Discussion and Perspectives

7.1. Detonation Signature

The preceding results lead to the identification of combinations of physical quantities that admit limits on the detonation front. Indeed, Theorem 2 shows that the following combinations of the unknowns tend to a finite limit at the singularity (as $\tau \to 0$):

$$\rho_1 - \mathbf{u}_1 \cdot \nabla\psi \quad \to \quad \sigma_0 - \mathbf{w}_0 \cdot \nabla\psi \tag{26a}$$

$$\rho_1 / T_1 \quad \to \quad k \tag{26b}$$

$$\mathbf{u}_1 - \frac{\nabla\psi}{\alpha B} T_1 \quad \to \quad \mathbf{w}_0 + \frac{\nabla\psi}{\alpha B} \ln\alpha \tag{26c}$$

$$|\nabla\psi|^2 T_1 - (\gamma - |\nabla\psi|^2)\rho_1 \quad \to \quad -|\nabla\psi|^2 \ln\alpha - (\gamma - |\nabla\psi|^2)\sigma_0 \tag{26d}$$

$$(\mathscr{D} - 1)\exp T_1 \quad \to \quad \varphi_1/\alpha. \tag{26e}$$

where the quantities α, B and φ_1 are determined in terms of the arbitrary functions (see Theorems 1 and 2). In particular, when $\nabla\psi = 0$ — that is, on the hot spot in the inertial frame at hand —*the leading order infinities vanish* in the expressions for the density and Eulerian velocities and we obtain the simple result

$$\rho_1 \quad \to \quad \sigma_0 \tag{27a}$$

$$\mathbf{u}_1 \quad \to \quad \mathbf{w}_0 \tag{27b}$$

$$(\mathscr{D} - 1)\exp T_1 \quad \to \quad \varphi_1/\alpha. \tag{27c}$$

These limiting behaviors, (26) and (27), give a characteristic signature of this detonation mechanism, that might be tested against measurements.

The fact that the first hot spot is not the cause of ignition is reflected in the fact that ψ is not a constant. The fact that the blow-up pattern is curved in general is reflected in the dependence of the coefficient σ_1 (in the expansion of the density) on the second-order derivatives of ψ, see (12a) and (13).

A particularly simple limiting behavior is $\rho_1/T_1 \to k$, where, we recall (see (9) and (6)), k only depends on the ratio of specific heats and the length of the gradient of ψ, that is, on the normal velocity in (S) of the detonation front. An overview of measurement techniques may be found in the second chapter of [32]; see also recent papers, such as [6,34].

7.2. Other Asymptotics

In this paper, we assume that the three spatial variables are scaled using the same characteristic length ℓ_0. It would be interesting to introduce different scalings, since some measurements are performed on thin domains, that are essentially two-dimensional [33,34]. The gap width plays the role of a second small parameter, in addition to the inverse activation energy. The detonation signature could exhibit a marked dependence on gap width, which would be consistent with the results of [33,34].

An important application of much current interest is the possibility of introducing ammonia rather than methane in fuel composition, since the former does not contain carbon. The literature is extensive [5–7]. This would have obvious advantages from the environmental point of view. It would be of interest to determine the non-dimensional parameters for different fuel compositions.

7.3. Relativistic Effects

The fact that the first hot spot for one observer may not be the first for another is a consequence of the kinematics of special relativity, irrespective of the importance of relativistic effects on the chemistry. This being said, there are two ways to introduce further relativistic effects in this problem. The first is to consider situations in which relativistic effects are significant, as in astrophysics. The second would be to repeat measurements such as those of [33,34] in a moving frame. For instance, one could set the cell used in these measurements in rapid rotation and perform measurements in the (stationary) laboratory frame. One could similarly consider a circular shock tube. Such measurements are perhaps feasible; their interpretation would depend on the importance of inertial effects.

8. Conclusions

(1) When a singularity is formed along a smooth hypersurface of Minkowski spacetime, with an equation of the form $\tau := t - \psi(x, y, z) = 0$, the spacetime location of the first hot spot is not a Lorentz invariant. This is a consequence of the Lorentz transformation between observers in special relativity and is independent of the size of the relativistic effects in the modeling of the physical situation that led to singularity-formation. However, the set of all singularities seen by all observers is a well-defined geometric object—the blow-up set.

(2) These ideas apply to the weak detonation problem. We have solved the appropriate system of PDEs, in the limit of high activation energy, by integrating them from singularity data given on the blow-up set, or detonation locus, and obtained a general solution of the equations. It contains the maximum number of arbitrary functions, namely, five. This solution improves earlier results in three respects: (i) It provides a description of the solution that is valid when it is large, but not infinite; in the weak detonation problem, the temperature never becomes actually infinite. (ii) It identifies which combinations of physical quantities remain finite at blow-up. (iii) It takes into account the kinematics of special relativity.

(3) In addition, the arbitrary functions $(\sigma_0, \mathbf{w}_0, \psi)$ in the general solution admit a physical interpretation in terms of the behavior of density, velocity and temperature at the singularity. This provides a signature for this type of ignition mechanism.

(4) Perspectives include the following: (i) a similar asymptotic study for nearly two-dimensional situations used in some measurements; (ii) the measurement of a detonation signature on the basis of the limiting behavior of the physical quantities, including the curvature of the blow-up pattern; (iii) the inclusion of the relativistic effects in the chemistry; (iv) the impact of fuel composition, especially the inclusion of hydrogen or ammonia in the mix, on the non-dimensional parameters, therefore on the domain of validity of this ignition mechanism.

Funding: This research was partially funded by CNRS through its research group UMR9008 in Reims. The APC was waived by MDPI.

Institutional Review Board Statement: Not applicable.

Informed Consent Statement: Not applicable.

Acknowledgments: The author thanks Leila Zhang at MDPI for encouraging him to prepare and submit this paper, and the referees for their kind comments.

Conflicts of Interest: The author declares no conflict of interest. The funders had no role in the design of the study; in the collection, analyses, or interpretation of data; in the writing of the manuscript, or in the decision to publish the results.

Appendix A. Derivation of the Basic System

Let us consider a reactive fluid, with reactant mass fraction y (i.e., one gram of fluid contains y grams of reactant and $1-y$ grams of reaction products). The overall reaction is modeled by a one-step irreversible reaction in the form $A_1 \longrightarrow A_2$, where the reactant A_1 and the product A_2 have the same specific heats c_p and c_v, as well as the same molar mass μ. This makes it possible to consider A_1 and A_2 as forming a single perfect fluid, with density ρ, specific volume $v = 1/\rho$, Eulerian velocity $\mathbf{u} = (u_1, u_2, u_3)$, pressure p and temperature T, that all vary with in space and time. Therefore, the equation of state is

$$\frac{p}{\rho} = pv = \frac{R}{\mu}T,$$

where R is the perfect gas constant, $\gamma = c_p/c_v$, and

$$\frac{R}{\mu} = c_p - c_v = (\gamma - 1)c_v.$$

The specific internal energy is

$$\varepsilon = c_v T + qy = \frac{p/\rho}{\gamma - 1} + qy,$$

where the constant q is the heat release rate. The total specific energy is

$$e = \varepsilon + \frac{1}{2}u^2,$$

where $u^2 = \mathbf{u} \cdot \mathbf{u}$. The reaction rate is given by the Arrhenius law

$$\frac{Dy}{Dt} = r(y,t) := -Ay\exp\left(-\frac{E}{RT}\right). \tag{A1}$$

where A and E are constants.

To write the conservation laws, let us introduce the material derivative operator

$$\frac{D}{Dt} = \frac{\partial}{\partial t} + \sum_{i=1}^{3} u_i \frac{\partial}{\partial x_i}.$$

Mass, momentum and energy conservation read

$$\frac{D\rho}{Dt} + \rho \operatorname{div} \mathbf{u} = 0 \tag{A2a}$$

$$\frac{D\mathbf{u}}{Dt} + \frac{1}{\rho}\nabla p = 0 \tag{A2b}$$

$$\rho\frac{De}{Dt} + \operatorname{div}(p\mathbf{u}) = 0. \tag{A2c}$$

By (A2a), $\rho_t + \operatorname{div}(\rho\mathbf{u}) = 0$. Therefore,

$$\begin{aligned}\operatorname{div}(p\mathbf{u}) &= p\operatorname{div}\mathbf{u} + \mathbf{u}\cdot\nabla p\\ &= -\frac{p}{\rho}\frac{D\rho}{Dt} - \rho\mathbf{u}\frac{D\mathbf{u}}{Dt}\\ &= \rho\frac{D}{Dt}\left[\frac{p}{\rho} - \frac{1}{2}u^2\right] - \frac{Dp}{Dt}.\end{aligned}$$

In addition, $p/\rho = (c_p - c_v)T$. On the other hand,

$$\rho\frac{De}{Dt} = \rho\frac{D}{Dt}\left[c_vT + qy + \frac{1}{2}u^2\right].$$

Therefore, Equation (A2c) takes the form

$$\rho\frac{De}{Dt} + \operatorname{div}(p\mathbf{u}) = \rho\frac{D}{Dt}\left\{[c_vT + qy + \frac{1}{2}u^2] + [(c_p - c_v)T - \frac{1}{2}u^2]\right\} - \frac{Dp}{Dt}.$$

Using the Arrhenius law,

$$\rho c_p\frac{DT}{Dt} - \frac{Dp}{Dt} = -\rho q\frac{Dy}{Dt} = -\rho qr = A\rho qy\exp(-\frac{E}{RT}). \tag{A3}$$

To sum up, we have to solve the following system:

$$p = (\gamma-1)c_v\rho T \tag{A4a}$$
$$\frac{D\rho}{Dt} + \rho\operatorname{div}\mathbf{u} = 0 \tag{A4b}$$
$$\frac{D\mathbf{u}}{Dt} + \frac{1}{\rho}\nabla p = 0 \tag{A4c}$$
$$\rho c_p\frac{DT}{Dt} - \frac{Dp}{Dt} = A\rho qy\exp(-\frac{E}{RT}) \tag{A4d}$$
$$\frac{Dy}{Dt} = -Ay\exp(-\frac{E}{RT}). \tag{A4e}$$

The objective is to solve this system in the limit when the activation energy E is large.

Appendix B. Non-Dimensionalization

Appendix B.1. Non-Dimensional Variables

Let us introduce a reference state characterized by the values p_0, ρ_0, T_0 and y_0, a reference length ℓ_0 and reference time t_0. They determine $u_0 = \ell_0/t_0$. We introduce scaled variables by

$$x^* = \frac{x}{\ell_0};\quad t^* = \frac{t}{t_0};\quad p^* = \frac{p}{p_0};\quad \mathbf{u}^* = \frac{\mathbf{u}}{u_0};\quad \rho^* = \frac{\rho}{\rho_0};\quad y^* = \frac{y}{y_0};\quad T^* = \frac{T}{T_0}.$$

We take $y_0 = 1$ and assume the equation of state holds for the reference state $p_0 = (\gamma-1)c_v\rho_0T_0$. The velocity $c_0 = \sqrt{\gamma p_0/\rho_0}$ determines the acoustic time $t_a = \frac{\ell_0}{c_0}$. The main non-dimensional parameters for the present analysis are

$$\begin{aligned}
\theta &= E/RT_0 \quad \text{(non-dimensional activation parameter)} \\
t_r &= A^{-1}\exp\theta \quad \text{(initial reaction time)} \\
c_0 &= \sqrt{\gamma p_0/\rho_0} \\
M &= \frac{t_a}{t_0} = \frac{u_0}{c_0} \quad \text{(Mach number)} \\
\beta &= \frac{q y_0}{c_p T_0} \quad \text{(heat release parameter).}
\end{aligned}$$

Substituting these relations into (A4), we obtain

$$p^* = \rho^* T^* \tag{A5a}$$

$$\frac{D\rho^*}{Dt^*} + \rho^* \operatorname{div}^* \mathbf{u}^* = 0 \tag{A5b}$$

$$\rho^* \frac{D\mathbf{u}^*}{Dt^*} + \frac{1}{\gamma M^2}\nabla^* p^* = 0 \tag{A5c}$$

$$\rho^* \frac{DT^*}{Dt^*} - \frac{\gamma-1}{\gamma}\frac{Dp^*}{Dt^*} = \frac{t_0}{t_r}\beta\rho^* y^* \exp[\theta - \frac{\theta}{T^*}] \tag{A5d}$$

$$\frac{Dy^*}{Dt^*} = -\frac{t_0}{t_r} y^* \exp[\theta - \frac{\theta}{T^*}]. \tag{A5e}$$

Appendix B.2. Choice of Time and Length Scales

The modeling assumptions (A2–A4) translate into the following choices:

- Fix the reference time t_0 by $t_0/t_r = 1/(\beta\theta)$.
- Fix the reference length so that $M = 1$. That means $t_a = t_0$, or $\ell_0 = c_0 t_0$.

To express the compatibility of these assumptions, we relate M and θ. First,

$$M = \frac{t_a}{t_0} = \frac{t_a}{t_r}\beta\theta = \frac{t_a}{A^{-1}\exp\theta}\frac{q y_0}{c_p T_0}\theta.$$

Therefore, since $t_a = t_0$,

$$M = 1 = A\beta\theta t_a e^{-\theta}.$$

The assumptions $M = 1$ and $\theta \gg 1$ are compatible if $A\beta$ is large; this is consistent with the assumption that the reaction is strongly exothermic. Additionally, since $t_0 \ll t_r$ for $\beta\theta \gg 1$, the reaction proceeds slowly with respect to the reference time scale t_0.

With these assumptions, the dimensionless equations become

$$p^* = \rho^* T^* \tag{A6a}$$

$$\frac{D\rho^*}{Dt^*} + \rho^* \operatorname{div}^* \mathbf{u}^* = 0 \tag{A6b}$$

$$\rho^* \frac{D\mathbf{u}^*}{Dt^*} + \frac{1}{\gamma}\nabla^* p^* = 0 \tag{A6c}$$

$$\rho^* \frac{DT^*}{Dt^*} - \frac{\gamma-1}{\gamma}\frac{Dp^*}{Dt^*} = \frac{1}{\theta}\rho^* y^* \exp[\theta - \frac{\theta}{T^*}] \tag{A6d}$$

$$\frac{Dy^*}{Dt^*} = -\frac{1}{\beta\theta} y^* \exp[\theta - \frac{\theta}{T^*}]. \tag{A6e}$$

Multiplying (A6e) by $\beta\rho^*$ and adding the result to (A6d) yields

$$(\rho^* - 1)\frac{D}{Dt^*}[T^* + \beta y^*] + \frac{D}{Dt^*}\left[T^* - \frac{\gamma-1}{\gamma}p^* + \beta y^*\right] = 0. \tag{A7}$$

References

1. Lee, J.H.S. *The Detonation Phenomenon*; Cambridge University Press: New York, NY, USA, 2008.
2. Sivashinsky, G.I. Some developments in premixed combustion modeling. *Proc. Combust. Inst.* **2002**, *29*, 1737–1761. [CrossRef]
3. Urtiew, P.A.; Oppenheim, A.K. Experimental observations of the transition to detonation in an explosive gas. *Proc. R. Soc. Lond. A* **1966**, *295*, 13–28.
4. Williams, F.A. *Combustion Theory*; Addison-Wesley: Reading, MA, USA, 1985.
5. Hasan, M.H.; Mahlia, T.M.I.; Mofijur, M.; Rizwanul Fattah, I.M.; Handayani, F.; Ong, H.C.; Silitonga, A.S. A comprehensive review on the recent development of ammonia as a renewable energy carrier. *Energies* **2021**, *14*, 3732. [CrossRef]
6. Tang, G.; Jin, P.; Bao, Y.; Chai, W.S.; Zhou, L. Experimental investigation of premixed combustion limits of hydrogen and methane additives in ammonia. *Int. J. Hydrogen Energy* **2021**, *46* 20765–20776. [CrossRef]
7. Valera-Medina, A.; Xiao, H.; Owen-Jones, M.; David, W.I.F.; Bowen, P.J. Ammonia for power. *Prog. Energy Combust. Sci.* **2018**, *69*, 63–102. [CrossRef]
8. Kichenassamy, S. *Fuchsian Reduction: Applications to Geometry, Cosmology and Mathematical Physics*; Birkhäuser: Boston, MA, USA, 2007.
9. Kichenassamy, S. Stability of blow-up patterns for nonlinear wave equations. In *Nonlinear PDEs, Dynamics and Continuum Physics*; Bona, J., Saxton, K.R., Eds.; AMS: Providence, RI, USA, 2000; pp. 139–162.
10. Kichenassamy, S.; Littman, W. Blow-up surfaces for nonlinear wave equations, I & II. *Commun. Partial Differ. Equ.* **1993**, *18*, 431–452 and 1869–1899.
11. Kichenassamy, S. Control of blow-up singularities for nonlinear wave equations, *Evol. Equ. Control Theory* **2013**, *2*, 667–677.
12. Cabart, G.; Kichenassamy, S. Explosion et normes L^p pour l'équation des ondes non linéaire cubique. *Comptes Rendus Math.* **2002**, *335*, 903–908.
13. Kapila, A.K.; Jackson, T.L. Dynamics of hot-spot evolution in a reactive, compressible flow. In *Computational Fluid Dynamics and Reacting Gas Flows*; IMA Volumes in Mathematics and its Applcations 12; Engquist, B., Luskin, M., Majda, A., Eds.; Springer: Boston, MA, USA, 1988; pp. 123–150.
14. Blythe, P.A.; Crighton, D.G. Shock-generated ignition: the induction zone. *Proc. R. Soc. Lond. A* **1989**, *426*, 189–209.
15. Clarke, J.F.; Cant, R.S. Non-steady gasdynamic effects in the induction domain behind a strong shock wave. In *Dynamics of Flames and Reactive Systems*; Progress in Astronautics and Aeronautics 95; Bowen, J.R., Manson, N., Oppenheim, A.K., Eds.; American Institute of Aeronautics: New York, NY, USA, 1984; pp. 142–163.
16. Jackson, T.L.; Kapila, A.K.; Stewart, D.S. Evolution of a reaction center in an explosive material. *SIAM J. Appl. Math.* **1989**, *49*, 432–458. [CrossRef]
17. Campbell, A.W.; Davis, W.C.; Travis, J.R. Shock initiation of detonation in liquid explosives. *Phys. Fluids* **1961**, *4*, 498–510. [CrossRef]
18. Clarke, J.F. Propagation of gas dynamic disturbance in an explosive atmosphere. *Prog. Astronaut. Aeronaut.* **1981**, *76*, 383–402.
19. Dold, J.W.; Kapila, A.K.; Short, M. Theoretical description of direct initiation of detonation for one-step chemistry. In *Dynamic Structure of Detonation in Gaseous and Dispersed Media*; Borisov, A.A., Ed.; Kluwer: Dordrecht, The Netherlands, 1991; pp. 245–256.
20. Dremin, A.N. *Toward Detonation Theory*; Springer: Berlin, Germany, 1999.
21. Fickett, W.; Davis, W.C. *Detonation: Theory and Experiment*; Dover: Mineola, NY, USA, 2000.
22. Jackson, T.L.; Kapila, A.K. Shock-induced thermal runaway. *SIAM J. Appl. Math.* **1985**, *45*, 130–137. [CrossRef]
23. Kapila, A.K.; Dold, J.W. A theoretical picture of shock-to-detonation transition in a homogeneous explosive. In *9th Symposium (International) on Detonation*; OCNR 113291-7; Office of Naval Research: Washington, DC, USA, 1989; pp. 219–227.
24. Kassoy, D.R. Extremely rapid transient phenomona in combustion, ignition and explosion. In *Asymptotic Methods and Singular Perturbations*; SIAM-AMS Proc; O'Malley, R.E. Ed.; AMS: Providence, RI, USA, 1976; Volume 10, pp. 61–72.
25. Meyer, J.W.; Oppenheim, A.K. Dynamic response of a plane-symmetrical exothermic reaction center. *AIAA J. (Publ. Am. Inst. Aeronaut. Astronaut.)* **1972**, *10*, 1509–1513. [CrossRef]
26. Poland, J.; Kassoy, D.R. The induction period of a thermal explosion in a gas between infinite parallel plates. *Combust. Flame* **1983**, *50*, 259–274. [CrossRef]
27. Shepherd, J.E. Detonation in gases. *Proc. Combust. Inst.* **2009**, *32*, 83–98. [CrossRef]
28. Short, M. On the critical conditions for the initiation of a detonation in a nonuniformly perturbed reactive fluid. *SIAM J. Appl. Math.* **1997**, *57*, 1242–1280. [CrossRef]
29. Short M.; Dold J.W. Weak detonations, their paths and transition to strong detonation. *Combust. Theory Modeling* **2002**, *6*, 279–296. [CrossRef]
30. Zajac, L.J.; Oppenheim, A.K. Dynamics of an explosive reaction center. *AIAA J. (Publ. Am. Inst. of Aeronaut. Astronaut.)* **1971**, *9*, 545–553. [CrossRef]
31. Kichenassamy, S. The blow-up problem for exponential nonlinearities. *Commun. Partial Differ. Equ.* **1996**, *21*, 125–162. [CrossRef]
32. Warnatz, J.; Maas, U.; Dibble, R.W. *Combustion—Physical and Chemical Foundations, Modeling and Simulation, Experiments, Pollutant Formation*, 3rd ed.; Springer: Berlin, Germany, 2001.
33. Wu, M.-H.; Kuo, W.-C. Accelerative expansion and DDT of stoichiometric ethylene/oxygen flame rings in microgaps. *Proc. Combust. Inst.* **2013**, *34*, 2017–2024. [CrossRef]
34. Ssu, H.-W.; Wu, M.-H. Formation and characteristics of composite reaction—Shock clusters in narrow channels. *Proc. Combust. Inst.* **2021**, *28* 3473–3480. [CrossRef]

axioms

MDPI

Article

Bäcklund Transformations for Liouville Equations with Exponential Nonlinearity

Tatyana V. Redkina [1,2], Robert G. Zakinyan [2,3,*], Arthur R. Zakinyan [2,3] and Olga V. Novikova [2,4]

1 Faculty of Mathematics and Computer Sciences Named after Professor N. I. Chervyakov, North-Caucasus Federal University, 1 Pushkin Street, 355017 Stavropol, Russia; TVR59@mail.ru
2 North-Caucasus Center for Mathematical Research, 1 Pushkin Street, 355017 Stavropol, Russia; zakinyan.a.r@mail.ru (A.R.Z.); oly-novikova@yandex.ru (O.V.N.)
3 Physical-Technical Faculty, North-Caucasus Federal University, 1 Pushkin Street, 355017 Stavropol, Russia
4 Institute for Digital Development, North-Caucasus Federal University, 2 Kulakov Avenue, 355017 Stavropol, Russia
* Correspondence: zakinyan@mail.ru; Tel.: +7-918-7788-675

Abstract: This work aims to obtain new transformations and auto-Bäcklund transformations for generalized Liouville equations with exponential nonlinearity having a factor depending on the first derivatives. This paper discusses the construction of Bäcklund transformations for nonlinear partial second-order derivatives of the soliton type with logarithmic nonlinearity and hyperbolic linear parts. The construction of transformations is based on the method proposed by Clairin for second-order equations of the Monge–Ampere type. For the equations studied in the article, using the Bäcklund transformations, new equations are found, which make it possible to find solutions to the original nonlinear equations and reveal the internal connections between various integrable equations.

Keywords: nonlinear equations in partial derivatives; hyperbolic equations; Bäcklund transformations; Clairin's method; differential relationships; the Liouville equation

Citation: Redkina, T.V.; Zakinyan, R.G.; Zakinyan, A.R.; Novikova, O.V. Bäcklund Transformations for Liouville Equations with Exponential Nonlinearity. *Axioms* **2021**, *10*, 337. https://doi.org/10.3390/axioms10040337

Academic Editor: Hans J. Haubold

Received: 31 October 2021
Accepted: 7 December 2021
Published: 9 December 2021

Publisher's Note: MDPI stays neutral with regard to jurisdictional claims in published maps and institutional affiliations.

Copyright: © 2021 by the authors. Licensee MDPI, Basel, Switzerland. This article is an open access article distributed under the terms and conditions of the Creative Commons Attribution (CC BY) license (https://creativecommons.org/licenses/by/4.0/).

1. Introduction

The study of Bäcklund transformations is one of the current topics in the theory of partial differential equations. Such transformations are used to find solutions to nonlinear differential equations. Due to the complexity of various nonlinear equations, there is no single method for solving them. For integrable systems, effective methods have been developed, such as the inverse scattering method [1,2], the Hirota method [3–5], the Painleve method [6,7], Bäcklund transformations [8–11] and the mapping and deformation method [2].

Bäcklund transformations are an example of differential geometric structures generated by differential equations. They make it possible to obtain not only pairs of equations but also a solution to one of them if the solution to the other is known. These transformations play an important role in integrable systems since they reveal internal connections between various properties, such as the definition of symmetries [12,13] and the presence of a Hamiltonian structure [14–16]. More recently, many studies have been carried out in this area [11,17–19].

This article is a presentation of new results on transformations and auto-Bäcklund transformations for equations of the Klein–Gordon type, using the method of constructing transformations for the Liouville equation. The paper considers special cases of equations with exponential–power nonlinearity having a factor depending on the first derivatives. The construction of transformations is based on Clairin's method [20].

2. Methods

We consider the following nonlinear equation of the hyperbolic form:

$$v_{\xi\eta} = f(v, v_\xi, v_\eta). \quad (1)$$

The method developed by Clairin to construct Bäcklund transformations of a general form is applicable when the functions z and v satisfy different partial differential equations. The technique of constructing Bäcklund transformations is general to any hyperbolic equation and completely repeats the construction for the Liouville equation.

Differential equations of the second order of the form

$$f_1(\xi,\eta,z,z_\xi,z_\eta)\, z_{\xi\xi} + f_2(\xi,\eta,z,z_\xi,z_\eta)\, z_{\xi\eta} + f_3(\xi,\eta,z,z_\xi,z_\eta)\, z_{\eta\eta} + f_4(\xi,\eta,z,z_\xi,z_\eta) = 0.$$

are called Monge–Ampere equations [21]. The Bäcklund transformation linking two such second-order equations for the v and z functions is given by a pair of first-order differential equations:

$$\frac{\partial z}{\partial \xi} = F_1\left(z, v, \frac{\partial v}{\partial \xi}, \frac{\partial v}{\partial \eta}\right). \quad (2)$$

$$\frac{\partial z}{\partial \eta} = F_2\left(z, v, \frac{\partial v}{\partial \xi}, \frac{\partial v}{\partial \eta}\right). \quad (3)$$

To define an explicit transformation type, it is necessary to find the functions F_1 and F_2. The integrability condition (the equality of the mixed second derivatives) requires that the functions (2), (3) satisfy the relation

$$\frac{\partial^2 z}{\partial \eta \partial \xi} - \frac{\partial^2 z}{\partial \xi \partial \eta} = 0.$$

Each of the variables $z,\ z_\xi,\ z_\eta$ and, respectively, $v,\ v_\xi,\ v_\eta$, depends on ξ and η. Given the equality (2), we obtain

$$\frac{\partial^2 z}{\partial \eta \partial \xi} = \frac{\partial F_1}{\partial \eta} = \frac{\partial F_1}{\partial z} z_\eta + \frac{\partial F_1}{\partial v} v_\eta + \frac{\partial F_1}{\partial v_\xi} v_{\xi\eta} + \frac{\partial F_1}{\partial v_\eta} v_{\eta\eta} \quad (4)$$

$$\frac{\partial^2 z}{\partial \xi \partial \eta} = \frac{\partial F_2}{\partial \xi} = \frac{\partial F_2}{\partial z} z_\xi + \frac{\partial F_2}{\partial v} v_\xi + \frac{\partial F_2}{\partial v_\xi} v_{\xi\xi} + \frac{\partial F_2}{\partial v_\eta} v_{\eta\xi} \quad (5)$$

Using Formulas (2), (3) to exclude z_ξ and z_η, finally, we obtain the condition of compatibility as

$$\left(-\frac{\partial F_2}{\partial v_\xi}\right) v_{\xi\xi} + \left(\frac{\partial F_1}{\partial v_\xi} - \frac{\partial F_2}{\partial v_\eta}\right) v_{\xi\eta} + \frac{\partial F_1}{\partial v_\eta} v_{\eta\eta} - \frac{\partial F_2}{\partial v} v_\xi + \frac{\partial F_1}{\partial v} v_\eta + F_2 \frac{\partial F_1}{\partial z} - F_1 \frac{\partial F_2}{\partial z} = 0 \quad (6)$$

We consider the function z as a solution to some simple equation, the form of which is defined below. Then, while at least one of the coefficients,

$$\frac{\partial F_1}{\partial v_\eta},\ \frac{\partial F_2}{\partial v_\xi} \text{ or } \left(\frac{\partial F_1}{\partial v_\xi} - \frac{\partial F_2}{\partial v_\eta}\right),$$

is not zero, Equation (6) is a partial differential equation for the function v.

Since Equation (1) contains $v_{\xi\eta}$, but not $v_{\xi\xi}$ or $v_{\eta\eta}$, from the condition of compatibility (6), we expect that

$$\frac{\partial F_2}{\partial v_\xi} = 0,\ \frac{\partial F_1}{\partial v_\eta} = 0,\ \frac{\partial F_1}{\partial v_\xi} - \frac{\partial F_2}{\partial v_\eta} \neq 0$$

Then, we must assume

$$\frac{\partial z}{\partial \xi} = F_1\left(z, v, \frac{\partial v}{\partial \xi}\right), \quad (7)$$

$$\frac{\partial z}{\partial \eta} = F_2\left(z, v, \frac{\partial v}{\partial \eta}\right). \tag{8}$$

Therefore, Equation (6) takes the form

$$\left(\frac{\partial F_1}{\partial v_\xi} - \frac{\partial F_2}{\partial v_\eta}\right)v_{\xi,\eta} - \frac{\partial F_2}{\partial v}v_\xi + \frac{\partial F_1}{\partial v}v_\eta + F_2\frac{\partial F_1}{\partial z} - F_1\frac{\partial F_2}{\partial z} = 0$$

The η-derivative of (7) is

$$\frac{\partial^2 z}{\partial \eta \partial \xi} = \frac{\partial F_1}{\partial z}z_\eta + \frac{\partial F_1}{\partial \tilde{z}}v_\eta + \frac{\partial F_1}{\partial v_\xi}v_{\xi,\eta}. \tag{9}$$

Further reasoning depends on the type of equation under consideration. Let us consider the following equations:

$$v_{\eta\xi} = (a + bv)e^v v_\xi - v_\xi v_\eta, \tag{10}$$

$$v_{\eta\xi} = \frac{\alpha_{32}\alpha_{21}}{8\alpha_{11}}e^v(1+2v)v_\xi - v_\xi v_\eta, \tag{11}$$

$$v_{\eta\xi} = \frac{\alpha_{21}}{8\alpha_{11}^2}e^v(v_\eta - v_\xi), \tag{12}$$

$$v_{\eta\xi} = e^v v_\eta - e^{-v}v_\xi. \tag{13}$$

These equations have a hyperbolic linear form on the left side and a nonlinear right side depending on the function and the first derivatives to variables η and ξ, wherein the derivatives v_η, v_ξ are included in equations only in the first degree, so the general form of these equations is rewritten as

$$v_{\eta\xi} = \Omega(v, v_\xi^1, v_\eta^1),$$

Here, a one in the exponent indicates that these variables are included in this equality only to the first degree.

We assume that the Bäcklund transformations make it possible to move to the simplest hyperbolic equation $z_{\xi\eta} = 0$.

Using Equations (8)–(10), we obtain

$$z_{\xi\eta} = \frac{\partial F_1}{\partial z}F_2 + \frac{\partial F_1}{\partial v}v_\eta + \frac{\partial F_1}{\partial v_\xi}\Omega(v, v_\xi^1, v_\eta^1) = 0. \tag{14}$$

Take from (14) the derivative to v_η. Then, $\frac{\partial\Omega(v, v_\xi^1, v_\eta^1)}{\partial v_\eta}$ does not depend on v_η, since v_η comes into equality only in the first degree

$$\frac{\partial^2 F_1}{\partial z \partial v_\eta}F_2 + \frac{\partial F_1}{\partial z}\frac{\partial F_2}{\partial v_\eta} + \frac{\partial^2 F_1}{\partial v \partial v_\eta}v_\eta + \frac{\partial F_1}{\partial v} + \frac{\partial^2 F_1}{\partial v_\xi \partial v_\eta}\Omega(v, v_\xi^1, v_\eta^1) + \frac{\partial F_1}{\partial v_\xi}\frac{\partial\Omega(v, v_\xi^1, v_\eta^1)}{\partial v_\eta} = 0.$$

Taking into account equalities (7), (8), we have $\frac{\partial F_1}{\partial v_\eta} = 0$, $\frac{\partial F_2}{\partial v_\xi} = 0$ and then $\frac{\partial^2 F_1}{\partial z \partial v_\eta} = 0$, and equality remains

$$\frac{\partial F_1}{\partial z}\frac{\partial F_2}{\partial v_\eta} + \frac{\partial F_1}{\partial v} + \frac{\partial F_1}{\partial v_\xi}\frac{\partial\Omega(v, v_\xi^1, v_\eta^1)}{\partial v_\eta} = 0.$$

Having performed re-differentiation to v_η, we have

$$\frac{\partial^2\Omega(v, v_\xi^1, v_\eta^1)}{\partial v_\eta^2} = 0,$$

$$\frac{\partial^2 F_1}{\partial z \partial v_\eta}\frac{\partial F_2}{\partial v_\eta}+\frac{\partial F_1}{\partial z}\frac{\partial^2 F_2}{\partial v_\eta^2}+\frac{\partial^2 F_1}{\partial v \partial v_\eta}+\frac{\partial^2 F_1}{\partial v_\xi \partial v_\eta}\frac{\partial \Omega(v, v_\xi^1, v_\eta^1)}{\partial v_\eta}=0.$$

Considering $\frac{\partial F_1}{\partial v_\eta}=0$, we obtain

$$\frac{\partial F_1}{\partial z}\frac{\partial^2 F_2}{\partial v_\eta^2}=0.$$

We conduct similar actions with equality (3). Differentiating to v_ξ twice, we obtain

$$\frac{\partial F_2}{\partial v}\frac{\partial^2 F_1}{\partial v_\xi^2}=0.$$

Therefore, the functions F_1 and F_2 have a linear form to v_η and v_ξ, respectively. Then, we have

$$\frac{\partial z}{\partial \xi}=f_1(z,v)+p_1(z,v)\frac{\partial v}{\partial \xi}, \tag{15}$$

$$\frac{\partial z}{\partial \eta}=f_2(z,v)+p_2(z,v)\frac{\partial v}{\partial \eta}. \tag{16}$$

We write the compatibility condition of Equation (6) with the new conditions (15) and (16):

$$\begin{gathered}(p_1-p_2)\Omega(v, v_\xi^1, v_\eta^1)-\frac{\partial(f_2+p_2 v_\eta)}{\partial v}v_\xi\\ +\frac{\partial(f_1+p_1 v_\xi)}{\partial v}v_\eta+(f_2+p_2 v_\eta)\frac{\partial(f_1+p_1 v_\xi)}{\partial z}-(f_1+p_1 v_\xi)\frac{\partial(f_2+p_2 v_\eta)}{\partial z}=0\end{gathered} \tag{17}$$

After differentiating this expression to variable v_η and v_ξ, we proceed to the analysis of the equation

$$(p_1-p_2)\frac{\partial^2 \Omega(v, v_\xi^1, v_\eta^1)}{\partial v_\xi \partial v_\eta}-\frac{\partial p_2}{\partial v}+\frac{\partial p_1}{\partial v}+p_2\frac{\partial p_1}{\partial z}-p_1\frac{\partial p_2}{\partial z}=0. \tag{18}$$

Further studies depend significantly on $\frac{\partial^2 \Omega(v, v_\xi^1, v_\eta^1)}{\partial v_\xi \partial v_\eta}$, so let us move on to a detailed analysis of equality (18) for each equation studied separately.

3. Results

3.1. *Bäcklund Transformations for Nonlinear Equation*

Let us perform the Bäcklund transformation for nonlinear Equation (10).

Equation (18) for (4), (5), considering that $\frac{\partial^2 \Omega(v, v_\xi^1, v_\eta^1)}{\partial v_\xi \partial v_\eta}=-1$ takes the form

$$\frac{\partial p_1}{\partial v}+p_2\frac{\partial p_1}{\partial z}-p_1=p_1\frac{\partial p_2}{\partial z}+\frac{\partial p_2}{\partial v}-p_2. \tag{19}$$

It can be assumed that $p_1 \neq p_2$, and we define the relationship between the functions $p_1(z,v)$ and $p_2(z,v)$. We convert (19) to the following form:

$$\frac{\partial(p_1-p_2)}{\partial v}-(p_1-p_2)=p_1^2\frac{\partial}{\partial z}\frac{p_2}{p_1},$$

then, if we assume $p_1-p_2=e^v\varphi(z)$, then $p_1=p_2+e^v\varphi(z)$, and for the function p_2, we have the equation

$$[p_2+e^v\varphi(z)]^2\frac{\partial}{\partial z}\frac{p_2}{p_2+e^v\varphi(z)}=0.$$

Obviously, if $p_2 + e^v \varphi(z) = 0$, then $p_1 = 0$. This option could be considered but with only one undefined function $\varphi(z)$ remaining, which reduces the possibility of varying the unknowns in further reasoning, so calculate $p_2 + e^v \varphi(z) \neq 0$, and then

$$\frac{\partial}{\partial z} \frac{p_2}{p_2 + e^v \varphi(z)} = 0.$$

This leads to the dependence $\frac{p_2}{p_2+e^v \varphi(z)} = \psi(v)$ and the definition of functions p_2 and p_1 in the form

$$p_2 = \frac{\psi(v)}{1-\psi(v)} e^v \varphi(z), \quad p_1 = \frac{1}{1-\psi(v)} e^v \varphi(z).$$

Now, Equation (17) will take the form

$$\left[(a+bv)e^{2v}\varphi(z) - \tfrac{\partial f_2}{\partial v} + \tfrac{1}{1-\psi(v)} e^v \left(f_2 \tfrac{\partial \varphi(z)}{\partial z} - \varphi(z) \tfrac{\partial f_2}{\partial z} \right) \right] v_\xi$$
$$+ f_2 \tfrac{\partial f_1}{\partial z} - f_1 \tfrac{\partial f_2}{\partial z} + \left[\tfrac{\psi(v)}{1-\psi(v)} e^v \left(\varphi(z) \tfrac{\partial f_1}{\partial z} - f_1 \tfrac{\partial \varphi(z)}{\partial z} \right) + \tfrac{\partial f_1}{\partial v} \right] v_\eta = 0.$$

We differentiate the last equation to the variable v_η and the same expression to the variable v_ξ; as a result, we obtain the system

$$(a+bv)e^{2v}\varphi(z) - \frac{\partial f_2}{\partial v} + \frac{1}{1-\psi(v)} e^v \left(f_2 \frac{\partial \varphi(z)}{\partial z} - \varphi(z) \frac{\partial f_2}{\partial z} \right) = 0, \tag{20}$$

$$f_2 \frac{\partial f_1}{\partial z} - f_1 \frac{\partial f_2}{\partial z} = 0, \tag{21}$$

$$\frac{\psi(v)}{1-\psi(v)} e^v \left(\varphi(z) \frac{\partial f_1}{\partial z} - f_1 \frac{\partial \varphi(z)}{\partial z} \right) + \frac{\partial f_1}{\partial v} = 0. \tag{22}$$

We look for functions $f_1(z,v)$, $f_2(z,v)$ in the following form

$$f_1(z,v) = \psi_1(v) g_1(z), \quad f_2(z,v) = \psi_2(v) g_2(z).$$

We substitute these equations in the system (20)–(22) and isolate the logarithmic derivatives $\ln g_2$, $\ln g_1$, $\ln \varphi$:

$$(a+bv)e^{2v}\varphi(z) - g_2(z) \frac{\partial \psi_2(v)}{\partial v} + \frac{\psi_2(v)}{1-\psi(v)} e^v g_2(z) \varphi(z) \frac{\partial}{\partial z} \left(\ln \frac{\varphi(z)}{g_2(z)} \right) = 0, \tag{23}$$

$$\psi_2(v)\psi_1(v) g_2(z) g_1(z) \frac{\partial}{\partial z} \left(\ln \frac{g_1(z)}{g_2(z)} \right) = 0, \tag{24}$$

$$g_1(z) \frac{\partial \psi_1}{\partial v} - \frac{\psi(v)\psi_1(v)}{1-\psi(v)} e^v \varphi(z) g_1(z) \frac{\partial}{\partial z} \left(\ln \frac{\varphi(z)}{g_1(z)} \right) = 0. \tag{25}$$

One can choose a special form of functions $g_j(z)$, $\varphi(z)$ so that the system takes a simpler form; then, the differential equations can be explicitly integrated. We calculate

$$g_1(z) = g_2(z) = k_1 z, \qquad \varphi(z) = k_2 z, \qquad k_1, k_2 = \text{const},$$

and the system (23)–(25) will take the form

$$k_2 (a+bv) e^{2v} z - k_1 \frac{\partial \psi_2(v)}{\partial v} = 0,$$

$$k_1 z \frac{\partial \psi_1}{\partial v} = 0.$$

As a result, simple differential equations are obtained for functions $\psi_2(v)$, $\psi_1(v)$. Let us define them:

$$\psi_2 = \frac{k_2}{k_1}\int (a+bv)e^{2v}dv = \frac{k_2}{4k_1}(2a-b+2bv)e^{2v} + C_1, \quad \psi_1 = k = \text{const}.$$

Now, the transformations (2), (3) take the form

$$\frac{\partial z}{\partial \xi} = kk_1 z + \frac{k_2}{1-\psi(v)} e^v z \frac{\partial v}{\partial \xi}, \tag{26}$$

$$\frac{\partial z}{\partial \eta} = z\left[\frac{k_2}{4}(2a-b+2bv)e^{2v} + C_1\right] + k_2 e^v z \frac{\psi(v)}{1-\psi(v)}\frac{\partial v}{\partial \eta}. \tag{27}$$

Thus, the Bäcklund transformation is obtained in the form (26), (27). The system (26), (27) is combined with any function $\psi(v)$. We consider the following option: $\psi(v) = 2$, $C_1 = 0, \quad k = k_1 = 1, \quad k_2 = -2$, and then the relations (26), (27) will take the form

$$\frac{\partial z}{\partial \xi} = z + 2e^v z\frac{\partial v}{\partial \xi}, \ \frac{\partial z}{\partial \eta} = 4e^v z\frac{\partial v}{\partial \eta} - \left(a - \frac{1}{2}b + bv\right) z e^{2v}. \tag{28}$$

Let us check whether it is possible to obtain Equation (10) from the system (28).

If we differentiate the first equality of the system (28) to the variable η and the second to the variable ξ, we obtain

$$z_{\xi\eta} = z_\eta + 2e^v z v_\eta v_\xi + 2e^v z_\eta v_\xi + 2e^v z v_{\eta\xi},$$

$$z_{\xi\eta} = 4e^v z v_\xi v_\eta + 4e^v z_\xi v_\eta + 4e^v z v_{\xi\eta} - 2(a+bv)ze^{2v}v_\xi - \left(a - \frac{1}{2}b + bv\right) z_\xi e^{2v}.$$

We subtract the upper equality from the lower one and collect similar terms:

$$(1+2e^v v_\xi)z_\eta = 2e^v z v_{\xi\eta} + 2e^v z v_\xi v_\eta - 2(a+bv)ze^{2v}v_\xi + \left[4v_\eta - \left(a - \frac{1}{2}b + bv\right)e^v\right]e^v z_\xi.$$

We eliminate the derivatives z_η, z_ξ, using the relations (28). Canceling by non-zero functions, we obtain Equation (10).

Let us see which equation goes to the original Equation (10) using transformations (28). To do this, we convert Equation (28) to the form

$$\frac{\partial \ln z}{\partial \xi} = 1 + 2\frac{\partial e^v}{\partial \xi}, \ \frac{\partial \ln z}{\partial \eta} = 2\frac{\partial e^v}{\partial \eta} - \left(a - \frac{1}{2}b + bv\right)e^{2v}. \tag{29}$$

Let us try to identify how the functions $z(\xi,\eta)$ and $v(\xi,\eta)$ are related, taking into account that the functions satisfy Equation (10). Let us differentiate the second equality (29) to the variable ξ:

$$(\ln z)_{\eta\xi} = 4(e^v)_{\eta\xi} - 2(a+bv)e^{2v}v_\xi$$

and replace the expression $(a+bv)e^v v_\xi$ with the terms of Equation (10), then

$$(\ln z)_{\eta\xi} = 4(e^v)_{\eta\xi} - 2(v_{\eta\xi} + v_\xi v_\eta)e^v = 2(e^v)_{\eta\xi}.$$

Taking into account the first differential constraint (29), the derivatives can be omitted up to a constant

$$\ln z = 2e^v + \xi.$$

Then, the function $v(\xi,\eta)$ is expressed through $z(\xi,\eta)$

$$v = \ln\left[\frac{1}{2}(\ln z - \xi)\right]. \tag{30}$$

Denoting $\ln z = w(\xi, \eta)$ and substituting (30) in (10), we obtain

$$w_{\eta\xi} = \frac{1}{2}\left(a + b\ln\left[\frac{1}{2}(w - \xi)\right]\right)(w - \xi)(w_\xi - 1). \tag{31}$$

Theorem 1. *Bäcklund transformations*

$$\frac{\partial w}{\partial \eta} = 2e^v \frac{\partial v}{\partial \eta} - \left(a - \frac{1}{2}b + bv\right)e^{2v}, \ \frac{\partial w}{\partial \xi} = 1 + 2e^v \frac{\partial v}{\partial \xi}, \tag{32}$$

link Equations (10)–(31).

Equation (11) is a special case of (10), so the following conclusion can be formulated for this equation.

Corollary 1. *Bäcklund transformations*

$$\frac{\partial w}{\partial \xi} = 1 + 2e^v \frac{\partial v}{\partial \xi}, \ \frac{\partial w}{\partial \eta} = 4e^v \frac{\partial v}{\partial \eta} - \frac{\alpha_{32}\alpha_{21}}{4\alpha_{11}} v e^{2v}, \tag{33}$$

link Equation (11) to the following equation:

$$\frac{\alpha_{32}\alpha_{21}}{8\alpha_{11}}(w_\xi - 1)(w - \xi)\left(\frac{1}{2} + \ln\left[\frac{1}{2}(w - \xi)\right]\right) - w_{\xi\eta} = 0. \tag{34}$$

Theorem 2. *For Equation (11), there is a Bäcklund auto-transformation of the form*

$$e^g \frac{\partial g}{\partial \xi} = e^v \frac{\partial v}{\partial \xi}, \ e^g \frac{\partial g}{\partial \eta} = 2e^v \frac{\partial v}{\partial \eta} - \frac{\alpha_{32}\alpha_{21}}{8\alpha_{11}} v e^{2v}. \tag{35}$$

Proof of Theorem 2. Let us write the equality (35) in the following form:

$$\frac{\partial e^g}{\partial \xi} = e^v \frac{\partial v}{\partial \xi}, \qquad \frac{\partial e^g}{\partial \eta} = 2e^v \frac{\partial v}{\partial \eta} - \frac{\alpha_{32}\alpha_{21}}{8\alpha_{11}} v e^{2v},$$

and cross-differentiate. Equalizing the left parts gives

$$e^v \frac{\partial v}{\partial \xi}\frac{\partial v}{\partial \eta} + e^v \frac{\partial^2 v}{\partial \xi \partial \eta} - \frac{\alpha_{32}\alpha_{21}}{8\alpha_{11}} \frac{\partial v}{\partial \xi} e^{2v} - \frac{\alpha_{32}\alpha_{21}}{4\alpha_{11}} \frac{\partial v}{\partial \xi} v e^{2v} = 0$$

or Equation (11). □

Now, we rewrite the second equality (35) in the form

$$\frac{\partial e^g}{\partial \eta} = 2\frac{\partial e^v}{\partial \eta} - \frac{\alpha_{32}\alpha_{21}}{8\alpha_{11}} v e^{2v},$$

and differentiate by ξ,

$$\frac{\partial^2 e^g}{\partial \eta \partial \xi} = 2\frac{\partial^2 e^v}{\partial \eta \partial \xi} - \frac{\alpha_{32}\alpha_{21}}{8\alpha_{11}} e^{2v}(1 - 2v)\frac{\partial v}{\partial \xi}.$$

We replace the term $\frac{\alpha_{32}\alpha_{21}}{4\alpha_{11}} v_\xi e^v(1 + 2v)$ in the last equality with the remaining terms from (11), and then

$$\frac{\partial^2 e^g}{\partial \eta \partial \xi} = 2\frac{\partial^2 e^v}{\partial \eta \partial \xi} - e^v\left(\frac{\partial^2 v}{\partial \eta \partial \xi} + \frac{\partial v}{\partial \xi}\frac{\partial v}{\partial \eta}\right),$$

which leads to the equality

$$\frac{\partial^2 e^g}{\partial\eta\partial\xi} = \frac{\partial^2 e^v}{\partial\eta\partial\xi}.$$

This means that the functions e^g and e^v can differ only by arbitrary terms of the form $\varphi(\xi) + \psi(\eta)$, so

$$e^g + \varphi(\xi) + \psi(\eta) = e^v. \tag{36}$$

If $\varphi(\xi) = \psi(\eta) = 0$, then $g = v$, and from equalities (35), we obtain Equation (11).

Let us determine what happens if $\varphi(\xi) \neq 0, \quad \psi(\eta) \neq 0$. Let us perform substitution (33) in Equation (11); then, we obtain

$$e^g_{\eta\xi} = \frac{\alpha_{32}\alpha_{21}}{8\alpha_{11}}([e^g + \varphi(\xi) + \psi(\eta)]^2 \ln[e^g + \varphi(\xi) + \psi(\eta)])_\xi.$$

Let us perform differentiation

$$\begin{aligned} e^g(g_{\eta\xi} + g_\xi g_\eta) = \tfrac{\alpha_{32}\alpha_{21}}{8\alpha_{11}} 2[e^g + \varphi(\xi) + \psi(\eta)] \ln[e^g + \varphi(\xi) + \psi(\eta)](e^g g_\xi + \varphi'(\xi)) \\ + \tfrac{\alpha_{32}\alpha_{21}}{8\alpha_{11}}[e^g + \varphi(\xi) + \psi(\eta)](e^g g_\xi + \varphi'(\xi)) \end{aligned}$$

and group the terms with a common derivative; we obtain

$$e^g(g_{\eta\xi} + g_\xi g_\eta) = \frac{\alpha_{32}\alpha_{21}}{8\alpha_{11}}[e^g + \varphi(\xi) + \psi(\eta)](2\ln[e^g + \varphi(\xi) + \psi(\eta)] + 1)(e^g g_\xi + \varphi'(\xi)),$$

here, $\varphi(\xi)$, $\psi(\eta)$ are arbitrary functions.

Corollary 2. *Bäcklund transformations*

$$\frac{\partial q}{\partial\xi} + \varphi'(\xi) = e^v \frac{\partial v}{\partial\xi},$$

$$\frac{\partial q}{\partial\eta} + \psi'(\eta) = 2e^v \frac{\partial v}{\partial\eta} - \frac{\alpha_{32}\alpha_{21}}{8\alpha_{11}} v e^{2v},$$

associate Equation (11) with the equation

$$q_{\eta\xi} = \frac{\alpha_{32}\alpha_{21}}{8\alpha_{11}}[q + \varphi(\xi) + \psi(\eta)](2\ln[q + \varphi(\xi) + \psi(\eta)] + 1)(q_\xi + \varphi'(\xi)),$$

here, $\varphi(\xi)$, $\psi(\eta)$ are arbitrary functions.

Similarly, starting the transformation with a detailed analysis of equality (18), in each case for the remaining studied Equations (12) and (13), the following theorems are proved:

Theorem 3. *Bäcklund transformations of the form*

$$w_{\xi\xi} = w_\xi \frac{\partial v}{\partial\xi} - \frac{\alpha_{21}}{16\alpha_{11}^2} e^v w_\xi, \tag{37}$$

$$w_{\xi\eta} = \frac{w_\xi}{2}\frac{\partial v}{\partial\eta} - \frac{\alpha_{21}}{16\alpha_{11}^2} e^v w_\xi, \tag{38}$$

connect Equation (12) with the equation

$$(w^2)_{\xi\eta} = 4w_\xi^2. \tag{39}$$

Theorem 4. *Bäcklund transformations of the form*

$$\frac{\partial w}{\partial \xi} = \frac{\partial v}{\partial \xi} - e^{v}, \ \frac{\partial w}{\partial \eta} = e^{-v}$$

connect Equation (13) with the equation

$$w_{\eta\xi} + w_{\xi} w_{\eta} = -1.$$

3.2. Applying Differential Couplings to Obtain Exact Solutions

Theorem 5. *If Equation (39) has a solution*

$$w = 2\eta + \xi, \tag{40}$$

then Equation (12) has a solution:

$$v = -\ln\left[C - \frac{\alpha_{21}}{16\alpha_{11}^2}(\xi + \eta)\right], \quad C = \text{const.} \tag{41}$$

Proof of Theorem 5. We use the found transformations (37), (38) and substitute the known solution (40) in them, and then system (37), (38) takes the form

$$\frac{\partial v}{\partial \xi} = \frac{\alpha_{21}}{16\alpha_{11}^2} e^{v}, \qquad \frac{\partial v}{\partial \eta} = \frac{\alpha_{21}}{16\alpha_{11}^2} e^{v},$$

from here, we find

$$e^{-v} = C - \frac{\alpha_{21}}{16\alpha_{11}^2}(\xi + \eta),$$

here, C is an arbitrary constant. As a result, the solution (41) of Equation (12) was found. □

Let us perform some transformations in Equation (39), multiplying both sides by w^2:

$$w^2 (w^2)_{\xi\eta} - \left[(w^2)_{\xi}\right]^2 = 0,$$

and, using the Fourier method of separation of variables, we obtain a solution to Equation (39) in the form

$$w = e^{\frac{\lambda}{2}(\eta + \xi)}, \quad \lambda = \text{const.}$$

Theorem 6. *If Equation (39) has a solution*

$$w = e^{\frac{\lambda}{2}(\eta + \xi)}, \quad \lambda = \text{const} \tag{42}$$

then Equation (12) has a solution

$$v = \ln\lambda + \lambda\left(\eta + \frac{1}{2}\xi\right) - \ln\left|1 - \frac{\alpha_{21}}{8\alpha_{11}^2} e^{\lambda[\eta + \frac{1}{2}\xi]}\right|. \tag{43}$$

Proof of Theorem 6. Using the found transformations (37), (38), we substitute the known solution (42) into it, and then system (37), (38), after cancellation by $\frac{\lambda}{2}e^{\frac{\lambda}{2}(\eta+\xi)}$, takes the form

$$\frac{\lambda}{2} = \frac{\partial v}{\partial \xi} - \frac{\alpha_{21}}{16\alpha_{11}^2}e^v, \qquad \lambda = \frac{\partial v}{\partial \eta} - \frac{\alpha_{21}}{8\alpha_{11}^2}e^v, \tag{44}$$

from the first linear partial differential equation, we find

$$v - \ln\left[\frac{\lambda}{2} + \frac{\alpha_{21}}{16\alpha_{11}^2}e^v\right] = \frac{\lambda}{2}\xi + \varphi(\eta),$$

where $\varphi(\eta)$ is an arbitrary function, and from the second equation of system (44), we determine the form of the function $\varphi(\eta)$: $\varphi(\eta) = 2e^{\lambda\eta}$. □

Expressing the function v explicitly, we obtain the solution (43) of Equation (12).

Theorem 7. *If Equation (12) has a solution*

$$v = a(\eta + \xi),$$

then Equation (39) has a solution

$$w = -\frac{16\alpha_{11}^2}{\alpha_{21}}\exp\left[-\frac{\alpha_{21}}{16\alpha\alpha_{11}^2}e^{\alpha(\eta+\xi)} - \frac{\alpha}{2}\eta\right]. \tag{45}$$

Proof of Theorem 7. Using the found transformations (37), (38), we substitute the known solution $v = a(\eta + \xi)$, and then we obtain the system of equations

$$(\ln w_\xi)_\xi = \alpha - \frac{\alpha_{21}}{16\alpha_{11}^2}e^{\alpha(\eta+\xi)},$$

$$(\ln w_\xi)_\eta = \frac{\alpha}{2} - \frac{\alpha_{21}}{16\alpha_{11}^2}e^{\alpha(\eta+\xi)},$$

which can be easily integrated over the corresponding variables

$$\ln w_\xi = \alpha\xi - \frac{\alpha_{21}}{16\alpha\alpha_{11}^2}e^{\alpha(\eta+\xi)} + \varphi(\eta), \tag{46}$$

$$\ln w_\xi = \frac{\alpha}{2}\eta - \frac{\alpha_{21}}{16\alpha\alpha_{11}^2}e^{\alpha(\eta+\xi)} + \psi(\xi), \tag{47}$$

where $\varphi(\eta)$, $\psi(\xi)$ are the constants of integration (arbitrary functions). □

Let us extend the definition of the functions $\varphi(\eta)$ and $\psi(\xi)$ so that the obtained values of the right-hand sides of the system (46), (47) coincide as follows:

$$\varphi(\eta) = \frac{\alpha}{2}\eta, \quad \psi(\xi) = \alpha\xi.$$

As a result, an expression for the function w_ξ is defined:

$$w_\xi = e^{\alpha(\xi+\frac{1}{2}\eta) - \frac{\alpha_{21}}{16\alpha\alpha_{11}^2}e^{\alpha(\eta+\xi)}}.$$

We perform integration over ξ, and we obtain the unknown function $w(\xi,\eta)$ (45). The form of function (45) is shown from two angles in Figure 1 (for $\alpha = 1$, $\frac{\alpha_{21}}{\alpha_{11}^2} = -32$).

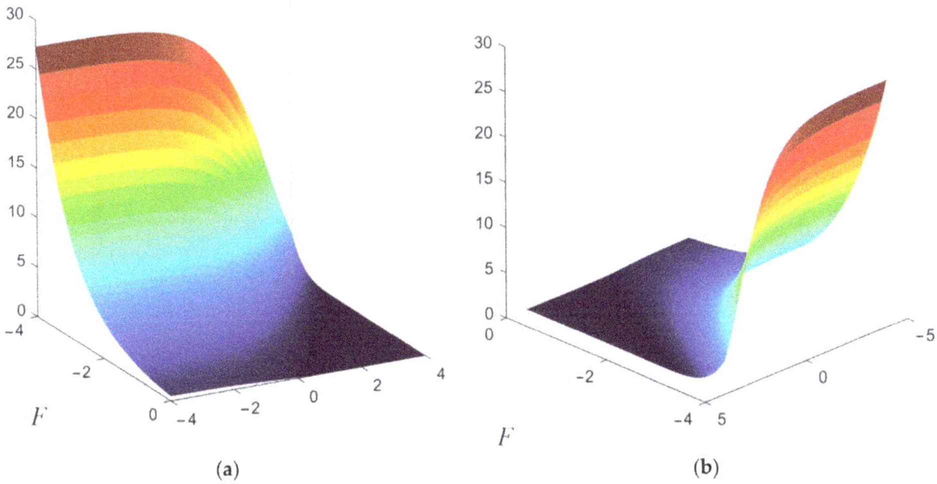

Figure 1. A 3D graph of the function (45) shown from two angles (**a**) and (**b**). Here, $F \equiv w(\xi, \eta)$.

Theorem 8. *Equation (31) has a solution implicitly given in the form of a series*

$$\ln\left|\ln\left[\frac{1}{2}(w-\xi)\right]\right| + \sum_{n=1}^{\infty}\frac{(-1)^n}{n\cdot n!}\ln^n\left[\frac{1}{2}(w-\xi)\right] = \frac{\alpha_{32}\alpha_{21}}{8\alpha_{11}}(\gamma\xi+\eta)+C_2,$$

where constants γ, C_2 *are arbitrary constants.*

The proof is carried out by simple verification.

Theorem 9. *If (10) has a solution* $v = a$, *then (31) has a solution*

$$w = \xi - \frac{\alpha_{32}\alpha_{21}}{4\alpha_{11}}ae^{2a}\eta. \tag{48}$$

Proof of Theorem 9. We substitute $v = a$ into the found differential links (34) and integrate each equality.

$$w = \xi + \varphi(\eta),$$
$$w = -\frac{\alpha_{32}\alpha_{21}}{4\alpha_{11}}ae^{2a}\eta + \psi(\xi).$$

We equate the obtained expressions for the function w and redefine arbitrary functions $\varphi(\eta)$, $\psi(\xi)$. As a result, we obtain (48). □

Theorem 10. *If Equation (11) has a solution* $v = \eta$, *then Equation (34) has a solution*

$$w = 4e^{\eta} - \frac{\alpha_{32}\alpha_{21}}{16\alpha_{11}}(2\eta-1)e^{2\eta} + \xi. \tag{49}$$

Proof of Theorem 10. Substitute $v = \eta$ into the found differential constraints (33) and integrate each equality.

$$w = \xi + \varphi(\eta),$$

$$w = 4e^{\eta} - \frac{\alpha_{32}\alpha_{21}}{16\alpha_{11}}(2\eta - 1)e^{2\eta} + \psi(\xi).$$

We equate the obtained expressions for the function w and redefine arbitrary functions $\varphi(\eta)$, $\psi(\xi)$. As a result, we obtain (49) (Figure 2). □

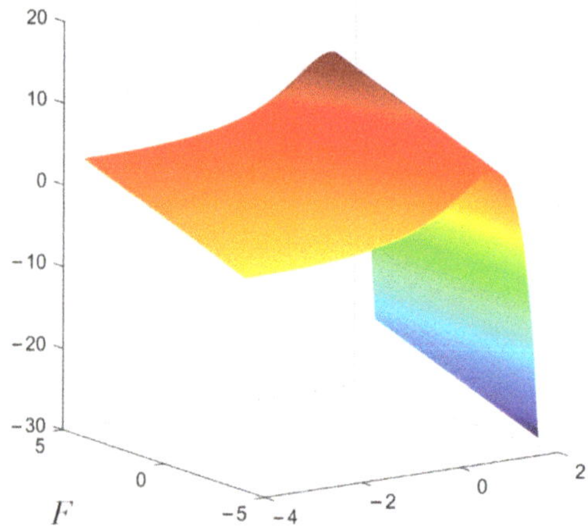

Figure 2. The plot is according to Formula (49) at $\frac{\alpha_{32}\alpha_{21}}{16\alpha_{11}} = \frac{1}{3}$. Here, $F \equiv w(\xi, \eta)$.

Theorem 11. *Equation (11) has a solution implicitly given in the form of a series*

$$\ln|v| + \sum_{n=1}^{\infty} \frac{(-v)^n}{n \cdot n!} = \frac{\alpha_{32}\alpha_{21}}{8\alpha_{11}}(\gamma\xi + \eta) + C_2, \quad (50)$$

where constants γ, C_2 are arbitrary constants.

The proof is carried out by simple verification.

Solution (50) is a cylindrical surface with a guide shown in Figure 3.

Let us use the found auto-Bäcklund transformations (35) for Equation (11) and solution (50). If we assume that $g(\xi, \eta) = v(\xi, \eta)$, then using (35), we can find a new solution to Equation (11). Substitute expression (50) into the left-hand side of (35):

$$\frac{\alpha_{32}\alpha_{21}}{8\alpha_{11}} g = \frac{1}{\gamma} e^{v} \frac{\partial v}{\partial \xi},$$

$$\frac{\alpha_{32}\alpha_{21}}{8\alpha_{11}} g = 2e^{v} \frac{\partial v}{\partial \eta} - \frac{\alpha_{32}\alpha_{21}}{8\alpha_{11}} v e^{2v}.$$

Equating the left-hand sides, we obtain a linear first-order equation for the function $v(\xi, \eta)$

$$2\frac{\partial v}{\partial \eta} - \frac{1}{\gamma}\frac{\partial v}{\partial \xi} = \frac{\alpha_{32}\alpha_{21}}{8\alpha_{11}} v e^{v}, \quad (51)$$

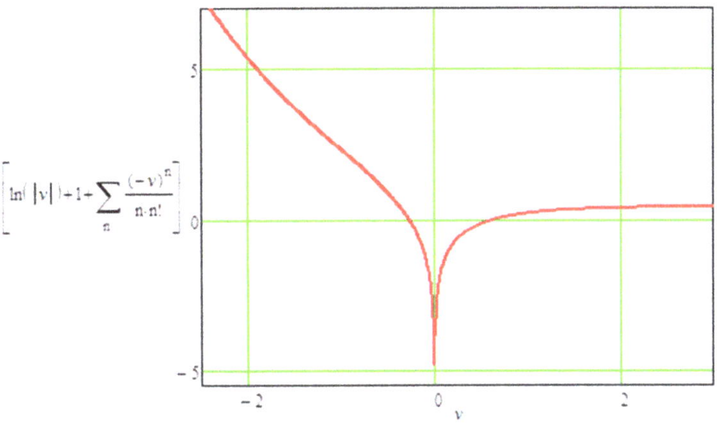

Figure 3. Cylindrical surface guide (50), where $n = 1, 2, \ldots, 100$.

To find the general solution of this equation, we find the first integrals of the system

$$2\gamma\xi + \eta = C_1, \quad \frac{8\alpha_{11}}{\gamma\alpha_{32}\alpha_{21}}\left[\ln|v| + \sum_{n=1}^{\infty}\frac{(-v)^n}{n\cdot n!}\right] + \xi = C_2. \tag{52}$$

The general solution to equation (51) has the form

$$F\left(2\gamma\xi + \eta,\ \frac{8\alpha_{11}}{\gamma\alpha_{32}\alpha_{21}}\left[\ln|v| + \sum_{n=1}^{\infty}\frac{(-v)^n}{n\cdot n!}\right] + \xi\right) = C,$$

Here, F is any function.

Equation (11) is nonlinear; therefore, it is necessary to clarify the form of the function F. Let us substitute expression (52) into Equation (11)

$$\frac{16\alpha_{11}}{\alpha_{32}\alpha_{21}}\left[2\frac{F_1}{F_2^2}F_{12} - \frac{F_1^2}{F_2^3}F_{22} - \frac{F_{11}}{F_2}\right] = -\left(\frac{1}{\gamma} + \frac{F_1}{F_2}\right)\left(2\gamma\frac{F_1}{F_2} + 1\right)e^v(1 + 2v). \tag{53}$$

Here, F_1 is the derivative of F by the first component, and F_2 is the derivative of F by the second component.

As one can see, equality (53) is not identically fulfilled; therefore, it is necessary to require that one of the systems is fulfilled:

$$\begin{cases} \frac{1}{\gamma} + \frac{F_1}{F_2} = 0, \\ 2\frac{F_1}{F_2^2}F_{12} - \frac{F_1^2}{F_2^3}F_{22} - \frac{F_{11}}{F_2} = 0, \end{cases}, \text{ or } \begin{cases} \frac{F_1}{F_2} = -\frac{1}{2\gamma}, \\ 2\frac{F_1}{F_2^2}F_{12} - \frac{F_1^2}{F_2^3}F_{22} - \frac{F_{11}}{F_2} = 0 \end{cases}. \tag{54}$$

All the terms of the equalities are homogeneous, so we use the technique that allows us to separate the arguments of the function. We represent F in the form

$$F = X(C_1)Y(C_2),$$

where X depends on the first component C_1 of the function F, and Y on the second component C_2 (52), and then the first equality of system (54) takes the following form (due to the similarity of the first equalities of the systems, the result of the substitution for the second system is written in parentheses):

$$\frac{X'}{X} = -\frac{1}{\gamma}\frac{Y'}{Y} = \lambda, \quad \left(\frac{X'}{X} = -\frac{1}{2\gamma}\frac{Y'}{Y} = \lambda\right),$$

13. Veerakumar, V.; Daniel, M. Modified Kadomtsev-Petviashvili (MKP) equation and electromagnetic soliton. *Math. Comput. Simul.* **2003**, *62*, 163–169. [CrossRef]
14. Song, J.F.; Hu, Y.H.; Ma, Z.Y. Bäcklund transformation and CRE solvability for the negative-order modified KdV equation. *Nonlinear Dyn.* **2017**, *90*, 575–580. [CrossRef]
15. Zakharov, V.E.; Kuznetsov, V.A. Hamiltonian formalism for nonlinear waves. *Adv. Phys. Sci.* **1997**, *167*, 1137–1168. (In Russian)
16. Gulenko, V.V.; Guschin, V.V. Hamiltonov formulation of new dynamic equations. *Rep. Acad. Sci. Ukr.* **1994**, 3, 73–77. (In Russian)
17. Cheng, J. Miura and auto-Bäcklund transformations for the q-deformed KP and q-deformed modified KP hierarchies. *J. Nonlinear Math. Phys.* **2017**, *24*, 7–19. [CrossRef]
18. Zabrodin, A.V. Bäcklund transformations for the difference Hirota equation and the supersymmetric Bethe ansatz. *Theor. Math. Phys.* **2008**, *155*, 74–93. [CrossRef]
19. Tsiganov, A.V. Bäcklund transformations and divisor doubling. *J. Geom. Phys.* **2018**, *126*, 148–158. [CrossRef]
20. Lamb, G.L. *Elements of Soliton Theory*; John Wiley & Sons: New York, NY, USA, 1980.
21. Pogorelov, A.V. *Multivariate Monge-Ampere Equation*; Science: Moscow, Russia, 1988. (In Russian)

axioms

MDPI

Article

A New q-Hypergeometric Symbolic Calculus in the Spirit of Horn, Borngässer, Debiard and Gaveau

Thomas Ernst

Department of Mathematics, Uppsala University, P.O. Box 480, SE-751 06 Uppsala, Sweden; thomas@math.uu.se

Abstract: The purpose of this article is to introduce a new complete multiple q-hypergeometric symbolic calculus, which leads to q-Euler integrals and a very similar canonical system of q-difference equations for multiple q-hypergeometric functions. q-analogues of recurrence formulas in Horns paper and Borngässers thesis lead to a more exact way to find these Frobenius solutions. To find the right formulas, the parameters in q-shifted factorials can be changed to negative integers, which give no extra q-factors. In proving these q-formulas, in the limit $q \to 1$ we obtain versions of the paper by Debiard and Gaveau for the solution of differential or q-difference equations. The paper is also a correction of some of the statements in the paper by Debiard and Gaveau, e.g., the Euler integrals and other solutions to differential equations for Appell functions, also without references to page numbers in the standard work of Appell and Kampé de Fériet. Sometimes the q-binomial theorem is used to simplify q-integral formulas. By the Horn method, we find another solution to the Appell Φ_1 function partial differential equation, which was not mentioned in the thesis by Le Vavasseur 1893.

Keywords: symbolic calculus; canonical system of q-difference equations; q-Euler integral

MSC: Primary 33D05; Secondary 33C65; 33D70; 33D60

Citation: Ernst, T. A New q-Hypergeometric Symbolic Calculus in the Spirit of Horn, Borngässer, Debiard and Gaveau. *Axioms* **2022**, *11*, 64. https://doi.org/10.3390/axioms11020064

Academic Editors: Hans J. Haubold and Anna Maria Fino

Received: 28 December 2021
Accepted: 30 January 2022
Published: 4 February 2022

Publisher's Note: MDPI stays neutral with regard to jurisdictional claims in published maps and institutional affiliations.

Copyright: © 2022 by the author. Licensee MDPI, Basel, Switzerland. This article is an open access article distributed under the terms and conditions of the Creative Commons Attribution (CC BY) license (https://creativecommons.org/licenses/by/4.0/).

1. Introduction

We refer to our standard work [1] and to the paper on multiple q-hypergeometric functions [2]. The pathbreaking paper [3] by Debiard and Gaveau on a new umbral calculus led to the automatic solutions of differential equations for multiple hypergeometric functions according to Frobenius and Horn. In this paper, we generalize this method to the q-case and slightly change the notation for a better overview. As examples, the exponent in method of Frobenius is changed from α to λ and the Euler operator $x\frac{d}{dx}$ is changed to $\theta_{(q)}$ as in [1]. Our umbral calculus simply means that a $\theta_{q,1} \vee \theta_{q,2}$ before a double power series is replaced by the exponents of $x \vee y$. The same goes for additive arguments in the Γ_q function.

A proper notation is extremely important in papers on special functions, since long computations often occur and the origin of the variables is crucial for the understanding of the formulas. The notation and especially the computations in [3] are sometimes erroneous, one example is the notation on the top of page 789, where small a, α_i and α occur, together with a misprint. For operators, we mention the spaces of formal power series in their definitions. We also remember that in Horns paper ([4], p. 387) and in Borngässers thesis [5], recurrence formulas for the determination of the other solutions in the method of Frobenius were given, which was missed in [3].

The paper is organized as follows: In Section 1 we define all q-functions. In Section 2 we present Horns and Borngässers recurrence formulas for the coefficients in the method of Frobenius, which have a very similar form as before. In Section 3 we introduce the general symbolic calculus. In Section 4 we find bases for the spaces of solutions by the Frobenius method for the first q-Appell function. In Sections 5–8 we consider the q-Appell functions Φ_1, Φ_2, Φ_3, Φ_4.

Let $\delta > 0$ be an arbitrary small number. We will always use the following branch of the logarithm: $-\pi + \delta < \text{Im}(\log q) \leq \pi + \delta$. This defines a simply connected space in the complex plane.

The power function is defined by

$$q^a \equiv e^{a\log(q)}. \tag{1}$$

A q-analogue of a complex number is also a complex number.

Definition 1. *The q-analogue of a complex number a is defined as follows:*

$$\{a\}_q \equiv \frac{1-q^a}{1-q},\; q \in \mathbb{C}\setminus\{0,1\}, \tag{2}$$

The q-shifted factorial is defined by

$$\langle a;q\rangle_n \equiv \prod_{m=0}^{n-1}(1-q^{a+m}), \tag{3}$$

The q-derivative is defined by

$$(D_q\varphi)(x) \equiv \begin{cases} \frac{\varphi(x)-\varphi(qx)}{(1-q)x}, & \text{when } q \in \mathbb{C}\setminus\{1\},\ x \neq 0; \\ \frac{d\varphi}{dx}(x), & \text{when } q = 1; \\ \frac{d\varphi}{dx}(0), & \text{when } x = 0. \end{cases} \tag{4}$$

Definition 2. *The following operator will also be useful.*

$$\theta_{q,j} \equiv x_j D_{q,x_j}. \tag{5}$$

Definition 3. *[1]. The q-analogues of the Appell functions are*

$$\Phi_1(a;b,b';c|q;x_1,x_2) \equiv \sum_{m_1,m_2=0}^{\infty} \frac{\langle a;q\rangle_{m_1+m_2}\langle b;q\rangle_{m_1}\langle b';q\rangle_{m_2}}{\langle 1;q\rangle_{m_1}\langle 1;q\rangle_{m_2}\langle c;q\rangle_{m_1+m_2}} x_1^{m_1}x_2^{m_2}, \quad max(|x_1|,|x_2|) < 1. \tag{6}$$

$$\Phi_2(a;b,b';c,c'|q;x_1,x_2) \equiv \sum_{m_1,m_2=0}^{\infty} \frac{\langle a;q\rangle_{m_1+m_2}\langle b;q\rangle_{m_1}\langle b';q\rangle_{m_2}}{\langle 1;q\rangle_{m_1}\langle 1;q\rangle_{m_2}\langle c;q\rangle_{m_1}\langle c';q\rangle_{m_2}} x_1^{m_1}x_2^{m_2}, \quad |x_1| \oplus_q |x_2| < 1. \tag{7}$$

$$\Phi_3(a,a';b,b';c|q;x_1,x_2) \equiv \sum_{m_1,m_2=0}^{\infty} \frac{\langle a;q\rangle_{m_1}\langle a';q\rangle_{m_2}\langle b;q\rangle_{m_1}\langle b';q\rangle_{m_2}}{\langle 1;q\rangle_{m_1}\langle 1;q\rangle_{m_2}\langle c;q\rangle_{m_1+m_2}} x_1^{m_1}x_2^{m_2}, \quad max(|x_1|,|x_2|) < 1. \tag{8}$$

$$\Phi_4(a;b;c,c'|q;x_1,x_2) \equiv \sum_{m_1,m_2=0}^{\infty} \frac{\langle a;q\rangle_{m_1+m_2}\langle b;q\rangle_{m_1+m_2}}{\langle 1;q\rangle_{m_1}\langle 1;q\rangle_{m_2}\langle c;q\rangle_{m_1}\langle c';q\rangle_{m_2}} x_1^{m_1}x_2^{m_2}, \quad |\sqrt{x_1}| \oplus_q |\sqrt{x_2}| < 1. \tag{9}$$

2. q-Analogues of Horns and Borngässers Recurrence Formulas

The purpose of this section is to introduce q-analogues of Horns and Borngässers recurrence formulas ([4], p. 387), ([5], p. 26 ff) for double series. We just state the formulas, the proofs are simple; in the process we slightly improve the notation. We start with the double q-hypergeometric series

$$z \equiv H(x,y) = \sum_{m,n=0}^{\infty} A_{mn} x^m y^n, \tag{10}$$

where the two quotients

$$f(m,n) \equiv \frac{A_{m+1,n}}{A_{mn}},\ g(m,n) \equiv \frac{A_{m,n+1}}{A_{mn}} \tag{11}$$

are rational functions in q-analogues of m,n. Now put

$$f(m,n) \equiv \frac{F(m,n)}{F'(m,n)},\ g(m,n) \equiv \frac{G(m,n)}{G'(m,n)}, \tag{12}$$

where $F(mn),\ G(mn),\ F'(mn),\ G'(mn)$ are entire products of q-analogues in m,n of maximal second order in m,n. We assume that $F'(mn)$ has the factor $\{1+m\}_q$, $G'(mn)$ has the factor $\{1+n\}_q$.

All these q-functions are q-analogues of the Appell, confluent Humbert, etc., and Horn functions.

We just state a q-analogue of a generalization of the Euler operator ([4], p. 387). Assume that $\alpha,\ \beta \in \mathbb{R}[\theta_{q,1},\theta_{q,2}]$. $\alpha,\ \beta$ are linear functions of $\theta_{q,1}$ and $\theta_{q,2}$ with coefficients $\in \mathbb{Z}$, and z is defined by (10).

$$(1-q)^2\{\alpha\}_q\{\beta\}_q z = \sum_{m,n=0}^{\infty} \langle\alpha;q\rangle_1\langle\beta;q\rangle_1 A_{mn} x^m y^n. \tag{13}$$

A q-analogue of an improved version of ([4], p.387), where we have skipped the sums $\sum_{\alpha,\beta}$. Assume that $\alpha,\ \beta,\ \gamma,\ \delta, \alpha',\ \beta',\ \gamma',\ \delta' \in \mathbb{R}[\theta_{q,1},\theta_{q,2}]$ are linear functions of $\theta_{q,1}$ and $\theta_{q,2}$ with coefficients $\in \mathbb{Z}$. Furthermore, the function z in (10) satisfies the system of q-difference equations

$$\begin{aligned} &(x\{\alpha\}_q\{\beta\}_q - \{\alpha'\}_q\{\beta'\}_q)z = 0, \\ &(y\{\gamma\}_q\{\delta\}_q - \{\gamma'\}_q\{\delta'\}_q)z = 0, \end{aligned} \tag{14}$$

with convenient boundary values.

Case I. Assume instead that

$$z = \sum_{m,n=0}^{\infty} C_{mn} x^{m+\rho} y^{n+\sigma}, \tag{15}$$

where ρ and σ are unknown real constants. In the previous case, C_{mn} become A_{mn}. We now have the recurrence formulas

$$\begin{cases} F'(m+\rho,n+\sigma)C_{m+1,n} = F(m+\rho,n+\sigma)C_{m,n} \\ G'(m+\rho,n+\sigma)C_{m,n+1} = G(m+\rho,n+\sigma)C_{m,n}, \end{cases} \tag{16}$$

which follow from the previous recurrence formulas for A_{mn}.

By comparing the coefficients of

$$x^{m+\rho}y^{n+\sigma},\ m=-1; n \geq 0 \tag{17}$$

in the first recurrence, and the coefficients of

$$x^{m+\rho}y^{n+\sigma},\ m \geq 0,\ n=-1 \tag{18}$$

in the second recurrence, we obtain the equations ([5], p. 27), ([4], p. 388) for the determination of the exponents ρ and σ

$$\begin{cases} F'(\rho-1,\sigma)=0, \\ G'(\rho,\sigma-1)=0 \end{cases} \tag{19}$$

Case II. For the determination of the solutions in the vicinity of the point (∞,∞), we look at series of the form

$$z=\sum_{m,n=0}^{\infty} C_{mn}x^{\rho-m}y^{\sigma-n}, \tag{20}$$

where ρ and σ are unknown real constants. We now have the recurrence formulas

$$\begin{cases} F(\rho-m-1,\sigma-n)C_{m+1,n}=F'(\rho-m-1,\sigma-n)C_{m,n} \\ G(\rho-m,\sigma-n-1)C_{m,n+1}=G'(\rho-m,\sigma-n-1)C_{m,n}. \end{cases} \tag{21}$$

By comparing the coefficients of

$$x^{\rho-m}y^{\sigma-n}, \tag{22}$$

we obtain the equations ([5], p. 28), ([4], p. 388) for the determination of the exponents ρ and σ

$$\begin{cases} F(\rho,\sigma)=0, \\ G(\rho,\sigma)=0 \end{cases} \tag{23}$$

Case III. For the determination of the solutions in the vicinity of the point $(0,\infty)$, we look at series of the form

$$z=\sum_{m,n=0}^{\infty} C_{mn}x^{\rho+m}y^{\sigma-n}, \tag{24}$$

which leads to the recurrence formulas

$$\begin{cases} F'(\rho+m,\sigma-n)C_{m+1,n}=F(\rho+m,\sigma-n)C_{m,n} \\ G(\rho+m,\sigma-n-1)C_{m,n+1}=G'(\rho+m,\sigma-n-1)C_{m,n}. \end{cases} \tag{25}$$

By comparing the coefficients of

$$x^{\rho+m}y^{\sigma-n}, \tag{26}$$

we obtain the equations ([5], p. 29), ([4], p. 388) for the determination of the exponents ρ and σ

$$\begin{cases} F'(\rho-1,\sigma)=0, \\ G(\rho,\sigma)=0. \end{cases} \tag{27}$$

Case IV. Finally, for the determination of the solutions in the vicinity of the point $(\infty,0)$, we look at series of the form

$$z=\sum_{m,n=0}^{\infty} C_{mn}x^{\rho-m}y^{\sigma+n}, \tag{28}$$

which leads to the recurrence formulas

$$\begin{cases} F(\rho-m-1,\sigma+n)C_{m+1,n}=F'(\rho-m-1,\sigma+n)C_{m,n} \\ G'(\rho-m,\sigma+n)C_{m,n+1}=G(\rho-m,\sigma+n)C_{m,n}. \end{cases} \tag{29}$$

By comparing the coefficients of

$$x^{\rho-m}y^{\sigma+n}, \tag{30}$$

we obtain the equations ([5], p. 29), ([4], p. 388) for the determination of the exponents ρ and σ

$$\begin{cases} F(\rho,\sigma) = 0, \\ G'(\rho,\sigma-1) = 0 \end{cases} \tag{31}$$

3. General Symbolic Calculus

The purpose of this section is to introduce the general symbolic calculus for double series.

Definition 4. *We put*

$$H_q(a,b,c,x) \equiv \{\theta_{q,1}+c\}_q D_{q,x} - \{\theta_{q,1}+a\}_q\{\theta_{q,1}+b\}_q. \tag{32}$$

Then we have

$$H_q(a,b,c,x)\ {}_2\phi_1\left[\begin{matrix} a,b \\ c \end{matrix}\middle| q;x\right] = 0. \tag{33}$$

We are always interested in solutions to the equation

$$H_q(a,b,c,x)(f(x)) = 0. \tag{34}$$

Around $x=0$ another solution, apart from y_1 in (33) is (6.186 [1])

$$y_2 = x^{1-c}\ {}_2\phi_1\left[\begin{matrix} a-c+1, b-c+1 \\ 2-c \end{matrix}\middle| q;x\right]. \tag{35}$$

The purpose of the next definition is to keep the powers of the variables in the operator.

Definition 5. *Let A, B, C be three operators $\mathbb{R}[[x]] \to \mathbb{R}[[x]]$, which are linear in $\theta_{q,1}$, $\theta_{q,2}$. Then we define*

$$H_q(A,B,C,x) \equiv \{\theta_{q,1}+C\}_q D_{q,x} - \{\theta_{q,1}+A\}_q\{\theta_{q,1}+B\}_q. \tag{36}$$

Lemma 1. *Compare with ([3], p. 777). Let $\mathscr{F}(a,b,c,x)$ be a solution of*

$$H_q(a,b,c,x)\mathscr{F} = 0. \tag{37}$$

Then

$$\mathscr{F}(A,B,C,x) \tag{38}$$

is a solution of

$$H_q(A,B,C,x)\mathscr{F}(A,B,C,x) = 0. \tag{39}$$

Definition 6. *Compare with ([3], p. 777). Assume $x>0$, C, $\lambda \in \mathbb{R}$, $\psi(y) \times y^{-\lambda} \in \mathbb{R}[[y]]$. Then, in the umbral sense,*

$$x^{C\theta_{q,2}}\psi(y) \ddot{=} \psi(x^C y). \tag{40}$$

Assume that

$$\phi(\vec{a},\vec{b},y) \equiv y^\lambda \sum_{k=0}^{\infty} \frac{\langle \vec{a};q\rangle_k}{\langle 1,\vec{b};q\rangle_k} y^k. \tag{41}$$

Then, in the umbral sense,

$$\phi(\vec{a}+\vec{a'}\theta_{q,2},\vec{b}+\vec{b'}\theta_{q,2},y) \equiv y^{\lambda}\sum_{k=0}^{\infty}\frac{\langle \vec{a}+\vec{a'}(\lambda+k);q\rangle_k}{\langle 1,\vec{b}+\vec{b'}(\lambda+k);q\rangle_k}y^k. \tag{42}$$

Furthermore, for the Γ_q-function:

$$\Gamma_q(\alpha+\theta_{q,2})y^{\lambda} \doteq \Gamma_q(\alpha+\lambda)y^{\lambda}. \tag{43}$$

We can generalize this to many variables.

Definition 7. *Compare with (3.9 [3]). Let $H_q(x,y)$ be an operator $\mathbb{R}[[x]] \to \mathbb{R}[[x]]$:*

$$H_q(x,y) \equiv \{\theta_{q,1}+\gamma\theta_{q,2}+c\}_q D_{q,x} - \{\theta_{q,1}+\alpha\theta_{q,2}+a\}_q\{\theta_{q,1}+\beta\theta_{q,2}+b\}_q. \tag{44}$$

The parameters in $H_q(x,y)$ will always be the same.

Remark 1. *The function (44) generalizes the basic definition (32) and is a special case of the more general definition (36). The notation in ([3], 3.9) is slightly misleading.*

Theorem 1. *Compare with ([3], (3.12) p. 779). Let $\mathscr{F}_j(a,b,c,x),\ j=1,2$ of the form (41) be two independent solutions of*

$$H_q(a,b,c,x)\mathscr{F}_j = 0. \tag{45}$$

Furthermore, let $\psi_j(y) \in \mathbb{R}[[y]],\ j=1,2$ be q-hypergeometric series, with suitable convergence radii. Then the general solution of the equation

$$H_q(x,y)f = 0 \tag{46}$$

in the umbral form (42) is given by

$$f(x,y) = \sum_{j=1}^{2}\mathscr{F}_j(\alpha\theta_{q,2}+a,\beta\theta_{q,2}+b,\gamma\theta_{q,2}+c,x)\phi_j(y). \tag{47}$$

Proof. This follows from (39). □

Theorem 2. *Compare with ([3], (4.5) p. 780). Let $\psi(y) \in \mathbb{R}[[y]]$ be a q-hypergeometric series, with suitable convergence radius. Then the series*

$$F(x,y:q) \equiv {}_2\phi_1(\alpha\theta_{q,2}+a,\beta\theta_{q,2}+b,\gamma\theta_{q,2}+c|q;x)\psi(y) \tag{48}$$

is a double q-hypergeometric series, convergent in the vicinity of $(0,0)$.

Proof. Similar to ([3], p. 780). □

Definition 8. *Compare with (5.1 [3]). Introduce the two general operators $\mathbb{R}[[x,y]] \to \mathbb{R}[[x,y]]$:*

$$\begin{aligned} H_{1;q}(x,y) &\equiv \{\gamma_1\theta_{q,2}+c_1\}_q D_{q,x} - \{\alpha_1\theta_{q,2}+a_1\}_q\{\beta_1\theta_{q,2}+b_1\}_q, \\ H_{2;q}(y,x) &\equiv \{\gamma_2\theta_{q,1}+c_2\}_q D_{q,y} - \{\alpha_2\theta_{q,1}+a_2\}_q\{\beta_2\theta_{q,1}+b_2\}_q. \end{aligned} \tag{49}$$

We wish to study the system of q-difference equations

$$\begin{cases} H_{1;q}(x,y)f(x,y) = 0 \\ H_{2;q}(y,x)f(x,y) = 0. \end{cases} \tag{50}$$

The system (50) is called q-compatible if it has common solutions $f(x,y)$.

Theorem 3. *Compare with ([3], p. 785). The system (50) is q-compatible if the following two products of q-analogues are equal,*

$$P_1(m,n;q) = P_2(m,n;q), \tag{51}$$

where

$$\begin{aligned} &P_1(m,n;q) \equiv \{m+\alpha_1 n + a_1\}_q\{m+\beta_1 n + b_1\}_q\{m+\gamma_1(n+1)+c_1\}_q \\ &\{n+\alpha_2(m+1)+a_2\}_q\{n+\beta_2(m+1)+b_2\}_q\{n+\gamma_2 m + c_2\}_q, \\ &P_2(m,n;q) \equiv \{m+\alpha_1(n+1)+a_1\}_q\{m+\beta_1(n+1)+b_1\}_q\{m+\gamma_1 n + c_1\}_q \\ &\{n+\alpha_2 m + a_2\}_q\{m+\beta_2 n + b_2\}_q\{n+\gamma_2(m+1)+c_2\}_q. \end{aligned} \tag{52}$$

For the following proof, compare with ([3], p. 786).

Proof. We put

$$f(x,y) = \sum_{m,n=0}^{\infty} A_{mn} x^m y^n. \tag{53}$$

We first calculate the following operator formulas.

$$\{\theta_{q,1}+\gamma_1\theta_{q,2}+c_1\}_q \mathrm{D}_{q,x} f = \sum_{m,n=0}^{\infty} a_{m+1,n}\frac{\{m+\gamma_1 n + c_1\}_q}{\langle 1;q\rangle_m \langle 1;q\rangle_n} x^m y^n \tag{54}$$

$$\begin{aligned} &\{\theta_{q,1}+\alpha_1\theta_{q,2}+a_1\}_q\{\theta_{q,1}+\beta_1\theta_{q,2}+b_1\}_q f \\ &= \sum_{m,n=0}^{\infty} a_{mn}\frac{\{m+\alpha_1 n + a_1\}_q\{m+\beta_1 n + b_1\}_q}{\langle 1;q\rangle_m \langle 1;q\rangle_n} x^m y^n, \end{aligned} \tag{55}$$

where

$$A_{mn} \equiv \frac{a_{mn}}{\langle 1;q\rangle_m \langle 1;q\rangle_n}. \tag{56}$$

The first Equation (50) is satisfied when

$$\frac{a_{m+1,n}}{a_{mn}} = \frac{\{m+\alpha_1 n + a_1\}_q\{m+\beta_1 n + b_1\}_q}{\{m+\gamma_1 n + c_1\}_q}. \tag{57}$$

The second Equation (50) is satisfied when

$$\frac{a_{m,n+1}}{a_{mn}} = \frac{\{n+\alpha_2 m + a_2\}_q\{n+\beta_2 m + b_2\}_q}{\{n+\gamma_2 m + c_2\}_q}. \tag{58}$$

Using Horn's notation, we have

$$f(m,n) \equiv \frac{A_{m+1,n}}{A_{mn}} = \frac{\{m+\alpha_1 n + a_1\}_q\{m+\beta_1 n + b_1\}_q}{\{m+\gamma_1 n + c_1\}_q\{m+1\}_q}, \tag{59}$$

$$g(m,n) \equiv \frac{A_{m,n+1}}{A_{mn}} = \frac{\{n+\alpha_2 m + a_2\}_q\{n+\beta_2 m + b_2\}_q}{\{n+\gamma_2 m + c_2\}_q\{n+1\}_q}. \tag{60}$$

Now (51) follows from the compatibility condition

$$f(m,n)g(m+1,n) = f(m,n+1)g(m,n). \tag{61}$$

□

Similarly, we find that the q-hypergeometric functions defined by (6)–(9), after rescaling, satisfy the systems (50).

4. Bases for the Spaces of Solutions by the Frobenius Method

Assuming that our system (50) is q-compatible, by using Lemma 1, we construct a basis for its solutions. Like in ([3], p. 788), the parameter $c_i \ni -\mathbb{Z}$. Many of these solutions are obtained simply by the same formula after q-deformation like in [1]. We note that other solutions are obtained by permutation of the variables.

Solutions in the Vicinity of $(0,0)$

Now we assume that

$$\psi(y) \equiv y^{\lambda} \sum_{k=0}^{\infty} a_n \frac{y^k}{\langle 1;q\rangle_k}. \tag{62}$$

Consider the basis $y_1(x),\ y_2(x)$ in (33) and (35). In all these assumptions, we put the first coefficient $a_0 = 1$. Note that this was not mentioned in [3]. In order to obtain solutions of our system of q-difference equations, define (49)

$$f(x,y;q) \equiv {}_2\phi_1(\alpha_1\theta_{q,2} + a_1, \beta_1\theta_{q,2} + b_1, \gamma_1\theta_{q,2} + c_1 | q; x)\psi(y). \tag{63}$$

This implies

$$f(x,y;q) = \sum_{m,n=0}^{\infty} a_n \frac{\langle \alpha_1(n+\lambda) + a_1, \beta_1(n+\lambda) + b_1; q\rangle_m}{\langle 1, \gamma_1(n+\lambda) + c_1; q\rangle_m \langle 1;q\rangle_n} x^m y^{n+\lambda} \tag{64}$$

$$\begin{aligned}(\theta_{q,2} + \gamma_2\theta_{q,1} + c_2)\mathrm{D}_{q,y} f = y^{\lambda} \sum_{m,n=0}^{\infty} a_n x^m y^{n-1} \{n+\lambda\}_q \\ \frac{\langle \alpha_1(n+\lambda) + a_1, \beta_1(n+\lambda) + b_1; q\rangle_m \{n+\lambda-1+\gamma_2 m + c_2\}_q}{\langle 1, \gamma_1(n+\lambda) + c_1; q\rangle_m \langle 1;q\rangle_n}\end{aligned} \tag{65}$$

$$\begin{aligned}&(\theta_{q,2} + \alpha_2\theta_{q,1} + a_2)(\theta_{q,2} + \beta_2\theta_{q,1} + b_2) f \\ &= y^{\lambda} \sum_{m,n=0}^{\infty} a_n x^m y^n \{n+\lambda+\alpha_2 m + a_2\}_q \\ &\frac{\langle \alpha_1(n+\lambda) + a_1, \beta_1(n+\lambda) + b_1; q\rangle_m \{n+\lambda+\beta_2 m + b_2\}_q}{\langle 1, \gamma_1(n+\lambda) + c_1; q\rangle_m \langle 1;q\rangle_n}\end{aligned} \tag{66}$$

By equating the last two formulas for $n = 0$, we obtain the indicial equation

$$\{\lambda\}_q \{\lambda - 1 + \gamma_2 m + c_2\}_q = 0,\ m \geq 0, \tag{67}$$

which implies

$$\begin{aligned}&\lambda = 0,\ \forall \gamma_2, \\ &\lambda = 0 \vee \lambda = 1 - c_2 \text{ if } \gamma_2 = 0.\end{aligned} \tag{68}$$

5. First q-Appell Function

We now apply the general method from the previous section to the first q-Appell function. Put $\gamma_i = \alpha_i = 1,\ c_i = c,\ a_i = a$ in (49). Like before the system is denoted by $(H_i f(x,y;q))_{i=1}^2$.

$$\begin{cases} \left[\{\theta_{q,1} + \theta_{q,2} + c\}_q \mathrm{D}_{q,x} - \{\theta_{q,1} + \theta_{q,2} + a\}_q \{\theta_{q,1} + b_1\}_q\right] f(x,y;q) = 0 \\ \left[\{\theta_{q,1} + \theta_{q,2} + c\}_q \mathrm{D}_{q,y} - \{\theta_{q,1} + \theta_{q,2} + a\}_q \{\theta_{q,2} + b_2\}_q\right] f(x,y;q) = 0. \end{cases} \tag{69}$$

With (62), $a_0 = 1$ and y_1 in (33), we get the first q-Appell function. Next consider the function y_2 in (35).

We find that the following equation can be rewritten by (40) and (43) as

$$\begin{aligned} &f_D(x,y;q)\\ &\equiv x^{1-c-\theta_{q,2}}{}_2\phi_1(a+1-c,b_1+1-c-\theta_{q,2};2-c-\theta_{q,2}|q;x)\psi(y)\\ &=x^{1-c}\sum_{m,n=0}^{\infty} a_n\frac{\langle a+1-c,b_1+1-c-n-\lambda;q\rangle_m}{\langle 1,2-c-n-\lambda+c_1;q\rangle_m\langle 1;q\rangle_n}x^{m-n-\lambda}y^{n+\lambda}. \end{aligned} \tag{70}$$

Lemma 2.

$$\frac{\langle b_1+1-c-n;q\rangle_m\langle c-1;q\rangle_n\langle 2-c;q\rangle_{m-n}}{\langle 2-c-n;q\rangle_m\langle c-b_1;q\rangle_n\langle b_1+1-c;q\rangle_{m-n}}=q^{n(b_1-1)}. \tag{71}$$

This lemma is used in the proof (77). Similar to ([3], p. 793) we find that

$$\begin{aligned} &\left[\{\theta_{q,1}+\theta_{q,2}+c\}_q \mathrm{D}_{q,y}\right](x^{m-n-\lambda+1-c}y^{n+\lambda})\\ &=\{m\}_q\{n+\lambda\}_q x^{m-n-\lambda+1-c}y^{n+\lambda-1}. \end{aligned} \tag{72}$$

Again, $\lambda=0$, and we have

$$\begin{aligned} H_2f_D(x,y;q)&=x^{1-c}\left[\sum_{m,n=1}^{\infty} a_n\frac{\langle a+1-c,b_1+1-c-n;q\rangle_m}{\langle 2-c-n;q\rangle_m\langle 1;q\rangle_{m-1}\langle 1;q\rangle_{n-1}}x^{m-n}y^{n-1}\right.\\ &\left.-\sum_{m,n=0}^{\infty} a_n\langle n+b_2,1+m+a-c;q\rangle_1\frac{\langle a+1-c,b_1+1-c-n;q\rangle_m}{\langle 1,2-c-n;q\rangle_m\langle 1;q\rangle_n}x^{m-n}y^n\right]\\ &=x^{1-c}\sum_{m,n=0}^{\infty}\left[a_{n+1}\frac{\langle a+1-c,b_1+1-c-n;q\rangle_{m+1}}{\langle 1-c-n;q\rangle_{m+1}\langle 1;q\rangle_m\langle 1;q\rangle_n}\right.\\ &\left.-a_n\langle n+b_2,1+m+a-c;q\rangle_1\frac{\langle a+1-c,b_1+1-c-n;q\rangle_m}{\langle 1,2-c-n;q\rangle_m\langle 1;q\rangle_n}\right]x^{m-n}y^n. \end{aligned} \tag{73}$$

By the condition $H_2f=0$ we obtain

$$\frac{a_{n+1}}{a_n}=\frac{\langle n+b_2,1-c-n;q\rangle_1}{\langle b_1-c-n;q\rangle_1}. \tag{74}$$

This implies

$$a_n=\frac{\langle b_2,c-1;q\rangle_n}{\langle c-b_1;q\rangle_n}, \tag{75}$$

$$\psi(y)={}_2\phi_1(b_2,c-1;c-b_1|q;y). \tag{76}$$

Then, we can induce by (71)

$$\begin{aligned} &f_D(x,y;q)\\ &=x^{1-c}\sum_{m,n=0}^{\infty}\frac{\langle a+1-c,b_1+1-c-n;q\rangle_m\langle b_2,c-1;q\rangle_n}{\langle 1,2-c-n;q\rangle_m\langle 1,c-b_1;q\rangle_n}x^{m-n}y^nq^{n(1-b_1)}\\ &=x^{1-c}\sum_{m,n=0}^{\infty}\frac{\langle a+1-c;q\rangle_m\langle b_1+1-c;q\rangle_{m-n}\langle b_2;q\rangle_n}{\langle 2-c;q\rangle_{m-n}\langle 1;q\rangle_m\langle 1;q\rangle_n}x^{m-n}y^n. \end{aligned} \tag{77}$$

We can rewrite this in the form of q-Horn function, convenient for convergence aspects.

$$\begin{aligned} &f_D(x,y;q)=x^{1-c}\sum_{m,n=0}^{\infty}(-1)^{m-n}\mathrm{QE}\left(-\binom{m-n}{2}+(m-n)(c-2)\right)\\ &\frac{\langle a+1-c;q\rangle_m\langle b_1+1-c;q\rangle_{m-n}\langle c-1;q\rangle_{n-m}\langle b_2;q\rangle_n}{\langle 1;q\rangle_m\langle 1;q\rangle_n}x^{m-n}y^n. \end{aligned} \tag{78}$$

The operator form is

$$f_D(x,y;q) = x^{1-c}{}_2\phi_1(a+1-c, b_1+1-c-\theta_{q,2}; 2-c-\theta_{q,2}|q;x)\\ {}_2\phi_1\Big(b_2, c-1; c-b_1|q;\frac{y}{x}\Big). \tag{79}$$

The series (78) converges in a slightly larger region than

$$|x| < 1,\ |y/x| < 1. \tag{80}$$

5.1. First Horn Recurrence Solution

The Horn recurrence (16) for $\rho = 1-c,\ \sigma = 0$ gives

$$\begin{cases} \frac{C_{m+1,n}}{C_{mn}} = \frac{\langle a+1-c+m+n, b_1+1-c+m;q\rangle_1}{\langle 1+m+n, 2-c+m;q\rangle_1} \\ \frac{C_{m,n+1}}{C_{mn}} = \frac{\langle a+1-c+m+n, b_2+n;q\rangle_1}{\langle 1+m+n, 1+n;q\rangle_1}. \end{cases} \tag{81}$$

The solution to this recurrence is

$$f_2(x,y;q) = x^{1-c}\sum_{m,n=0}^{\infty} \frac{\langle a+1-c;q\rangle_{m+n}\langle b_1+1-c\rangle_m\langle b_2\rangle_n}{\langle 2-c\rangle_m\langle 1;q\rangle_{m+n}\langle 1;q\rangle_n} x^m y^n. \tag{82}$$

This solution, not of usual q-hypergeometric type, was not given in the thesis by Le Vavasseur.

By symmetry, we get a third solution $f_3(x,y;q)$, the three functions $\{f_i(x,y;q)\}$ form a basis for the system Φ_1 around $(0,0)$.

5.2. Q-Integral Representations

We now turn to q-integral representations of solutions to the system for Φ_1. The operator form

$$\Phi_1(a;b_1,b_2;c|q;x,y)\\ = {}_2\phi_1(a+\theta_{q,2}, b_1; c+\theta_{q,2}|q;x)\ {}_2\phi_1(a,b_2;c|q;y) \tag{83}$$

together with the q-integral for ${}_2\phi_1$ (7.50 [1]) gives the q-Picard integral (10.104 [1]) for the first q-Appell function.

The operator form (79) together with (7.50 [1]) gives a q-analogue of (7.11 [3]).

Theorem 4.

$$f_D(x,y;q) \cong x^{1-c}\,\Gamma_q\begin{bmatrix} 2-c \\ a+1-c, 1-a \end{bmatrix}\int_0^1 t^{a-c}\frac{(qt;q)_{-a}}{(xt;q)_{b_1+1-c}}\\ {}_3\phi_2\begin{bmatrix} a, b_2 \\ c-b_1 \end{bmatrix}q; yq^{b_1-c+a}\Bigg\|\ \frac{((xt)^{-1}q^{c-b_1};q)_k}{(t^{-1}q^a;q)_k}\Bigg] d_q(t). \tag{84}$$

Proof. We can apply (43) for following deduction.

$$f_D(x,y;q) \cong x^{1-c}\,\Gamma_q\begin{bmatrix} 2-c-\theta_{q,2} \\ a+1-c, 1-a-\theta_{q,2} \end{bmatrix}\int_0^1 t^{a-c}\frac{(qt;q)_{-a-\theta_{q,2}}}{(xt;q)_{b_1+1-c-\theta_{q,2}}}\\ {}_2\phi_1\begin{bmatrix} c-1, b_2 \\ c-b_1 \end{bmatrix}q;\frac{y}{x}\Bigg] d_q(t) = \text{RHS}. \tag{85}$$

□

Similarly, we get an improved version of (7.12, [3]).

Theorem 5.

$$f_D(x,y) = x^{1-c}\,\Gamma\left[\begin{array}{c} 2-c \\ b_1+1-c, 1-b_1 \end{array}\right] \int_0^1 t^{b_1-c}(1-t)^{-b_1}(1-xt)^{c-a-1} \left(1-\frac{y}{x}\right)^{-b_2} dt. \tag{86}$$

Proof. Permute the parameters in the proof. □

A q-analogue of (86)

$$f_D(x,y;q) \cong x^{1-c}\,\Gamma_q\left[\begin{array}{c} 2-c \\ b_1+1-c, 1-b_1 \end{array}\right] \int_0^1 t^{b_1-c}\frac{(qt;q)_{-b_1}}{(xt;q)_{a+1-c}} \frac{1}{(\frac{y}{x}q^{b_1-1};q)_{b_2}}\, d_q(t). \tag{87}$$

5.3. Solutions Around $(0,\infty)$

From Ansatz III we obtain the equations

$$\begin{cases} \{\rho\}_q\{c+\sigma+\rho-1\}_q = 0, \\ \{b_2+\sigma\}_q\{a+\rho+\sigma)_q = 0. \end{cases} \tag{88}$$

This has the three solutions

$$\begin{aligned} &\rho = 0,\ \sigma = -b_2 \\ &\rho = 0,\ \sigma = -a \\ &\rho = b_2+1-c,\ \sigma = -b_2. \end{aligned} \tag{89}$$

Put

$$\psi(y) \equiv \sum_{n=0}^{\infty} a_n \frac{y^{-n-\lambda}}{\langle 1;q\rangle_n}. \tag{90}$$

According to (40) and (43), the condition $H_1 g_1(x,y;q) = 0$ gives,

$$\begin{aligned} &g_1(x,y;q) \\ &\equiv {}_2\phi_1(a+\theta_{q,2}, b_1; c+\theta_{q,2}|q;x)\psi(y) \\ &= \sum_{m,n=0}^{\infty} a_n \frac{\langle a-n-\lambda, b_1;q\rangle_m}{\langle 1, c-n-\lambda;q\rangle_m \langle 1;q\rangle_n} x^m y^{-n-\lambda}. \end{aligned} \tag{91}$$

Lemma 3.

$$\begin{aligned} &\frac{\langle a-b_2-n;q\rangle_m \langle b_2-c+1;q\rangle_n (-1)^{m-n}}{\langle c-b_2-n;q\rangle_m \langle b_2+1-a;q\rangle_n \langle a-b_2;q\rangle_{m-n} \langle b_2+1-c;q\rangle_{n-m}} \\ &= \mathrm{QE}\left(-\binom{m}{2} - \binom{n}{2} - m(c-b_2-n) - n(2b_2+1-c)\right). \end{aligned} \tag{92}$$

This lemma is used in the following proof. Similar to ([3], p. 796) we find that

$$\begin{aligned} &\left[\{\theta_{q,1}+\theta_{q,2}+c\}_q \mathrm{D}_{q,y}\right](x^m y^{-n-\lambda}) \\ &= \{m-n-\lambda-1+c\}_q\{-n-\lambda\}_q x^m y^{-n-\lambda-1}. \end{aligned} \tag{93}$$

We have

$$H_2 f(x,y;q) = -\sum_{m,n=0}^{\infty} a_n \frac{\langle a-n-\lambda, b_1;q\rangle_m}{\langle 1, c-n-\lambda;q\rangle_m \langle 1;q\rangle_n} \Big[\{m-n-\lambda-1+c\}_q \{-n-\lambda\}_q x^m y^{-n-\lambda-1} + \{b_2-n-\lambda\}_q \{a+m-n-\lambda\}_q x^m y^{-n-\lambda}\Big]. \tag{94}$$

For $n=0$, the condition $H_2 f = 0$ implies $\lambda = b_2$, and we have

$$H_2 f(x,y;q) = y^{-b_2} \sum_{m,n=0}^{\infty} \Bigg[-a_n \frac{\langle a-b_2-n, b_1;q\rangle_m \{b_2+n\}_q q^{-n-b_2}}{\langle c-b_2-n;q\rangle_{m-1} \langle 1;q\rangle_m \langle 1;q\rangle_n} x^{m-n} y^{n-1} + a_{n+1} \frac{\langle a-b_2-n-1;q\rangle_{m+1} \langle b_1;q\rangle_m}{\langle 1, c-b_2-n-1;q\rangle_m \langle 1;q\rangle_n} \Bigg] x^m y^{-n-1} q^{-n}. \tag{95}$$

By the condition $H_2 f = 0$ we obtain

$$\frac{a_{n+1}}{a_n} = \frac{\langle n+b_2, c-b_2-n-1;q\rangle_1}{\langle a-b_2-n-1;q\rangle_1} q^{-b_2}. \tag{96}$$

This implies

$$a_n = \frac{\langle b_2, b_2+1-c;q\rangle_n}{\langle b_2+1-a;q\rangle_n} q^{n(c-a-b_2)}, \tag{97}$$

$$\psi(y) = {}_2\phi_1(b_2, b_2+1-c; b_2+1-a | q; y q^{c-a-b_2}). \tag{98}$$

According to (92), we should have

$$\begin{aligned} &g_1(x,y;q) \\ &= y^{-b_2} \sum_{m,n=0}^{\infty} \frac{\langle a-b_2-n, b_1;q\rangle_m \langle b_2+1-c, b_2;q\rangle_n}{\langle 1, c-b_2-n;q\rangle_m \langle 1, b_2+1-a;q\rangle_n} x^m y^{-n} q^{n(c-a-b_2)} \\ &= y^{-b_2} \sum_{m,n=0}^{\infty} \frac{\langle b_1;q\rangle_m \langle a-b_2;q\rangle_{m-n} \langle b_2;q\rangle_n \langle b_2+1-c;q\rangle_{n-m}}{\langle 1;q\rangle_m \langle 1;q\rangle_n} \\ &(-x)^m (-y)^{-n} \mathrm{QE}\left(-\binom{m}{2} - \binom{n}{2} - m(c-b_2-n) + n(c-2b_2-1)\right). \end{aligned} \tag{99}$$

We can again rewrite this in the form of the q-Horn function. The operator form is

$$\begin{aligned} g_1(x,y;q) &= {}_2\phi_1(a+\theta_{q,2}, b_1; c+\theta_{q,2} | q; x) \\ &{}_2\phi_1(b_2, b_2+1-c; b_2+1-a | q; y q^{c-a-b_2}). \end{aligned} \tag{100}$$

Again put

$$\psi(y) \equiv y^{-\lambda} \sum_{n=0}^{\infty} a_n \frac{y^{-n}}{\langle 1;q\rangle_n}, \tag{101}$$

and use the other q-hypergeometric function solution around 0.

According to (40) and (43), we have

$$\begin{aligned} &g_2(x,y;q) \\ &\equiv x^{1-c-\theta_{q,2}} {}_2\phi_1(a+1-c, b_1+1-c-\theta_{q,2}; 2-c-\theta_{q,2} | q; x) \psi(y) \\ &= x^{1-c+\lambda} y^{-\lambda} \sum_{m,n=0}^{\infty} a_n \frac{\langle a+1-c, b_1+1-c+n+\lambda;q\rangle_m}{\langle 1, 2-c+n+\lambda;q\rangle_m \langle 1;q\rangle_n} x^{m+n} y^{-n}. \end{aligned} \tag{102}$$

Similar to ([3], p. 797) we find that

$$\begin{aligned}&\left[\{\theta_{q,1}+\theta_{q,2}+c\}_q \mathrm{D}_{q,y}\right](x^{m+n+\lambda+1-c}y^{-n-\lambda})\\&=\{m\}_q\{-n-\lambda\}_q x^{m+n+\lambda+1-c}y^{-n-\lambda-1}.\end{aligned} \tag{103}$$

Because of the factor $\{b_2-n-\lambda\}_q$, $\lambda=b_2$, and we have

$$\begin{aligned}&H_2g_2(x,y;q)(1-q)^2=x^{1-c+b_2}y^{-b_2}\\&\left[\sum_{m=1,n=0}^{\infty}-a_n\frac{\langle a+1-c,b_1+1-c+n+b_2;q\rangle_m\langle n+b_2;q\rangle_1}{\langle 2-c+n+b_2;q\rangle_m\langle 1;q\rangle_{m-1}\langle 1;q\rangle_n}x^{m+n}y^{-n-1}q^{-n-b_2}\right.\\&\left.+\sum_{m=0,n=1}^{\infty}a_n\langle 1+m+a-c;q\rangle_1\frac{\langle a+1-c,b_1+1-c+n+b_2;q\rangle_m}{\langle 1,2-c+n+b_2;q\rangle_m\langle 1;q\rangle_{n-1}}x^{m+n}y^{-n}q^{-n}\right]\\&=x^{2-c+b_2}y^{-b_2}\sum_{m,n=0}^{\infty}\left[-a_n\frac{\langle a+1-c,b_1+1-c+n+b_2;q\rangle_{m+1}}{\langle 2-c+n+b_2;q\rangle_{m+1}\langle 1;q\rangle_m\langle 1;q\rangle_n}\langle n+b_2;q\rangle_1 q^{-b_2}\right.\\&\left.+a_{n+1}\langle 1+m+a-c;q\rangle_1\frac{\langle a+1-c,b_1+2-c+n+b_2;q\rangle_m}{q\langle 1,3-c+n+b_2;q\rangle_m\langle 1;q\rangle_n}\right]x^{m+n}y^{-n-1}q^{-n}.\end{aligned} \tag{104}$$

By the condition $H_2g_2=0$ we obtain

$$a_n=\frac{\langle b_2,b_1+b_2+1-c;q\rangle_n}{\langle b_2+2-c;q\rangle_n}q^{n(1-b_2)}, \tag{105}$$

$$\psi(y)=y^{-b_2}{}_2\phi_1(b_2,b_1+b_2+1-c;b_2+2-c|q;y^{-1}q^{1-b_2}). \tag{106}$$

$$\begin{aligned}&g_2(x,y;q)=x^{b_2+1-c}y^{-b_2}\sum_{m,n=0}^{\infty}\\&\frac{\langle a+1-c,b_1+1-c-n;q\rangle_m\langle b_2,c-1;q\rangle_n}{\langle 1,2-c+n+b_2;q\rangle_m\langle 1,b_2+2-c;q\rangle_n}x^{m+n}y^{-n}q^{n(1-b_1)}\\&=x^{b_2+1-c}y^{-b_2}\Phi_1\left(b_1+b_2+1-c;a+1-c,b_2;b_2+2-c\middle|q;x,\frac{x}{y}q^{n(1-b_2)}\right).\end{aligned} \tag{107}$$

The operator form is

$$\begin{aligned}&g_2(x,y;q)=x^{1-c-\theta_{q,2}}{}_2\phi_1(a+1-c,b_1+1-c-\theta_{q,2};2-c-\theta_{q,2}|q;x)\\&y^{-b_2}{}_2\phi_1(b_2,b_1+b_2+1-c;b_2+2-c|q;y^{-1}q^{1-b_2}).\end{aligned} \tag{108}$$

Type B2. Use the same $\psi(x)$ and the function (6.187 [1]), according to (40) and (43)

$$\begin{aligned}&g_4(x,y;q)\\&\equiv y^{-a-\theta_{q,1}}{}_2\phi_1(a+\theta_{q,1},a+1-c;a+1-b_2+\theta_{q,1}|q;\frac{1}{y})\psi(x)\\&=y^{-a}\sum_{m,n=0}^{\infty}a_n\frac{\langle a+n+\lambda,a+1-c;q\rangle_m}{\langle 1,a+1-b_2+n+\lambda;q\rangle_m\langle 1;q\rangle_n}x^{n+\lambda}y^{-m-n-\lambda}.\end{aligned} \tag{109}$$

Similar to ([3], p. 798) we find that

$$\begin{aligned}&\left[\{\theta_{q,1}+\theta_{q,2}+c\}_q \mathrm{D}_{q,x}\right](x^{n+\lambda}y^{-m-n-a-\lambda})\\&=\{c-a-1-m\}_q\{n+\lambda\}_q x^{n+\lambda-1}y^{-m-n-a-\lambda}.\end{aligned} \tag{110}$$

Again, $\lambda = 0$, and we have

$$H_1 g_4(x,y;q) = y^{-a}$$
$$\left[\sum_{m=0,n=1}^{\infty} a_n \frac{\langle c-a-1-m;q\rangle_1 \langle a+n,a+1-c;q\rangle_m}{\langle 1,a+1+n-b_2;q\rangle_m \langle 1;q\rangle_{n-1}} x^{n-1} y^{-n-m}\right.$$
$$\left. - \sum_{m=1,n=0}^{\infty} a_n \langle b_1+n;q\rangle_1 \frac{\langle a+n,a+1-c;q\rangle_m}{\langle 1,2-c-n;q\rangle_m \langle 1;q\rangle_n} q^{-m} x^n y^{-m-n}\right] \tag{111}$$
$$= y^{-a} \sum_{m,n=0}^{\infty} \left[a_{n+1} \frac{\langle c-a-1-m;q\rangle_1 \langle a+n+1,a+1-c;q\rangle_m}{\langle 1,a+2+n-b_2;q\rangle_m \langle 1;q\rangle_n}\right.$$
$$\left. -a_n \langle n+b_1;q\rangle_1 \frac{\langle a+1-c,a+n;q\rangle_{m+1}}{\langle a+1+n-b_2;q\rangle_{m+1} \langle 1;q\rangle_m \langle 1;q\rangle_n} q^{-m}\right] x^n y^{-m-n}.$$

By the condition $H_1 g_4 = 0$ we obtain

$$\frac{a_{n+1}}{a_n} = -q^{a+1-c} \frac{\langle n+b_1,n+a;q\rangle_1}{\langle a+1+n-b_2;q\rangle_1}. \tag{112}$$

This implies

$$a_n = (-1)^n \frac{\langle a,b_1;q\rangle_n}{\langle a+1-b_2;q\rangle_n} q^{n(a+1-c)} \tag{113}$$

$$\psi(x) = {}_2\phi_1(a,b_1;a+1-b_2|q;-xq^{a+1-c}). \tag{114}$$

Finally, we obtain a q-analogue of the corrected version of Levavasseur.

$$g_4(x,y;q) = y^{-a-\theta_{q,1}} {}_2\phi_1\left(a+\theta_{q,1},a+1-c;a+1-b_2+\theta_{q,1}\middle|q;\frac{1}{y}\right)$$
$${}_2\phi_1(a,b_1;a+1-b_2|q;-xq^{a+1-c}) \tag{115}$$
$$= y^{-a}\Phi_1\left(a;b_1,a+1-c;a+1-b_2\middle|q;-\frac{x}{y}q^{a+1-c},\frac{1}{y}\right).$$

Theorem 6. *The second solution is*

$$h_2(x,y;q) = y^{-a} \sum_{m,n=0}^{\infty} \frac{\langle a+1-c;q\rangle_{n-m} \langle a;q\rangle_n \langle b_1;q\rangle_m}{\langle a+1-b_2;q\rangle_n \langle 1;q\rangle_m \langle 1;q\rangle_{n-m}} x^m y^{-n}$$
$$QE((c-a)(m-n)+(1-b_2)n). \tag{116}$$

This solution is a q-analogue of ([5], p. 31).

Proof. From the recurrences, using (6.14 [1]), we can find

$$h_2(x,y;q) = y^{-a} \sum_{m,n=0}^{\infty} \frac{\langle c-a+m-n;q\rangle_{n-m} \langle 1-a-n;q\rangle_n \langle b_1;q\rangle_m}{\langle -a-n+b_2;q\rangle_n \langle 1;q\rangle_m \langle m-n;q\rangle_{n-m}} \tag{117}$$
$$x^m y^{-n} = \text{RHS}.$$

□

Theorem 7. *The third solution is*

$$h_3(x,y;q) = x^{b_2+1-c} y^{-b_2}$$
$$\sum_{m,n=0}^{\infty} \frac{\langle a+1-c;q\rangle_{m-n} \langle b_2;q\rangle_n \langle b_1+b_2+1-c;q\rangle_m}{\langle 2+b_2-c;q\rangle_m \langle 1;q\rangle_n \langle 1;q\rangle_{m-n}} x^m y^{-n} \tag{118}$$
$$QE(-nb_2+n).$$

This solution is a q-analogue of ([5], p. 31).

Proof. For the case $\rho = b_2 + 1 - c,\ \sigma = -b_2$ we obtain the recurrence

$$\begin{cases} \frac{C_{m+1,n}}{C_{mn}} = \frac{\langle a+1-c+m-n, b_1+b_2+1-c+m;q\rangle_1}{\langle 1+m-n, 2+b_2-c+m;q\rangle_1} \\ \frac{C_{m,n+1}}{C_{mn}} = \frac{\langle 1+m-n, -b_2-n;q\rangle_1}{\langle a+m-n+1-c, -1-n;q\rangle_1}. \end{cases} \tag{119}$$

The solution to this recurrence, using (6.14 [1]), is

$$\begin{aligned} &h_3(x,y;q) = x^{b_2+1-c}y^{-b_2} \sum_{m,n=0}^{\infty} \langle a+1-c;q\rangle_{m-n} x^m y^{-n} \\ &\frac{\langle -b_2+1-n;q\rangle_n \langle b_1+b_2+1-c;q\rangle_m}{\langle 2+b_2-c;q\rangle_m \langle -n;q\rangle_n \langle 1;q\rangle_{m-n}} = \text{RHS}. \end{aligned} \tag{120}$$

□

6. Second q-Appell Function

Now we put $\gamma_i = 0, ,\ \alpha_i = \beta_i = 1,\ a_i = a$ in (49).

Theorem 8. *A q-analogue of ((1) [6]), ([7], p. 50). The q-difference equation for Φ_2 has the following four independent solutions in the vicinity of $(0,0)$.*

$$\begin{aligned} &f_1(x,y;q) \equiv \Phi_2(a;b_1,b_2;c_1,c_2|q;x,y), \\ &f_2(x,y;q) \equiv x^{1-c_1}\Phi_2(a-c_1+1;b_1-c_1+1,b_2;2-c_1,c_2|q;x,y), \\ &f_3(x,y;q) \equiv y^{1-c_2}\Phi_2(a-c_2+1;b_1,b_2-c_2+1;c_1,2-c_2|q;x,y), \\ &f_4(x,y;q) \equiv x^{1-c_1}y^{1-c_2} \\ &\Phi_2(a-c_1-c_2+2;b_1-c_1+1,b_2-c_2+1;2-c_1,2-c_2|q;x,y). \end{aligned} \tag{121}$$

Proof. According to (40) and (43), we find

$$\begin{aligned} &f_3(x,y;q) \\ &\equiv {}_2\phi_1(a+\theta_{q,2},b_1;c_1|q;x)y^{1-c_2}{}_2\phi_1(a+1-c_2,b_2+1-c_2;2-c_2|q;y) \\ &= \text{RHS}. \end{aligned} \tag{122}$$

$$\begin{aligned} &f_4(x,y;q) \\ &= x^{1-c_1}{}_2\phi_1(a+1-c_1+\theta_{q,2},b+1-c_1;2-c_1|q;x) \\ &y^{1-c_2}{}_2\phi_1(a+2-c_1-c_2,b+2-c_2+1;2-c_2|q;y) \\ &= \text{RHS}. \end{aligned} \tag{123}$$

□

Remark 2. *The asymmetric expressions for $f_4(x,y)$ in ([3], p. 804 f) are in error.*

A q-analogue of ([3], p. 804).

Theorem 9. *A q-integral representation of $\Phi_2(a;b_1,b_2;c_1,c_2|q;x,y)$*

$$\Phi_2(a;b_1,b_2;c_1,c_2|q;x,y) \cong \Gamma_q\left[\begin{array}{c} c_1 \\ a, c_1-a \end{array}\right] \int_0^1 t^{a-1}\frac{(qt;q)_{c_1-a-1}}{(xt;q)_{b_1}}$$
$$\sum_{m=0}^{\infty} \frac{\langle b_2, a+1-c_1;q\rangle_m}{\langle 1,c_2;q\rangle_m}(-ty)^m(tq^{c_1-a};q)_{-m} \tag{124}$$
$$\mathrm{QE}\left(-\binom{m}{2}+m(c_1-a-1)\right) d_q(t).$$

Proof. According to (7.50 [1]) we have

$$\begin{aligned} \text{LHS} &= {}_2\phi_1(a+\theta_{q,2},b_1;c_1|q;x)\,{}_2\phi_1(a,b_2;c_2|q;y) \\ &=\Gamma_q\left[\begin{array}{c} c_1 \\ a+\theta_{q,2}, c_1-a-\theta_{q,2} \end{array}\right] \int_0^1 t^{a+\theta_{q,2}-1}\frac{(qt;q)_{c_1-a-\theta_{q,2}-1}}{(xt;q)_{b_1}} \\ &\quad {}_2\phi_1(a,b_2;c_2|q;y)\, d_q(t) = \text{RHS}. \end{aligned} \tag{125}$$

□

Theorem 10. *The functions $f_3(x,y;q)$ and $f_4(x,y;q)$ have q-integral representations*

$$f_3(x,y;q) \cong x^{1-c_1}\,\Gamma_q\left[\begin{array}{c} c_2 \\ a, c_2-a \end{array}\right] \int_0^1 t^{a-1}\frac{(qt;q)_{c_2-a-1}}{(yt;q)_{b_2}}$$
$$\sum_{m=0}^{\infty} \frac{\langle a+1-c_1, a+1-c_2, b_1+1-c_1;q\rangle_m}{\langle 1,2-c_1,a;q\rangle_m}(-xt)^m(tq^{c_2-a};q)_{-m} \tag{126}$$
$$\mathrm{QE}\left(-\binom{m}{2}+m(c_2-a-1)\right) d_q(t).$$

$$f_4(x,y;q) \cong x^{1-c_1}y^{1-c_2}\,\Gamma_q\left[\begin{array}{c} 2-c_1 \\ a+1-c_1, 1-a \end{array}\right] \int_0^1 t^{a-c_1}\frac{(qt;q)_{-a}}{(xt;q)_{b_1+1-c_1}}$$
$$\sum_{m=0}^{\infty} \frac{\langle a+2-c_1-c_2, b_2-c_2+1, a;q\rangle_m}{\langle 1,2-c_2,a+1-c_1;q\rangle_m}(-yt)^m(tq^{1-a};q)_{-m} \tag{127}$$
$$\mathrm{QE}\left(-\binom{m}{2}-ma\right) d_q(t).$$

Proof. Use formulas (122) and (123). □

7. Third q-Appell Function

Let us put $\alpha_i=\beta_i=0,\ \gamma_i=1,\ c_i=c,\ a_i=a,\ b_i=b$ in (49).

Theorem 11. *A q-analogue of ([3], p. 805). The third q-Appell function has q-integral representation*

$$\begin{aligned} &\Phi_3(a_1,a_2;b_1,b_2;c|q;x,y) \cong \Gamma_q\left[\begin{array}{c} c \\ a_1, c-a_1 \end{array}\right] \int_0^1 t^{a_1-1}\frac{(qt;q)_{c-a_1-1}}{(xt;q)_{b_1}} \\ &\sum_{m=0}^{\infty} \frac{\langle a_2,b_2;q\rangle_m}{\langle 1,c-a_1;q\rangle_m} y^m (q^{c-a_1}t;q)_m\, d_q(t). \end{aligned} \tag{128}$$

Proof. Using (7.50 [1]), we have

$$\begin{aligned} &\text{LHS}=\Gamma_q\left[\begin{array}{c} c+\theta_{q,2} \\ a_1, c-a_1+\theta_{q,2} \end{array}\right] \int_0^1 t^{a_1-1}\frac{(qt;q)_{c-a_1-\theta_{q,2}-1}}{(xt;q)_{b_1}} \\ &{}_2\phi_1\left[\begin{array}{c} a_2,b_2 \\ c \end{array}\middle| q;y\right] \overset{\text{by}(43)}{=} \text{RHS}. \end{aligned} \tag{129}$$

□

8. Fourth q-Appell Function

Finally, we put $\gamma_i = 0,\ \alpha_i = \beta_i = 1,\ a_i = a,\ b_i = b$ in (49). We have

$$\Phi_4(a,b;c_1,c_2|q;x,y) = {}_2\phi_1(a+\theta_{q,2}, b+\theta_{q,2};c_1|q;x){}_2\phi_1(a,b;c_2|q;y). \tag{130}$$

Theorem 12. *A q-analogue of ([3], p. 807). The fourth q-Appell function has q-integral representation*

$$\begin{aligned}&\Phi_4(a,b;c_1,c_2|q;x,y) \cong \Gamma_q\left[\begin{matrix} c_1 \\ a, c_1-a \end{matrix}\right] \int_0^1 t^{a-1}\frac{(qt;q)_{c_1-a-1}}{(xt;q)_b} \\ &\sum_{m=0}^{\infty} \frac{\langle a+1-c_1,b;q\rangle_m}{\langle 1,c_2;q\rangle_m}(-yt)^m \frac{(tq^{c_1-a};q)_{-m}}{(xtq^b t;q)_m} \\ &\mathrm{QE}\left(-\binom{m}{2} - m(a+1-c_1)\right) d_q(t).\end{aligned} \tag{131}$$

Proof. Using (7.50 [1]) and (43), we have

$$\begin{aligned}&\mathrm{LHS}{=}\Gamma_q\left[\begin{matrix} c_1+\theta_{q,2} \\ a+\theta_{q,2}, c_1-a-\theta_{q,2} \end{matrix}\right] \int_0^1 t^{a+\theta_{q,2}-1}\frac{(qt;q)_{c_1-a-\theta_{q,2}-1}}{(xt;q)_{b+\theta_{q,2}}} \\ &{}_2\phi_1\left[\begin{matrix} a,b \\ c_2 \end{matrix}\middle|q;y\right]{=}\mathrm{RHS}.\end{aligned} \tag{132}$$

□

Theorem 13. *A q-analogue of ([3], pp. 807–808). The q-difference equation for Φ_4 has the following four independent solutions in the vicinity of $(0,0)$.*

$$\begin{aligned}&f_1(x,y;q) \equiv \Phi_4(a,b;c_1,c_2|q;x,y), \\ &f_2(x,y;q) \equiv y^{1-c_2}\Phi_4(a-c_2+1;b-c_2+1;2-c_2,c_1|q;x,y), \\ &f_3(x,y;q) \equiv x^{1-c_1}\Phi_4(a-c_1+1;b-c_1+1;2-c_1,c_2|q;x,y), \\ &f_4(x,y;q) \equiv x^{1-c_1}y^{1-c_2} \\ &\Phi_4(a-c_1-c_2+2,b_-c_1-c_2+2;2-c_1,2-c_2|q;x,y).\end{aligned} \tag{133}$$

Proof. According to (40) and (43), we find

$$\begin{aligned}&f_2(x,y;q) \\ &= {}_2\phi_1(a+\theta_{q,2},b+\theta_{q,2};c_1|q;x)y^{1-c_2}{}_2\phi_1(a+1-c_2,b+1-c_2;2-c_2|q;y) \\ &{=}\mathrm{RHS}.\end{aligned} \tag{134}$$

$$\begin{aligned}&f_4(x,y;q) \\ &= x^{1-c_1}{}_2\phi_1(a+1-c_1+\theta_{q,2},b+1-c_1+\theta_{q,2};2-c_1|q;x) \\ &y^{1-c_2}{}_2\phi_1(a+2-c_1-c_2,b+2-c_1-c_2;2-c_2|q;y) \\ &{=}\mathrm{RHS}.\end{aligned} \tag{135}$$

□

Theorem 14. *The functions $f_2(x,y;q)$ and $f_4(x,y;q)$ have q-integral representations*

$$
\begin{aligned}
&f_2(x,y;q) \cong y^{1-c_2}\,\Gamma_q\left[\begin{matrix} c_1 \\ a, c_1-a \end{matrix}\right]\int_0^1 t^{a-1}\frac{(qt;q)_{c_1-a-1}}{(xt;q)_b} \\
&\sum_{m=0}^{\infty}\frac{\langle a+1-c_1, a+1-c_2, b+1-c_2;q\rangle_m}{\langle 1, a, 2-c_2;q\rangle_m}(-yt)^m\frac{(tq^{c_1-a};q)_{-m}}{(xtq^b t;q)_m} \\
&\mathrm{QE}\left(-\binom{m}{2}-m(a+1-c_1)\right)d_q(t).
\end{aligned}
\tag{136}
$$

$$
\begin{aligned}
&f_4(x,y;q) \cong x^{1-c_1}y^{1-c_2}\Gamma_q\left[\begin{matrix} 2-c_1 \\ a+1-c_1, 1-a \end{matrix}\right]\int_0^1 t^{a-c_1}\frac{(qt;q)_{-a}}{(xt;q)_{b_1+1-c_1}} \\
&\sum_{m=0}^{\infty}\frac{\langle a+2-c_1-c_2, b_2-c_2+1, a;q\rangle_m}{\langle 1, 2-c_2, a+1-c_1;q\rangle_m}(-yt)^m\left(tq^{1-a};q\right)_{-m} \\
&\mathrm{QE}\left(-\binom{m}{2}-ma\right)d_q(t).
\end{aligned}
\tag{137}
$$

Proof. Use formulas (134) and (135). □

9. Conclusions

We have given the other solutions to the systems of q-difference equations in three forms

1. the factorized, umbral form
2. the series expansion, with convergence regions, q-analogues of [3]
3. possibly, a q-integral representation

These convergence regions are always larger than in the ordinary case, sometimes q-deformed cones arise. Our method leads to more direct computation of the other solutions of Appell differential and similar differential equations than the papers by Horn and Borngässer. We have illustrated the new symbolic calculus in the special case q-Appell functions, since more complex functions would lead to longer computations. These computations are similar to the solutions of differential equations by the Frobenius method. We started with the solutions in the vicinity of $(0,0)$ and obtained the usual indicial equation for the exponents. Then we found all solutions, which was treated by Borngässer [5]. With the help of a lemma, we found a recurrence for the unknown coefficients, and the unknown function was sometimes another q-Appell function and sometimes a q-Horn function.

Then, by the symbolic operator formulas, we found q-integral representations of the formulas in the basis. For the solutions around $(0,\infty)$ we found $\lambda = b_2$ and by using another lemma, we obtained another q-Horn function in the basis of solutions.

10. Discussion

The Frobenius method [8] for solutions of differential equations originates from papers by Thomae [9], who studied logarithmic solutions of the Euler equation and Thomé [10], who wrote about very general solutions of differential equations, convergent in disks around a point a.

Thanks to Debiard and Gaveau for their most interesting papers on multiple hypergeometric functions. We have retained their notation as much as possible. The Debiard–Gaveau umbral method was neither used in the thesis by Borngässer [5], nor in the papers by Horn [4]. However, the umbral q-difference equations for q-Appel functions in our book ([1], p. 436), in the spirit of Mellin [11] and Thomae [9], are equivalent to the q-difference equations in this paper. The paper is also interesting for the case $q = 1$, since Borngässer's thesis [5], in German, is almost unknown, and is now available, in part, in English. In a future paper, we will discuss the q-difference equations and q-integral representations of the corresponding q-Horn functions. Likewise, the confluent forms [12], as well as other multiple q-hypergeometric functions can be treated with this method.

Funding: This research received no external funding.

Institutional Review Board Statement: Not applicable.

Informed Consent Statement: Not applicable.

Data Availability Statement: Not applicable.

Acknowledgments: Not applicable.

Conflicts of Interest: The author declares no conflict of interest.

References

1. Ernst, T. *A Comprehensive Treatment of q-Calculus*; Birkhäuser: Basel, Switzerland, 2012.
2. Ernst, T. On the q-analogues of Srivastava's triple hypergeometric functions. *Axioms* **2013**, *2*, 85–99. [CrossRef]
3. Debiard, A.; Gaveau, B. Hypergeometric symbolic calculus. I: Systems of two symbolic hypergeometric equations. *Bull. Sci. Math.* **2002**, *126*, 773–829. [CrossRef]
4. Horn, J. Hypergeometric functions of two variables. (Hypergeometrische Funktionen zweier Veränderlichen). *Math. Ann.* **1931**, *105*, 381–407. [CrossRef]
5. Borngässer, L. Über hypergeometrische Funktionen zweier Veränderlichen. Ph.D. Thesis, Pfeffer & Balzer, Darmstadt, Germany, 1933.
6. Olsson, P.O.M. On the integration of the differential equations of five-parametric double-hypergeometric functions of second order. *J. Math. Phys.* **1977**, *18*, 1285. [CrossRef]
7. Appell, P.; Kampé de Fériet, J. *Fonctions Hypergéométriques et Hypersphériques*; Polynomes D'Hermite: Paris, France, 1926.
8. Frobenius, G. On the integration of linear differential equations by means of series. (Über die Integration der linearen Differentialgleichungen durch Reihen.) *Crelle* **1873**, *76*, 214–235.
9. Thomae, J. On the higher hypergeometric series, especially the series ....(Über die höheren hypergeometrischen Reihen, insbesondere die Reihe ....) *Math. Ann.* **1870**, *2*, 427–444. (In Germany) [CrossRef]
10. Thomé, L.W. On the theory of linear differential equations. (Zur Theorie der linearen Differentialgleichungen). *Crelle* **1872**, *74*, 193–217. (In Germany)
11. Mellin, H.J. On the connection between the linear differential and difference equations, Über den Zusammenhang zwischen den linearen Differential- und Differenzengleichungen. *Acta Math.* **1902**, *25*, 139–164. (In Germany) [CrossRef]
12. Srivastava, H.M.; Karlsson, P.W. *Multiple Gaussian Hypergeometric Series*; Ellis Horwood: New York, NY, USA, 1985.

axioms

MDPI

Article

Control of String Vibrations by Displacement of One End with the Other End Fixed, Given the Deflection Form at an Intermediate Moment of Time

Vanya Barseghyan [1] and Svetlana Solodusha [2,*]

[1] Institute of Mechanics of the National Academy of Sciences of the Republic of Armenia, Yerevan State University, Yerevan 0019, Armenia; barseghyan@sci.am
[2] Melentiev Energy Systems Institute of Siberian Branch of the Russian Academy of Sciences, 664033 Irkutsk, Russia
* Correspondence: solodusha@isem.irk.ru

Abstract: We consider a boundary control problem for the equation of string vibration with given initial and final conditions, given the deflection form at an intermediate moment of time. The control is carried out by displacement of one end with the other end fixed. The problem is reduced to the problem of a distributed action control with zero boundary conditions. We propose a constructive approach to constructing a boundary control action by the separation of variables and methods of the theory of control of finite-dimensional systems. The approach is applied to given functions. A computational experiment was carried out with the construction of the corresponding graphs and their comparative analysis. They confirm the obtained results.

Keywords: vibration control; boundary control; intermediate state control; separation of variables

Citation: Barseghyan, V.; Solodusha, S. Control of String Vibrations by Displacement of One End with the Other End Fixed, Given the Deflection Form at an Intermediate Moment of Time. *Axioms* **2022**, *11*, 157. https://doi.org/10.3390/axioms11040157

Academic Editor: Hans J. Haubold

Received: 21 January 2022
Accepted: 24 March 2022
Published: 28 March 2022

Publisher's Note: MDPI stays neutral with regard to jurisdictional claims in published maps and institutional affiliations.

Copyright: © 2022 by the authors. Licensee MDPI, Basel, Switzerland. This article is an open access article distributed under the terms and conditions of the Creative Commons Attribution (CC BY) license (https://creativecommons.org/licenses/by/4.0/).

1. Introduction

Mathematical modeling of various controlled physical and engineering processes associated with vibration systems leads to wave equations. Controlled vibration systems are widespread in various theoretical and applied fields of science. In practice, control problems often arise for both distributed and lumped systems, in particular, when forming a given (desired) form of motion that satisfies multipoint intermediate conditions. Multipoint boundary value problems of control and optimal control of dynamical systems given both the classical boundary (initial and final) and multipoint intermediate conditions have applied value and theoretical importance. Therefore, they require research. In the scientific literature, multipoint boundary value problems of control are considered for systems described both by ordinary differential equations and partial differential equations. Unlike control problems for systems described by ordinary differential equations, control problems for ones described by partial differential equations with multipoint intermediate conditions are little studied.

Many researchers study problems of (optimal) control of vibrational processes. As a rule, both distributed and boundary-concentrated impacts are considered [1–19]. Modeling and control of dynamic systems is currently an actual scientific direction. At the same time, mathematical models of dynamic systems use both ordinary differential equations and partial differential equations with intermediate conditions. Studies of the above problems are the subject of such research contributions as [4–9,20,21] and others.

In production processes associated with the longitudinal movement of materials (for example, a paper web), undesirable transverse perturbations arise, which, for a vertical section, is described by the wave equation of a longitudinally moving string [22]. As a result, statements associated with generating the desired oscillation arise, i.e., oscillation control problems over a finite time. One of the possible approaches designed to prevent the appearance of unwanted disturbances can be considered the control of oscillations with given multipoint intermediate conditions. These conditions can be interpreted as a driving force.

Control and optimal control problems for the string oscillation equation with given initial and final conditions and undivided values of string point velocities at intermediate times are considered in [5,6]. The presented work is close to these articles.

This study solves the problem of boundary control of vibrations of a homogeneous string with given initial and final conditions, with a given form of deflection at an intermediate moment of time. Control is implemented by displacing the left end with the right end fixed. The problem is reduced to a distributed action control problem with zero boundary conditions. Using the method of separation of variables and methods of the theory of control of finite-dimensional systems for the first n harmonics of vibrations, we construct the required boundary control, under the action of which the deflection function of the string takes a given (or close to a given) value at an intermediate moment of time. In the paper, we formulate the corresponding statement and theorem for the first n harmonics. The results obtained for the first n harmonics are illustrated for $n = 1$ and $n = 2$. The presented study is located at the intersection of several scientific fields. We use terminology and approaches from the fields of control of systems with distributed parameters and control of finite-dimensional dynamic systems.

This paper is organized as follows. Section 2 contains formulas necessary for the analytical construction of the solution. Further, in Section 3, using the method of separation of variables and methods of the theory of control of finite-dimensional systems, for the first n harmonics of vibrations, we construct the required boundary control and the corresponding string deflection function. The presented formulas are necessary for the constructiveness of constructing an analytical solution. The constructed analytical solution of the formulated problem is compactly presented in Sections 2 and 3 with the corresponding formulations of the obtained general results in the form of a statement and a theorem. Section 4 presents formulas for fixed $n = 1$ and $n = 2$. They are also used in the Section 5 of the paper. In Section 5, we realize a computational experiment, build corresponding graphs and make a comparative analysis. They confirm the results of the study. The conclusion summarizes the main results.

2. Problem Statement and Its Reduction to a Problem with Zero Boundary Conditions

Consider the small transverse vibrations of a taut homogeneous string described by the function $Q(x,t)$, $0 \le x \le l$, $0 \le t \le T$, which obeys the wave equation

$$\frac{\partial^2 Q}{\partial t^2} = a^2 \frac{\partial^2 Q}{\partial x^2}, \quad 0 < x < l, \quad t > 0, \tag{1}$$

subject to boundary conditions

$$Q(0,t) = u(t), \quad Q(l,t) = 0, \quad 0 \le t \le T. \tag{2}$$

In the Equation (1) $a^2 = \frac{T_0}{\rho}$, where T_0 is string tension, ρ is density of the homogeneous string, and the function $u(t)$ is a boundary control ($u(t)$ is unknown function).

Let the initial and final conditions be given as follows:

$$Q(x,0) = \varphi_0(x), \quad \left.\frac{\partial Q}{\partial t}\right|_{t=0} = \psi_0(x), \quad 0 \le x \le l, \tag{3}$$

$$Q(x,T) = \varphi_T(x) = \varphi_2(x), \quad \left.\frac{\partial Q}{\partial t}\right|_{t=T} = \psi_T(x) = \psi_2(x), \quad 0 \le x \le l, \tag{4}$$

where T is some given moment of time. It is assumed that the function $Q(x,t) \in C^2(\Omega_T)$, where the set $\Omega_T = \{(x,t) : \ x \in [0,l], \ t \in [0,T]\}$.

Let at some moment of time t_1 $(0 < t_1 < T)$ an intermediate state of points (deflection) of the string be given:

$$Q(x,t_1) = \varphi_1(x), \quad 0 \le x \le l. \tag{5}$$

Let us state the following problem of boundary control of string vibrations.

Among possible boundary controls $u(t)$, $0 \le t \le T$, (2), it is required to find such a control that would cause the vibrating motion of the system (1) to pass from the given initial state (3) to the final state (4), taking a given form of deflection (5) at an intermediate moment of time.

Let us assume that the functions $\varphi_i(x)$ $(i = \overline{0,2})$ belong to the space $C^2[0,l]$ and the functions $\psi_0(x)$ and $\psi_T(x)$ belong to the space $C^1[0,l]$. The function $u(t) \in C^2[0,\ T]$ is called an admissible control. It is also assumed that all functions are such that the consistency conditions below are satisfied.

Since the boundary conditions (2) are not homogeneous, we reduce the solution to the problem stated to a control problem with zero boundary conditions. Hence, following [23], we find the solution to the Equation (1) in the form of the sum

$$Q(x,t) = V(x,t) + W(x,t), \tag{6}$$

where $V(x,t)$ is an unknown function to be determined, with homogeneous boundary conditions

$$V(0,t) = V(l,t) = 0, \tag{7}$$

and the function $W(x,t)$ is the solution to the Equation (1) with non-homogeneous boundary conditions

$$W(0,t) = u(t), \ W(l,t) = 0.$$

The function $W(x,t)$ has the form

$$W(x,t) = \left(1 - \frac{x}{l}\right)u(t). \tag{8}$$

Substituting (6) into (1) and considering (8), we obtain the following equation for the determination of the function $V(x,t)$:

$$\frac{\partial^2 V}{\partial t^2} = a^2\frac{\partial^2 V}{\partial x^2} + F(x,t), \tag{9}$$

where

$$F(x,t) = \left(\frac{x}{l} - 1\right)u''(t). \tag{10}$$

The function $V(x,t)$ by virtue of conditions (2)–(5) must satisfy the initial conditions

$$V(x,0) = \varphi_0(x) + \left(\frac{x}{l} - 1\right)u(0), \ \left.\frac{\partial V}{\partial t}\right|_{t=0} = \psi_0(x) + \left(\frac{x}{l} - 1\right)u'(0), \tag{11}$$

the intermediate condition

$$V(x,t_1) = \varphi_1(x) + \left(\frac{x}{l} - 1\right)u(t_1) \tag{12}$$

and final conditions

$$V(x,T) = \varphi_T(x) + \left(\frac{x}{l} - 1\right)u(T), \ \left.\frac{\partial V}{\partial t}\right|_{t=T} = \psi_T(x) + \left(\frac{x}{l} - 1\right)u'(T). \tag{13}$$

It follows from the condition (7) that

$$V(0,t_i) = V(l,t_i) = 0, \ \left.\frac{\partial V(0,t)}{\partial t}\right|_{t=t_i} = \left.\frac{\partial V(l,t)}{\partial t}\right|_{t=t_i} = 0, \ i = \overline{0,\ 2}. \tag{14}$$

From the conditions (11), (12) and (13), given (14), we obtain the following consistency conditions:

$$u(0) = \varphi_0(0),\ u'(0) = \psi_0(0),\ \varphi_0(l) = \psi_0(l) = 0, \tag{15}$$

$$u(t_1) = \varphi_1(0),\ \varphi_1(l) = 0, \tag{16}$$

$$u(T) = \varphi_T(0),\ u'(T) = \psi_T(0),\ \varphi_T(l) = \psi_T(l) = 0. \tag{17}$$

Thus, taking into account the conditions (15)–(17), the conditions (11)–(13) are written as follows:

$$V(x,0) = \varphi_0(x) + \left(\frac{x}{l} - 1\right)\varphi_0(0),\ \left.\frac{\partial V}{\partial t}\right|_{t=0} = \psi_0(x) + \left(\frac{x}{l} - 1\right)\psi_0(0), \tag{18}$$

$$V(x,t_1) = \varphi_1(x) + \left(\frac{x}{l} - 1\right)\varphi_1(0), \tag{19}$$

$$V(x,T) = \varphi_T(x) + \left(\frac{x}{l} - 1\right)\varphi_T(0),\ \left.\frac{\partial V}{\partial t}\right|_{t=T} = \psi_T(x) + \left(\frac{x}{l} - 1\right)\psi_T(0). \tag{20}$$

Thus, the solution to the stated problem of boundary control of vibrations of a string with a given form of deflection at an intermediate moment of time is reduced to the problem of control of (9) with boundary conditions (7) and is stated as follows: to find such a control $u(t), 0 \le t \le T$, under which the vibratory motion (9) with boundary conditions (7) from the given initial state (18) through the intermediate state (19) passes to the final state (20).

3. Problem Solution

Given that the boundary conditions (7) are homogeneous and consistency conditions are satisfied, according to the Fourier series theory, we find the solution to the Equation (9) in the form

$$V(x,t) = \sum_{k=1}^{\infty} V_k(t) \sin\frac{\pi k}{l}x. \tag{21}$$

Let us represent the functions $F(x,t)$, $\varphi_i(x)\ \ (i = \overline{0,\ 2})$, $\psi_0(x)$ and $\psi_T(x)$ as Fourier series, and by substituting their values together with $V(x,t)$ in the Equations (9) and (10) and in the conditions (18)–(20), we obtain

$$\ddot{V}_k(t) + \lambda_k^2 V_k(t) = F_k(t),\ \lambda_k^2 = \left(\frac{a\pi k}{l}\right)^2,\ F_k(t) = -\frac{2a}{\lambda_k l}u''(t), \tag{22}$$

$$V_k(0) = \varphi_k^{(0)} - \frac{2a}{\lambda_k l}\varphi_0(0),\ \dot{V}_k(0) = \psi_k^{(0)} - \frac{2a}{\lambda_k l}\psi_0(0), \tag{23}$$

$$\dot{V}_k(0) = \psi_k^{(0)} - \frac{2a}{\lambda_k l}\psi_0(0), \tag{24}$$

$$V_k(T) = \varphi_k^{(T)} - \frac{2a}{\lambda_k l}\varphi_T(0),\ \dot{V}_k(T) = \psi_k^{(T)} - \frac{2a}{\lambda_k l}\psi_T(0), \tag{25}$$

where $F_k(t)$, $\varphi_k^{(i)}\ \ (i = \overline{0,\ 2})$, $\psi_k^{(0)}$ and $\psi_k^{(T)}$ denote the Fourier coefficients of the functions $F(x,t)$, $\varphi_i(x)\ \ (i = \overline{0,\ 2})$, $\psi_0(x)$ and $\psi_T(x)$, respectively.

The general solution to the Equation (22) with the initial conditions (23) is of the form

$$V_k(t) = V_k(0)\cos\lambda_k t + \frac{1}{\lambda_k}\dot{V}_k(0)\sin\lambda_k t + \frac{1}{\lambda_k}\int_0^t F_k(\tau)\sin\lambda_k(t-\tau)d\tau. \tag{26}$$

Now, given the intermediate (24) and final (25) conditions and the consistency conditions (15)–(17), using the approaches given in [8,9], we obtain from (26) that the function $u(\tau)$ for each k must satisfy the following integral relation:

$$\int_0^T \overline{H}_k(\tau)u(\tau)d\tau = C_k(t_1,\ T),\ k = 1,\ 2, \tag{27}$$

$$\overline{H}_k(\tau) = \begin{pmatrix} \sin\lambda_k(T-\tau) \\ \cos\lambda_k(T-\tau) \\ h_k^{(1)}(\tau) \end{pmatrix}, \; h_k^{(1)}(\tau) = \begin{cases} \sin\lambda_k(t_1-\tau), & 0 \le \tau \le t_1, \\ 0, & t_1 < \tau \le T, \end{cases}$$

$$C_k(t_1, T) = \begin{pmatrix} C_{1k}(T) \\ C_{2k}(T) \\ C_{1k}(t_1) \end{pmatrix}, \tag{28}$$

$$C_{1k}(T) = \frac{1}{\lambda_k^2}\left[\frac{\lambda_k l}{2a}\widetilde{C}_{1k}(T) + X_{1k}\right], \; \widetilde{C}_{1k}(T) = \lambda_k V_k(T) - \lambda_k V_k(0)\cos\lambda_k T - \dot{V}_k(0)\sin\lambda_k T,$$

$$C_{2k}(T) = \frac{1}{\lambda_k^2}\left[\frac{\lambda_k l}{2a}\widetilde{C}_{2k}(T) + X_{2k}\right], \; \widetilde{C}_{2k}(T) = \dot{V}_k(T) + \lambda_k V_k(0)\sin\lambda_k T - \dot{V}_k(0)\cos\lambda_k T, \tag{29}$$

$$C_{1k}(t_1) = \frac{1}{\lambda_k^2}\left[\frac{\lambda_k l}{2a}\widetilde{C}_{1k}(t_1) + X_{1k}^{(1)}\right], \; \widetilde{C}_{1k}(t_1) = \lambda_k V_k(t_1) - \lambda_k V_k(0)\cos\lambda_k t_1 - \dot{V}_k(0)\sin\lambda_k t_1,$$

$$\begin{aligned} X_{1k} &= \lambda_k\varphi_T(0) - \psi_0(0)\sin\lambda_k T - \lambda_k\varphi_0(0)\cos\lambda_k T, \\ X_{2k} &= \psi_T(0) - \psi_0(0)\cos\lambda_k T + \lambda_k\varphi_0(0)\sin\lambda_k T, \\ X_{1k}^{(1)} &= \lambda_k\varphi_1(0) - \psi_0(0)\sin\lambda_k t_1 - \lambda_k\varphi_0(0)\cos\lambda_k t_1. \end{aligned} \tag{30}$$

Thus, to find the function $u(\tau)$, $\tau \in [0, T]$, we obtain the infinite integral relations (27). In practice, the first n harmonics of vibrations are selected and the problem of control synthesis is solved using methods of the theory of control of finite-dimensional systems [8,9,24].

For the first n harmonics, let us introduce the following block vector notations:

$$H_n(\tau) = \begin{pmatrix} \overline{H}_1(\tau) \\ \overline{H}_2(\tau) \\ \vdots \\ \overline{H}_n(\tau) \end{pmatrix}, \; \eta_n = \begin{pmatrix} C_1(t_1, T) \\ C_2(t_1, T) \\ \vdots \\ C_n(t_1, T) \end{pmatrix}. \tag{31}$$

with the dimensionalities $H_n(\tau) - (3n \times 1)$ and $\eta_n - (3n \times 1)$. Consequently, for the first n harmonics, taking into account (31) from (27), we have

$$\int_0^T H_n(\tau)u_n(\tau)d\tau = \eta_n \tag{32}$$

(here and elsewhere, the designation of the letter "n" in the lower index will mean "for the first n harmonics").

The obtained relation (32) implies the validity of the following statement.

Statement. *For each n, the process described by equation (22) with conditions (23)–(25) is completely controllable if and only if, for any given vector η_n (31), the control $u_n(t)$, $t \in [0, T]$, can be found, satisfying condition (32).*

For arbitrary numbers of first harmonics, the boundary control action $u_n(t)$, satisfying the integral relation (32), has the form [8,9,24]:

$$u_n(t) = H_n^T(t)S_n^{-1}\eta_n + f_n(t), \tag{33}$$

where $H_n^T(t)$ is a transposed matrix and $f_n(t)$ is some vector function such that

$$\int_0^T H_n(t)f_n(t)dt = 0, \; S_n = \int_0^T H_n(t)H_n^T(t)dt. \tag{34}$$

Here, $H_n(t)H_n^T(t)$ is the outer product, S_n is a known matrix of dimensionality $(3n \times 3n)$ and it is assumed that $detS_n \neq 0$.

Thus, the following theorem is true.

Theorem 1 *When the initial data of the problem specified in* Section 1 *are matched and the complete controllability condition is fulfilled, problem (1)–(5) has a solution determined for each harmonic of motion by the formula (33).*

Substituting (33) into (22) and the expression obtained for $F_k(t)$ into (26), we obtain the function $V_k(t)$, $t \in [0, T]$. Then, from (21), we have

$$V_n(x,t) = \sum_{k=1}^{n} V_k(t) \sin \frac{\pi k}{l} x, \tag{35}$$

and from (6) for the first n harmonics, the string deflection function $Q_n(x,t)$ is written as

$$Q_n(x,t) = V_n(x,t) + W_n(x,t), \tag{36}$$

where

$$W_n(x,t) = \left(1 - \frac{x}{l}\right) u_n(t). \tag{37}$$

4. Solution Construction in the Cases When $n = 1$ and $n = 2$

Applying the above approach, we construct the boundary control given $n = 1$ and given $n = 2$ and the string deflection function, respectively.

4.1. Case $n = 1$

Given $n = 1$ (therefore, $k = 1$), according to (31), we have $H_1(\tau) = \overline{H}_1(\tau)$ and $\eta_1 = C_1(t_1, T)$, and from (34) we obtain

$$S_1 = \int_0^T H_1(\tau) H_1^T(\tau) d\tau = \begin{pmatrix} s_{11}^{(1)} & s_{12}^{(1)} & s_{13}^{(1)} \\ s_{21}^{(1)} & s_{22}^{(1)} & s_{23}^{(1)} \\ s_{31}^{(1)} & s_{32}^{(1)} & s_{33}^{(1)} \end{pmatrix}.$$

Elements of the matrix S_1, according to the notation (28), have the following form:

$$s_{11}^{(1)} = \frac{T}{2} - \frac{1}{4\lambda_1} \sin 2\lambda_1 T,\ s_{12}^{(1)} = s_{21}^{(1)} = \frac{1}{2\lambda_1} \sin^2 \lambda_1 T,\ s_{22}^{(1)} = \frac{T}{2} + \frac{1}{4\lambda_1} \sin 2\lambda_1 T,$$

$$s_{33}^{(1)} = \tfrac{t_1}{2} - \tfrac{1}{4\lambda_1} \sin 2\lambda_1 t_1,\ s_{13}^{(1)} = s_{31}^{(1)} = \tfrac{t_1}{2} \cos \lambda_1 (T - t_1) - \tfrac{1}{2\lambda_1} \sin \lambda_1 t_1 \cos \lambda_1 T,$$
$$s_{23}^{(1)} = s_{32}^{(1)} = \tfrac{1}{2\lambda_1} \sin \lambda_1 t_1 \sin \lambda_1 T - \tfrac{t_1}{2} \sin \lambda_1 (T - t_1),$$

and $\Delta = \det S_1 \neq 0$. Denote by S_1^{-1} the symmetric matrix of dimension (3×3) inverse to the matrix S_1.

From (33), it follows that $u_1(\tau) = H_1^T(\tau) S_1^{-1} \eta_1 + f_1(\tau)$. Assuming that $f_1(\tau) = 0$, we obtain, given $\tau \in [0, t_1]$,

$$\begin{aligned} u_1(\tau) = {} & \sin \lambda_1 (T - \tau) [\hat{s}_{11} C_{11}(T) + \hat{s}_{12} C_{21}(T) + \hat{s}_{13} C_{11}(t_1)] \\ & + \cos \lambda_1 (T - \tau) [\hat{s}_{21} C_{11}(T) + \hat{s}_{22} C_{21}(T) + \hat{s}_{23} C_{11}(t_1)] \\ & + \sin \lambda_1 (t_1 - \tau) [\hat{s}_{31} C_{11}(T) + \hat{s}_{32} C_{21}(T) + \hat{s}_{33} C_{11}(t_1)], \end{aligned} \tag{38}$$

and given $\tau \in (t_1, T]$,

$$\begin{aligned} u_1(\tau) = {} & \sin \lambda_1 (T - \tau) [\hat{s}_{11} C_{11}(T) + \hat{s}_{12} C_{21}(T) + \hat{s}_{13} C_{11}(t_1)] \\ & + \cos \lambda_1 (T - \tau) [\hat{s}_{21} C_{11}(T) + \hat{s}_{22} C_{21}(T) + \hat{s}_{23} C_{11}(t_1)]. \end{aligned} \tag{39}$$

Note that according to (36), we can write the expression for the function $Q_1(x,t)$. Assume that $t_1 = \frac{l}{a}$, $T = 2\frac{l}{a}$. Then, given $\lambda_1 = \frac{a\pi}{l}$, we obtain $t_1\lambda_1 = \pi$, $T\lambda_1 = 2\pi$ and $\lambda_1(T-t_1) = \pi$. For matrices S_1 and S_1^{-1}, we have:

$$S_1 = \begin{pmatrix} \frac{l}{a} & 0 & -\frac{l}{2a} \\ 0 & \frac{l}{a} & 0 \\ -\frac{l}{2a} & 0 & \frac{l}{2a} \end{pmatrix}, \quad S_1^{-1} = \begin{pmatrix} \frac{2a}{l} & 0 & \frac{2a}{l} \\ 0 & \frac{a}{l} & 0 \\ \frac{2a}{l} & 0 & \frac{4a}{l} \end{pmatrix},$$

and for the control from (38) and (39), we obtain

$$u_1(\tau) = \begin{cases} \frac{a}{l}\cos\lambda_1\tau\ C_{21}(T) + \frac{2a}{l}\sin\lambda_1\tau\ C_{11}(t_1), & \tau \in [0,t_1], \\ \frac{a}{l}\cos\lambda_1\tau\ C_{21}(T) - \frac{2a}{l}\sin\lambda_1\tau\,(C_{11}(T)+C_{11}(t_1)), & \tau \in (t_1,T]. \end{cases} \tag{40}$$

For the function $V_1(t)$ from (26), given (22), so that $F_1(t) = -\frac{2a}{\lambda_1 l}u''_1(t)$, we obtain, given $t \in [0,t_1]$,

$$V_1(t) = \left(V_1(0) - \frac{a\lambda_1 t\, C_{11}(t_1)}{\pi l}\right)\cos\lambda_1 t + \left(\frac{\dot V_1(0)}{\lambda_1} + \frac{a(2C_{11}(t_1)+\lambda_1 t\, C_{21}(T))}{2\pi l}\right)\sin\lambda_1 t,$$

and given $t \in (t_1,T]$,

$$V_1(t) = \left[V_1(0) + \frac{a(t-t_1)\lambda_1}{\pi l}C_{11}(T) + \frac{a(t-2t_1)\lambda_1}{\pi l}C_{11}(t_1)\right]\cos\lambda_1 t$$
$$+\left[\frac{\dot V_1(0)}{\lambda_1} + \frac{at\lambda_1}{2\pi l}C_{21}(T) - \frac{a(C_{11}(T)+C_{11}(t_1))}{\pi l}\right]\sin\lambda_1 t.$$

From (36), given (35) and (37), we have

$$Q_1(x,t) = V_1(t)\sin\frac{\pi}{l}x + \left(1-\frac{x}{l}\right)u_1(t). \tag{41}$$

4.2. Case $n = 2$

Given $n = 2$ (i.e., $k = 1,\ 2$) from (31), according to (28)–(30), we have

$$H_2(\tau) = \begin{pmatrix} \overline{H}_1(\tau) \\ \overline{H}_2(\tau) \end{pmatrix} = \begin{pmatrix} \sin\lambda_1(T-\tau) \\ \cos\lambda_1(T-\tau) \\ h_1^{(1)}(\tau) \\ \sin\lambda_2(T-\tau) \\ \cos\lambda_2(T-\tau) \\ h_2^{(1)}(\tau) \end{pmatrix}, \quad \eta_2 = \begin{pmatrix} C_1(t_1,T) \\ C_2(t_1,T) \end{pmatrix} = \begin{pmatrix} C_{11}(T) \\ C_{21}(T) \\ C_{11}(t_1) \\ C_{12}(T) \\ C_{22}(T) \\ C_{12}(t_1) \end{pmatrix},$$

where

$$h_1^{(1)}(\tau) = \begin{cases} \sin\lambda_1(t_1-\tau), & 0 \le \tau \le t_1, \\ 0, & t_1 < \tau \le T, \end{cases} \quad h_2^{(1)}(\tau) = \begin{cases} \sin\lambda_2(t_1-\tau), & 0 \le \tau \le t_1, \\ 0, & t_1 < \tau \le T. \end{cases}$$

The values $C_{11}(T), C_{21}(T), C_{11}(t_1), C_{12}(T), C_{22}(T)$ and $C_{12}(t_1)$ can be easily calculated using formulas (29) and (30). Their explicit form is omitted for brevity.

From (34), we obtain

$$S_2 = \int_0^T H_2(\tau)H_2^T(\tau)d\tau$$

where S_2 is a symmetric matrix of dimension (6×6) and its elements $s_{ij}^{(2)}$ are equal to ones of the matrix S_1, i.e.,

$$s_{ij}^{(2)} = s_{ij}^{(1)} \text{ for } i,j = \overline{1,3}; \quad s_{ij}^{(2)} = s_{ji}^{(2)} \text{ for } i,j = \overline{1,6},\ i \neq j.$$

Given $\lambda_2 = \frac{2a\pi}{l}$ and using the assumptions made in Section 4.1, for the matrix S_2, we obtain:

$$S_2 = \begin{pmatrix} \frac{l}{a} & 0 & -\frac{l}{2a} & 0 & 0 & 0 \\ 0 & \frac{l}{a} & 0 & 0 & 0 & -\frac{4l}{3a\pi} \\ -\frac{l}{2a} & 0 & \frac{l}{2a} & 0 & -\frac{2l}{3a\pi} & 0 \\ 0 & 0 & 0 & \frac{l}{a} & 0 & \frac{l}{2a} \\ 0 & 0 & -\frac{2l}{3a\pi} & 0 & \frac{l}{a} & 0 \\ 0 & -\frac{4l}{3a\pi} & 0 & \frac{l}{2a} & 0 & \frac{l}{2a} \end{pmatrix},$$

where the calculation takes into account the following ratios: $t_1\lambda_2 = 2\pi$, $T\lambda_2 = 4\pi$, $\lambda_2(T - t_1) = 2\pi$, $(\lambda_1 + \lambda_2)T = 6\pi$, $\lambda_1 + \lambda_2 = \frac{3a\pi}{l}$, $\lambda_1 - \lambda_2 = -\frac{a\pi}{l}$, $\lambda_1^2 - \lambda_2^2 = -3\left(\frac{a\pi}{l}\right)^2$, $\lambda_1 T + \lambda_2 t_1 = 4\pi$, $\lambda_1 T - \lambda_2 t_1 = 0$, $\lambda_2 T + \lambda_1 t_1 = 5\pi$ and $\lambda_2 T - \lambda_1 t_1 = 3\pi$. Let us note that $\det S_2 = \frac{l^6}{6^4 a^6 \pi^4 pq}$, where $p = \left(9\pi^2 - 64\right)^{-1}$, $q = \left(9\pi^2 - 16\right)^{-1}$. Having the matrix S_2, it is not difficult to calculate the matrix S_2^{-1}, the inverse to it.

From (33), it follows that $u_2(\tau) = H_2^T(\tau)S_2^{-1}\eta_2 + f_2(\tau)$. For simplicity, assuming that $f_2(\tau) = 0$, we obtain given $\tau \in [0, t_1]$,

$$\begin{aligned} u_2(\tau) = {} & \tfrac{2a}{l} q\left(9\pi^2 C_{11}(t_1) + 8C_{11}(T) + 6\pi C_{22}(T)\right) \sin\lambda_1\tau \\ & + \tfrac{3a\pi}{l} p\left(16C_{12}(t_1) - 8C_{12}(T) + 3\pi C_{21}(T)\right) \cos\lambda_1\tau \\ & + \tfrac{2a}{l} p\left(32C_{12}(T) - 9\pi^2 C_{12}(t_1) - 12\pi C_{21}(T)\right) \sin\lambda_2\tau \\ & + \tfrac{3a\pi}{l} q\left(8C_{11}(t_1) + 4C_{11}(T) + 3\pi C_{22}(T)\right) \cos\lambda_2\tau, \end{aligned} \tag{42}$$

and given $\tau \in (t_1, T]$,

$$\begin{aligned} u_2(\tau) = {} & \tfrac{2a}{l} q\left[8C_{11}(T) - 9\pi^2(C_{11}(t_1) + C_{11}(T)) - 6\pi C_{22}(T)\right] \sin\lambda_1\tau \\ & + \tfrac{3a\pi}{l} p\left(16C_{12}(t_1) - 8C_{12}(T) + 3\pi C_{21}(T)\right) \cos\lambda_1\tau \\ & + \tfrac{2a}{l} p\left[9\pi^2(C_{12}(t_1) - C_{12}(T)) + 32C_{12}(T) + 12\pi C_{21}(T)\right] \sin\lambda_2\tau \\ & + \tfrac{3a\pi}{l} q\left(8C_{11}(t_1) + 4C_{11}(T) + 3\pi C_{22}(T)\right) \cos\lambda_2\tau, \end{aligned} \tag{43}$$

where

$$\begin{aligned} C_{11}(T) &= \frac{l}{2a}(V_1(T) - V_1(0)) + \frac{\varphi_T(0) - \varphi_0(0)}{\lambda_1}, \\ C_{21}(T) &= \frac{l}{2a\lambda_1}\left(\dot{V}_1(T) - \dot{V}_1(0)\right) + \frac{\psi_T(0) - \psi_0(0)}{\lambda_1^2}, \\ C_{11}(t_1) &= \frac{l}{2a}(V_1(t_1) + V_1(0)) + \frac{\varphi_1(0) + \varphi_0(0)}{\lambda_1}, \\ C_{12}(T) &= \frac{l}{2a}(V_2(T) - V_2(0)) + \frac{\varphi_T(0) - \varphi_0(0)}{\lambda_2}, \\ C_{22}(T) &= \frac{l}{2a\lambda_2}\left(\dot{V}_2(T) - \dot{V}_2(0)\right) + \frac{\psi_T(0) - \psi_0(0)}{\lambda_2^2}, \\ C_{12}(t_1) &= \frac{l}{2a}(V_2(t_1) - V_2(0)) + \frac{\varphi_1(0) - \varphi_0(0)}{\lambda_2}. \end{aligned} \tag{44}$$

From (26), for $V_2(t)$ given (22), so that $F_2(t) = -\frac{2a}{\lambda_2 l} u''{}_2(t)$, we obtain, given $t \in [0, t_1]$

$$V_2(t) = \frac{\alpha_1}{\lambda_1^2-\lambda_2^2}\cos\lambda_1 t - \frac{\beta_1}{\lambda_1^2-\lambda_2^2}\sin\lambda_1 t + \left(V_2(0) - \frac{\beta_2 t}{2\lambda_2} + \frac{\alpha_1}{\lambda_1^2-\lambda_2^2}\right)\cos\lambda_2 t$$
$$+\left(\frac{\dot{V}_2(0)}{\lambda_2} + \frac{\alpha_2 t}{2\lambda_2} + \frac{\beta_1\lambda_1}{\lambda_2(\lambda_1^2-\lambda_2^2)} + \frac{\beta_2}{2\lambda_2^2}\right)\sin\lambda_2 t$$

and given $t \in (t_1, T]$,

$$V_2(t) = \frac{\gamma_1}{\lambda_1^2-\lambda_2^2}\sin\lambda_1 t - \frac{\alpha_1}{\lambda_1^2-\lambda_2^2}\cos\lambda_1 t$$
$$+\left(V_2(0) + \frac{\alpha_1}{\lambda_1^2-\lambda_2^2} - \frac{\gamma_2 t}{2\lambda_2} + \frac{\pi(\lambda_1-2\lambda_2)(\gamma_2-\beta_2)}{2\lambda_2(\lambda_1^2-\lambda_2^2)}\right)\cos\lambda_2 t$$
$$+\left(\frac{\dot{V}_2(0)}{\lambda_2} + \frac{\alpha_2 t}{2\lambda_2} + \frac{\gamma_2}{2\lambda_2^2} + \frac{\lambda_1(\gamma_1+2\beta_1)}{\lambda_2(\lambda_1^2-\lambda_2^2)}\right)\sin\lambda_2 t,$$

where

$$\alpha_1 = \frac{3a\lambda_1^2 p}{l}[8(2C_{12}(t_1) - C_{12}(T)) + 3\pi C_{21}(T)],$$

$$\alpha_2 = \frac{3a\lambda_2^2 q}{l}[4(2C_{11}(t_1) + C_{11}(T)) + 3\pi C_{22}(T)],$$

$$\beta_1 = \frac{2a\lambda_1^2 q}{\pi l}\left(9\pi^2 C_{11}(t_1) + 6\pi C_{22}(T) + 8C_{11}(T)\right),$$

$$\beta_2 = \frac{2a\lambda_2^2 p}{\pi l}\left(32C_{12}(T) - 12\pi C_{21}(T) - 9\pi^2 C_{12}(t_1)\right),$$

$$\gamma_1 = \frac{2a\lambda_1^2 q}{\pi l}\left[9\pi^2(C_{11}(t_1) + C_{11}(T)) + 6\pi C_{22}(T) - 8C_{11}(T)\right],$$

$$\gamma_2 = \frac{2a\lambda_2^2 p}{\pi l}\left[32C_{12}(T) + 12\pi C_{21}(T) - 9\pi^2(C_{12}(T) - C_{12}(t_1))\right].$$

From (36), given (35) and (37), we have

$$Q_2(x,t) = V_2(x,t) + W_2(x,t) = V_1(t)\sin\frac{\pi}{l}x + V_2(t)\sin\frac{2\pi}{l}x + \left(1 - \frac{x}{l}\right)u_2(t). \quad (45)$$

5. Computational Experiment

Applying the above approach, we construct the boundary control given $n = 1$ and $n = 2$ and the string deflection function, respectively. This section includes the initial data, the results obtained and a discussion of the methodology's effectiveness.

5.1. Initial Data

Let us present the results of a computational experiment for a given initial, intermediate and final state of the string given $n = 1$ and $n = 2$ assuming that $a = \frac{1}{3}$ and $l = 1$ and compare the behavior of the string deflection function with the given initial functions. Given the chosen values of a and l, we have

$$t_1 = \frac{l}{a} = 3, \; T = 2\frac{l}{a} = 6, \; \lambda_1 = \frac{\pi}{3}, \; \lambda_2 = \frac{2\pi}{3}.$$

The choice of an intermediate value $t_1 = \frac{T}{2}$ is due to practical recommendations [4].

We choose the specific initial functions from the functions class from the problem statement (Section 2) that satisfied the consistency conditions (15)–(17).

Let the following initial state be specified given $t = 0$:

$$\varphi_0(x) = \frac{1}{2}x^2 - \frac{2x}{5} - \frac{1}{10}, \; \psi_0(x) = -\frac{x^2}{3} + \frac{x}{3},$$

Given $t_1 = 3$, an intermediate state is specified as follows:

$$\varphi_1(x) = \frac{x^3}{3} - x^2 + \frac{2x}{3},$$

Moreover, given $T = 6$, the next end state is specified:

$$\varphi_T(x) = 0,\ \psi_T(x) = 0.$$

The proposed approach is applicable for any initial functions that meet the necessary requirements given in Section 2 so that the selected functions are some of them.

Note the choice of final zero values does not reflect the essence of the limitations of the technique and is made to simplify the final formulas. In addition, the problem of stabilizing string vibrations is relevant for damping transverse vibrations of a longitudinally moving string (for example, a paper web) in production [22].

The coefficients of the Fourier series for the functions $\varphi_0(x)$, $\psi_0(x)$, $\varphi_1(x)$, $\varphi_T(x)$ and $\psi_T(x)$ are equal, respectively, to:

$$\varphi_1^{(0)} = -\frac{4}{\pi^3} - \frac{1}{5\pi},\ \varphi_2^{(0)} = -\frac{1}{10\pi},\ \psi_1^{(0)} = \frac{8}{3\pi^3},\ \varphi_1^{(1)} = \frac{4}{\pi^3},\ \varphi_2^{(1)} = \frac{1}{2\pi^3},$$

$$\psi_2^{(0)} = \varphi_1^{(T)} = \varphi_2^{(T)} = \psi_1^{(T)} = \psi_2^{(T)} = 0.$$

The values of these functions at the ends of the string are as follows:

$$\varphi_0(0) = -\tfrac{1}{10},\ \varphi_1(0) = \varphi_T(0) = \psi_T(0) = \psi_0(0) = \varphi_0(1) = \varphi_1(1) = \varphi_T(1) = \psi_T(1) = \psi_0(1) = 0.$$

From (23)–(25), we have

$$V_1(0) = -\frac{4}{\pi^3},\ \dot{V}_1(0) = \frac{8}{3\pi^3},\ V_1(3) = \frac{4}{\pi^3},\ V_1(6) = 0,\ \dot{V}_1(6) = 0,$$

$$V_2(0) = 0,\ \dot{V}_2(0) = 0,\ V_2(3) = \frac{1}{2\pi^3},\ V_2(6) = 0,\ \dot{V}_2(6) = 0.$$

From (44), we have

$$C_{11}(6) = \frac{6}{\pi^3} + \frac{3}{10\pi},\ C_{21}(6) = -\frac{12}{\pi^4},\ C_{11}(3) = -\frac{3}{10\pi},$$

$$C_{12}(6) = \frac{3}{20\pi},\ C_{22}(6) = 0,\ C_{12}(3) = \frac{3}{4\pi^3} + \frac{3}{20\pi}.$$

5.2. Results

In this section, we present the calculation formulas obtained for the functions u_1, u_2, V_1, V_2, Q_1 and Q_2. From (40), (42) and (43), we have

$$u_1(t) = \begin{cases} -\frac{4}{\pi^4}\cos\frac{\pi}{3}t - \frac{1}{5\pi}\sin\frac{\pi}{3}t, & t \in [0,3], \\ -\frac{4}{\pi^4}\cos\frac{\pi}{3}t - \frac{4}{\pi^3}\sin\frac{\pi}{3}t, & t \in (3,6], \end{cases} \tag{46}$$

$$u_2(t) = \tfrac{6}{5\pi^2}(\pi^2 - 20)p\cos\tfrac{\pi}{3}t - \tfrac{6}{5\pi^2}(\pi^2 - 20)q\cos\tfrac{2\pi}{3}t - \tfrac{1}{5\pi^3}(9\pi^4 - 8\pi^2 - 160)q\sin\tfrac{\pi}{3}t - \tfrac{1}{10\pi^3}(9\pi^4 + 13\pi^2 - 960)p\sin\tfrac{2\pi}{3}t,\ t \in [0,3], \tag{47}$$

$$u_2(t) = \tfrac{6}{5\pi^2}(\pi^2 - 20)p\cos\tfrac{\pi}{3}t - \tfrac{6}{5\pi^2}(\pi^2 - 20)q\cos\tfrac{2\pi}{3}t - \tfrac{4}{5\pi^3}(43\pi^2 - 40)q\sin\tfrac{\pi}{3}t + \tfrac{1}{10\pi^3}(77\pi^2 - 960)p\sin\tfrac{2\pi}{3}t,\ t \in (3,6]. \tag{48}$$

Note that the following estimates are obtained for the functions $u_1(t)$ and $u_2(t)$:

$$\max_{0\le t\le 6}|u_1(t)|\approx 0.1354,\quad \max_{0\le t\le 6}|u_2(t)|\approx 0.1193.$$

We obtain the following explicit expressions for the functions $V_1(t)$ and $V_2(t)$:

$$V_1(t)=\begin{cases}\frac{t\pi^2-120}{30\pi^3}\cos\frac{\pi}{3}t-\frac{20t+3\pi^2-240}{30\pi^4}\sin\frac{\pi}{3}t, & t\in[0,3],\\ \frac{3\pi^2+20t-180}{30\pi^3}\cos\frac{\pi}{3}t-\frac{2(t-9)}{3\pi^4}\sin\frac{\pi}{3}t, & t\in(3,6],\end{cases}\tag{49}$$

$$\begin{gathered}V_2(t)=\tfrac{2}{5\pi^3}(\pi^2-20)p\cos\tfrac{\pi}{3}t-\tfrac{1}{15\pi^4}(9\pi^4-8\pi^2-160)q\sin\tfrac{\pi}{3}t\\ +\tfrac{1}{30\pi^3}((9\pi^4+13\pi^2-960)t-12\pi^2+240)p\cos\tfrac{2\pi}{3}t\\ -\left(\tfrac{2}{5\pi^2}(\pi^2-20)tq+\tfrac{1}{60\pi^4}(81\pi^6+1215\pi^4-24,688\pi^2+25,600)pq\right)\sin\tfrac{2\pi}{3}t,\\ t\in[0,3],\end{gathered}\tag{50}$$

$$\begin{gathered}V_2(t)=\tfrac{2}{5\pi^3}(\pi^2-20)p\cos\tfrac{\pi}{3}t-\tfrac{4}{15\pi^4}(43\pi^2-40)q\sin\tfrac{\pi}{3}t\\ +\tfrac{1}{30\pi^3}((960-77\pi^2)t+27\pi^4+258\pi^2-5520)p\cos\tfrac{2\pi}{3}t\\ -\left(\tfrac{2}{5\pi^2}(\pi^2-20)tq+\tfrac{1}{60\pi^4}(-324\pi^6+3609\pi^4+8432\pi^2-66,560)pq\right)\sin\tfrac{2\pi}{3}t,\\ t\in(3,6].\end{gathered}\tag{51}$$

Note that the following estimates take place for the functions $V_1(t)$ and $V_2(t)$:

$$\max_{0\le t\le 6}|V_1(t)|\approx 0.1165,\quad \max_{0\le t\le 6}|V_2(t)|\approx 0.0321.$$

This confirms that the absolute value of each subsequent summand of series (21) decreases.

From (41) and (45), given (46)–(51), we obtain the following explicit expressions for the functions $Q_1(x,t)$ and $Q_2(x,t)$:

given $t\in[0,3]$,

$$\begin{gathered}Q_1(t,x)=\left(\frac{t\pi^2-120}{30\pi^3}\cos\frac{\pi}{3}t-\frac{20t+3\pi^2-240}{30\pi^4}\sin\frac{\pi}{3}t\right)\sin\pi x\\ -\left(\frac{4}{\pi^4}\cos\frac{\pi}{3}t+\frac{1}{5\pi}\sin\frac{\pi}{3}t\right)(1-x),\end{gathered}$$

given $t\in(3,6]$,

$$\begin{gathered}Q_1(t,x)=\left(\frac{3\pi^2+20t-180}{30\pi^3}\cos\frac{\pi}{3}t-\frac{2(t-9)}{3\pi^4}\sin\frac{\pi}{3}t\right)\sin\pi x\\ -\left(\frac{4}{\pi^4}\cos\frac{\pi}{3}t+\frac{4}{\pi^3}\sin\frac{\pi}{3}t\right)(1-x),\end{gathered}$$

given $t\in[0,3]$,

$$\begin{gathered}Q_2(x,t)=\left(\frac{t\pi^2-120}{30\pi^3}\cos\frac{\pi}{3}t-\frac{20t+3\pi^2-240}{30\pi^4}\sin\frac{\pi}{3}t\right)\sin\pi x\\ +\left(\frac{2}{5\pi^3}\left(\pi^2-20\right)p\cos\frac{\pi}{3}t-\frac{1}{15\pi^4}\left(9\pi^4-8\pi^2-160\right)q\sin\frac{\pi}{3}t\right.\\ +\tfrac{1}{30\pi^3}((9\pi^4+13\pi^2-960)t-12\pi^2+240)p\cos\tfrac{2\pi}{3}t\\ -\left(\tfrac{2}{5\pi^2}(\pi^2-20)tq+\tfrac{1}{60\pi^4}(81\pi^6+1215\pi^4+24,688\pi^2+25,600)pq\right)\end{gathered}$$

$$\cdot \sin\frac{2\pi}{3}t\Big)\sin 2\pi x+\Big(\frac{6}{5\pi^2}\left(\pi^2-20\right)p\,\cos\frac{\pi}{3}t-\frac{6}{5\pi^2}\left(\pi^2-20\right)q\,\cos\frac{2\pi}{3}t$$

$$-\frac{4}{5\pi^3}\left(43\pi^2-40\right)q\,\sin\frac{\pi}{3}t+\frac{1}{10\pi^3}\left(77\pi^2-960\right)p\,\sin\frac{2\pi}{3}t\Big)(1-x),$$

and given $t \in (3,6]$,

$$Q_2(x,t)=\left(\frac{3\pi^2+20t-180}{30\pi^3}\cos\frac{\pi}{3}t-\frac{2(t-9)}{3\pi^4}\sin\frac{\pi}{3}t\right)\sin\pi x$$

$$+\Big(\frac{2}{5\pi^3}\left(\pi^2-20\right)p\cos\frac{\pi}{3}t-\frac{4}{15\pi^4}\left(43\pi^2-40\right)q\sin\frac{\pi}{3}t$$

$$+\frac{1}{30\pi^3}\left(\left(960-77\pi^2\right)t+27\pi^4+258\pi^2-5520\right)p\cos\frac{2\pi}{3}t$$

$$-\left(\frac{2}{5\pi^2}\left(\pi^2-20\right)tq+\frac{1}{60\pi^4}\left(-324\pi^6+3609\pi^4+8432\pi^2-66,560\right)pq\right)$$

$$\cdot \sin\frac{2\pi}{3}t\Big)\sin 2\pi x+\Big(\frac{6}{5\pi^2}\left(\pi^2-20\right)p\,\cos\frac{\pi}{3}t-\frac{6}{5\pi^2}\left(\pi^2-20\right)q\,\cos\frac{2\pi}{3}t$$

$$-\frac{4}{5\pi^3}\left(43\pi^2-40\right)q\,\sin\frac{\pi}{3}t+\frac{1}{10\pi^3}\left(77\pi^2-960\right)p\,\sin\frac{2\pi}{3}t\Big)(1-x).$$

At the moment of time $t=0$, the functions $Q_1(x,0)$ and $Q_2(x,0)$ are equal to:

$$Q_1(x,0)=-\frac{4}{\pi^3}\sin\pi x-\frac{4}{\pi^4}(1-x),$$

$$Q_2(x,0)=-\frac{4}{\pi^3}\sin\pi x+\frac{288}{5\pi^2}\left(\pi^2-20\right)pq(1-x).$$

Calculate

$$\left.\frac{\partial Q_1(x,t)}{\partial t}\right|_{t=0}=\dot{Q}_1(x,0)=\frac{8}{3\pi^3}\sin\pi x-\frac{1}{15}(1-x),$$

$$\left.\frac{\partial Q_2(x,t)}{\partial t}\right|_{t=0}=\dot{Q}_2(x,0)=\frac{8}{3\pi^3}\sin\pi x$$

$$-\frac{1}{15\pi^2}\left(162\pi^6-675\pi^4-9776\pi^2+25,600\right)pq(1-x).$$

We can check that the expression of the deflection functions $Q_1(x,3)$ and $Q_2(x,3)$ at the final moment of the segment $[0,3]$ coincides with the corresponding expression at the beginning of the next time interval, and the functions have the form:

$$Q_1(x,3)=\frac{40-\pi^2}{10\pi^3}\sin\pi x+\frac{4}{\pi^4}(1-x),$$

$$Q_2(x,3)=\frac{40-\pi^2}{10\pi^3}\sin\pi x+\frac{1}{10\pi^3}\left(5\pi^2+9\pi^4-800\right)p\sin 2\pi x$$

$$-\frac{12}{5\pi^2}\left(\pi^2-20\right)\left(9\pi^2-40\right)pq(1-x).$$

The deflection function and its derivative at the moment of time $t=6$ are equal, respectively, to:

$$Q_1(x,6)=\frac{\pi^2-20}{10\pi^3}\sin\pi x-\frac{4}{\pi^4}(1-x),$$

$$\left.\frac{\partial Q_1(x,t)}{\partial t}\right|_{t=6}=\dot{Q}_1(x,6)=\frac{4}{3\pi^3}\sin\pi x-\frac{4}{3\pi^2}(1-x),$$

$$Q_2(x,6) = \frac{\pi^2 - 20}{10\pi^3}\sin \pi x + \frac{1}{10\pi}\sin 2\pi x + \frac{288}{5\pi^2}\left(\pi^2 - 20\right)pq(1-x)$$

$$\left.\frac{\partial Q_2(x,t)}{\partial t}\right|_{t=6} = \dot{Q}_2(x,6) = \frac{4}{3\pi^3}\sin \pi x - \frac{6}{5\pi}\left(\pi^2 - 20\right)q \sin 2\pi x$$

$$- \frac{1}{15\pi^2}\left(855\pi^4 - 2576\pi^2 - 5120\right)pq(1-x).$$

5.3. Illustrative Material

Let us illustrate the obtained formulas on the graphs. The graphs of the functions $u_1(t)$ and $u_2(t)$ are given in Figure 1.

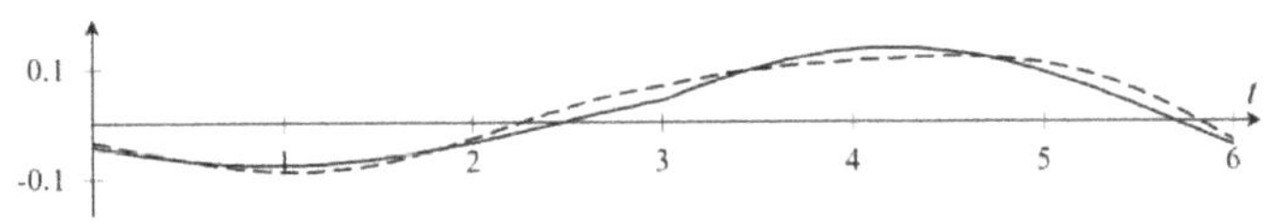

Figure 1. Graphs of $u_1(t)$ (solid line) and $u_2(t)$ (dashed line).

The graphs of the functions $V_1(t)$ and $V_2(t)$ are shown in Figure 2.

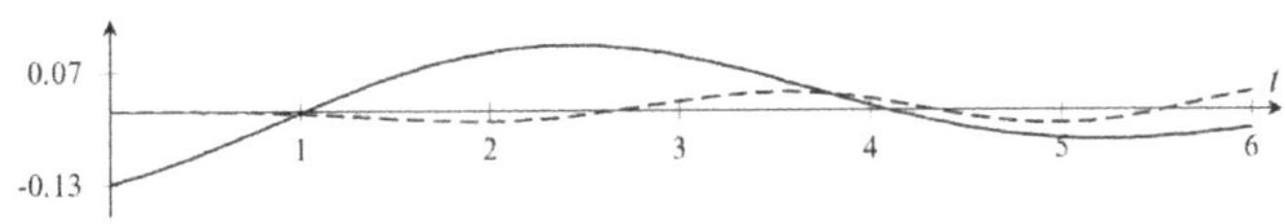

Figure 2. Graphs of the functions $V_1(t)$ (solid line) and $V_2(t)$ (dashed line).

The graphical representation of the functions $Q_1(x,0)$, $Q_2(x,0)$ and $\varphi_0(x)$ is illustrated in Figure 3.

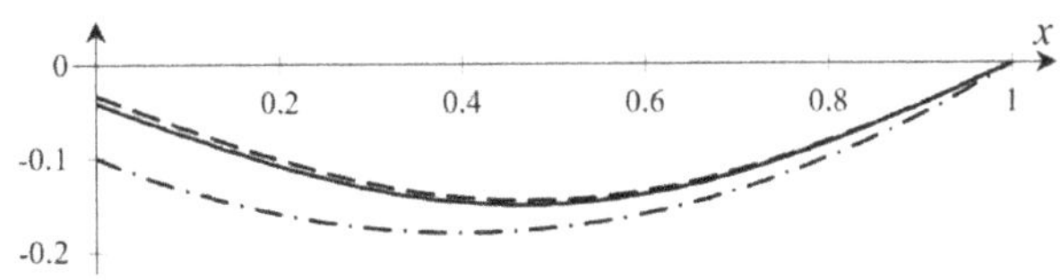

Figure 3. Graphs of $Q_1(x,0)$ (solid line), $Q_2(x,0)$ (dashed line) and $\varphi_0(x)$ (dash-dotted line).

The graphs of the functions $Q_1(x,3)$, $Q_2(x,3)$ and $\varphi_1(x)$ are shown in Figure 4, which illustrates the small discrepancies between these functions.

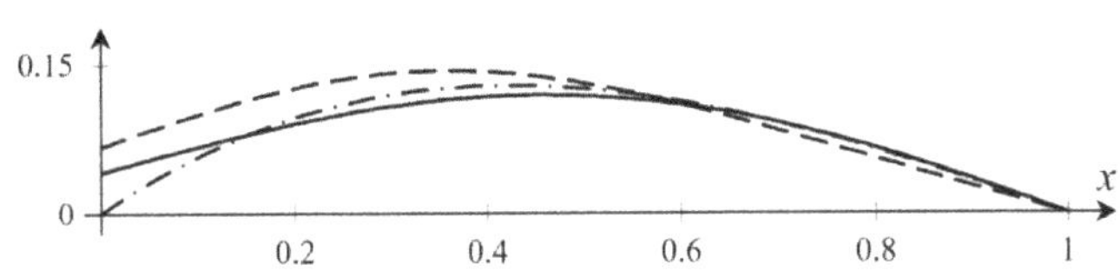

Figure 4. Graphs of $Q_1(x,3)$ (solid line), $Q_2(x,3)$ (dashed line) and $\varphi_1(x)$ (dash-dotted line).

Graphical representations of the functions $Q_1(x,6)$ and $Q_2(x,6)$ and $\dot{Q}_1(x,6)$ and $\dot{Q}_2(x,6)$ are shown in Figures 5 and 6, respectively.

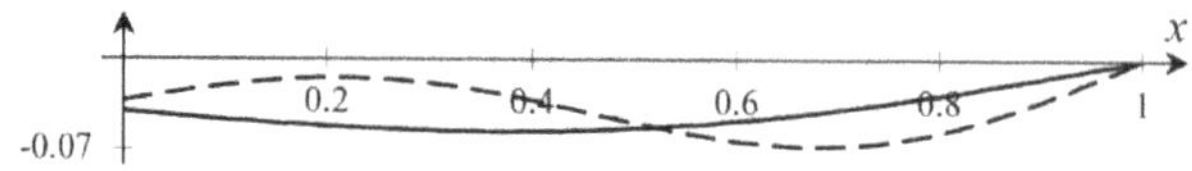

Figure 5. Graphs of $Q_1(x,6)$ (solid line) and $Q_2(x,6)$ (dashed line).

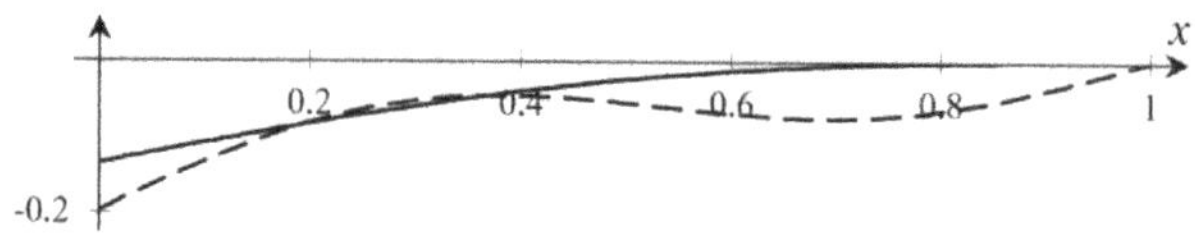

Figure 6. Graphs of $\dot{Q}_1(x,6)$ (solid line) and $\dot{Q}_2(x,6)$ (dashed line).

Figure 7 provides graphical illustrations of the dynamics of the behavior of the functions $Q_1(x,t)$ and $Q_2(x,t)$ given $t = 0, 1, 2, 3, 4, 5, 6$.

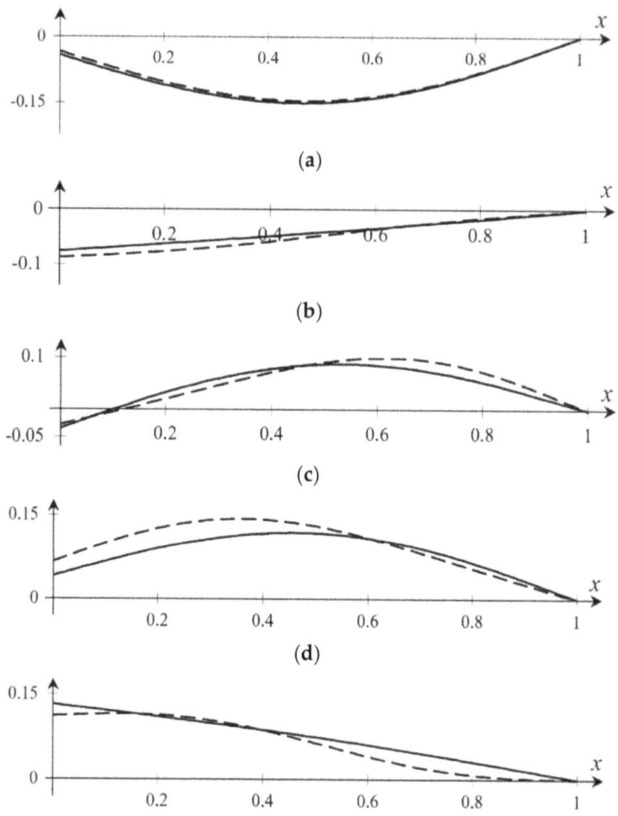

Figure 7. *Cont.*

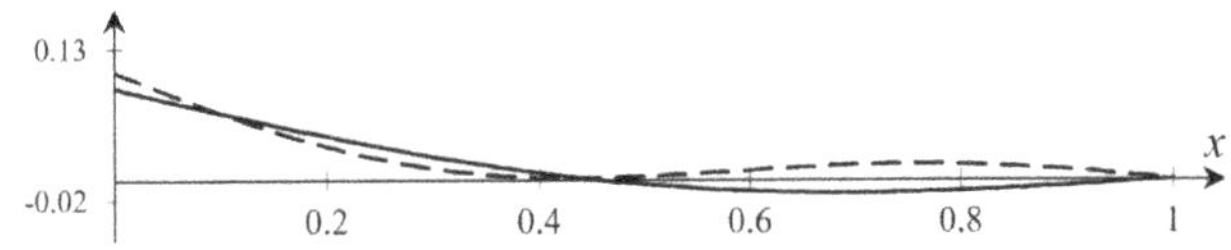

(f)

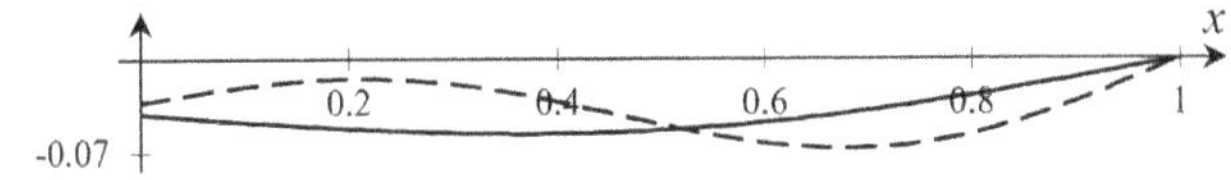

(g)

Figure 7. Graphs of the functions $Q_1(x,t)$ (solid line) and $Q_2(x,t)$ (dashed line) at fixed points in time t: **(a)** $t = 0$; **(b)** $t = 1$; **(c)** $t = 2$; **(d)** $t = 3$; **(e)** $t = 4$; **(f)** $t = 5$; **(g)** $t = 6$.

5.4. Discussion of Results

For a comparative analysis of the results obtained, we denote by $\varepsilon_n(x, t_j) = |Q_n(x,t_j) - \varphi_j(x)|$ and $\widetilde{\varepsilon}_n(x, t_m) = \left|\dot{Q}_n(x,t_j) - \psi_m(x)\right|$, $n = 1,2$, $m = 0,2$, $j = \overline{0, 2}$ (here, $m = j = 2$ corresponds to the moment of time $t_2 = T$), which illustrate the discrepancy between these functions.

The maximum values of residuals $\varepsilon_n(x, t_j), \widetilde{\varepsilon}_n(x, t_m), E_n(x, t_j) = \int_0^1 \varepsilon_n(x, t_j)dx$ and $\widetilde{E}_n(x, t_m) = \int_0^1 \widetilde{\varepsilon}_n(x, t_m)dx$ are given in the following table.

Tables 1 and 2 show that, under the constructed control, the behavior of the string deflection functions is quite close to that of the given initial ones. An illustration of the residuals at the initial and intermediate time points is shown in the following figures. The graphical representation of the functions $\varepsilon_n(x, 0)$, $n = 1,2$, is shown in Figure 8.

Table 1. Comparison of residuals for φ_j.

	t_0=0		t_1=3		t_2=6	
	n=1	n=2	n=1	n=2	n=1	n=2
$\max\limits_{0\le x\le 1} \varepsilon_n\left(x, t_j\right)$	0.0589	0.0673	0.0411	0.0665	0.0559	0.0669
$\max\limits_{0\le x\le 1} E_n\left(x, t_j\right)$	0.0307	0.0349	0.0068	0.0170	0.0413	0.0371

Table 2. Comparison of residuals for ψ_m.

	t_0=0		t_2=6	
	n=1	n=2	n=1	n=2
$\max\limits_{0\le x\le 1} \widetilde{\varepsilon}_n(x, t_m)$	0.0667	0.0714	0.1351	0.1970
$\max\limits_{0\le x\le 1} \widetilde{E}_n(x, t_m)$	0.0341	0.0365	0.0402	0.0711

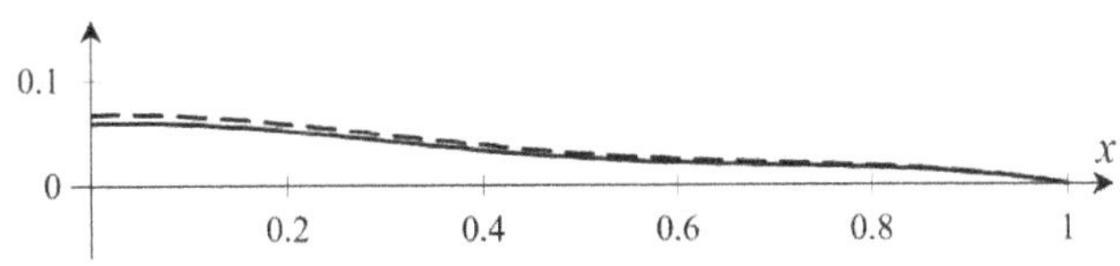

Figure 8. Graphs of the functions $\varepsilon_1(x, 0)$ (solid line) and $\varepsilon_2(x, 0)$ (dashed line).

The graphical representation of the functions $\varepsilon_n(x, 3)$, $n = 1,2$, is shown in Figure 9.

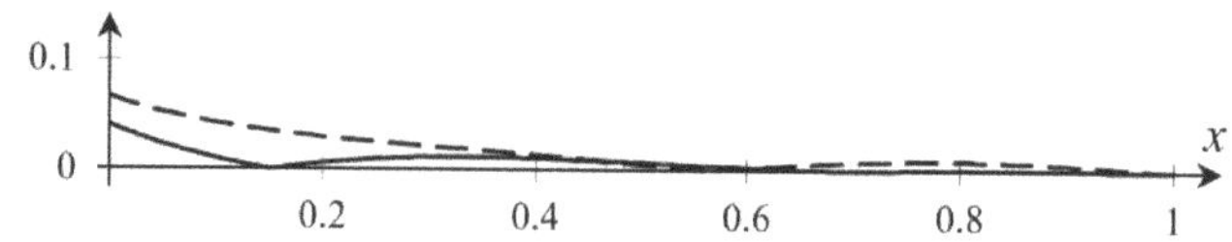

Figure 9. Graphs of the functions $\varepsilon_1(x, 3)$ (solid line) and $\varepsilon_2(x, 3)$ (dashed line).

The proposed analytical constructions are valid for any first n harmonics of string vibrations. Numerical calculations, illustrations of the results and their analysis were carried out with the help of the developed general approach for $n = 1, 2$. The series (21) is uniformly convergent for functions from the above classes. The behavior of the functions $V_1(t)$ and $V_2(t)$ shows it (see Figure 2).

Thus, given $n = 1$ and $n = 2$, we construct explicit expressions of the boundary control $u_1(t)$ and $u_2(t)$ and those of the string deflection functions $Q_1(x, t)$ and $Q_2(x, t)$.

6. Conclusions

We proposed a constructive method for constructing the control of vibrations of a homogeneous string with a given deflection shape at an intermediate moment. We also proposed a constructive method for constructing the control of homogeneous string vibrations with a given deflection shape at an intermediate moment. The control was carried out by shifting one end with the other end fixed. The construction scheme was as follows: We reduced the original problem to the control problem of distributed influences with zero boundary conditions. Further, we used the method of separation of variables and methods of control theory for finite-dimensional systems with multipoint intermediate conditions.

We formulated the corresponding statement and theorem for the first n harmonics. A specific example illustrated the obtained results. We realized a computational experiment, constructed the corresponding graphs and made a comparative analysis. They confirm the results of the study. The proposed method can be extended to other non-one-dimensional vibrational systems. The results presented in the paper can be used in the design of boundary control of vibration processes in physical and technological systems.

Author Contributions: Conceptualization, V.B. and S.S.; methodology, V.B.; software, S.S.; validation, V.B. and S.S.; formal analysis, V.B. and S.S.; investigation, V.B. and S.S.; resources, V.B. and S.S.; data curation, S.S.; writing—original draft preparation, V.B. and S.S.; writing—review and editing, V.B. and S.S.; visualization, S.S.; supervision, V.B. and S.S.; project administration, V.B. and S.S.; funding acquisition, V.B. and S.S. All authors have read and agreed to the published version of the manuscript.

Funding: Part of the research of S.S. was carried out under State Assignment Project no. FWEU-2021-0006, reg. number AAAA-A21-121012090034-3, of the Fundamental Research Program of Russian Federation 2021–2030 using the resources of the High-Temperature Circuit Multi-Access Research Center (Ministry of Science and Higher Education of the Russian Federation, project no 13.CKP.21.0038).

Institutional Review Board Statement: Not applicable.

Informed Consent Statement: Not applicable.

Data Availability Statement: Not applicable.

Conflicts of Interest: The authors declare no conflict of interest.

References

1. Abdukarimov, M.F. On the optimal boundary control of displacements of the forced vibration process at the two ends of the string. *Rep. Acad. Sci. Repub. Tajikistan* **2013**, *56*, 612–618.
2. Amara, J.B.; Beldi, E. Boundary controllability of two vibrating strings connected by a point mass with variable coefficients. *SIAM J. Control Optim.* **2019**, *57*, 3360–3387. [CrossRef]
3. Andreev, A.A.; Leksina, S.V. The problem of boundary control for a system of wave equations. *J. Samara State Tech. Univ. Ser. Phys. Math. Sci.* **2008**, *1*, 5–10.

4. Barseghyan, V.R.; Sahakyan, M.A. Optimal control of vibrations of a string with given states at intermediate moments of time. *Proc. Natl. Acad. Sci. Repub. Armen. Mekhanika* **2008**, *61*, 52–60.
5. Barseghyan, V.R. Control Problem of String Vibrations with Inseparable Multipoint Conditions at Intermediate Points in Time. *Mech. Solids* **2019**, *54*, 1216–1226. [CrossRef]
6. Barseghyan, V.R. The Problem of Optimal Control of String Vibrations. *Int. Appl. Mech.* **2020**, *56*, 471–480. [CrossRef]
7. Barseghyan, V.R. The problem of control of rod heating process with nonseparated conditions at intermediate moments of time. *Arch. Control. Sci.* **2021**, *31*, 481–493. [CrossRef]
8. Barseghyan, V.R. *Control of Composite Dynamical Systems and Systems with Multipoint Intermediate Conditions*; Nauka: Moscow, Russia, 2016.
9. Barseghyan, V.R.; Barseghyan, T.V. On an approach to the problems of control of dynamic systems with nonseparated multipoint intermediate conditions. *Automat. Remote. Control.* **2015**, *76*, 549–559. [CrossRef]
10. Butkovsky, A.G. *Methods for Control of Systems with Distributed Parameters*; Nauka: Moscow, Russia, 1975.
11. Gibkina, N.V.; Sidorov, M.V.; Stadnikova, A.V. Optimal boundary control of homogeneous string vibrations. *Radioelektron. Inform.* **2016**, *2*, 3–11.
12. Ilyin, V.A.; Moiseev, E.I. Optimization of boundary controls of string vibrations. *Uspekhi Mat. Nauk.* **2005**, *60*, 89–114. [CrossRef]
13. Kopets, M.M. The problem of optimal control of the string vibration process. In *Theory of Optimal Solutions: Collection of Research Contributions*; Glushkov Institute of Cybernetics of the NAS of Ukraine: Kyiv, Ukraine, 2014; pp. 32–38.
14. Moiseev, E.I.; Kholomeeva, A.A.; Frolov, A.A. Boundary displacement control for the oscillation process with boundary conditions of damping type for a time less than critical. *J. Math. Sci.* **2019**, *160*, 74–84. [CrossRef]
15. Zhang, S.; He, W.; Li, G. Boundary control design for a flexible string system with input backlash. In Proceedings of the 2016 American Control Conference (ACC), Boston, MA, USA, 6–8 July 2016; pp. 6698–6702. [CrossRef]
16. Zhao, Z. Boundary and distributed control for a vibrating string system. In Proceedings of the 2017 Chinese Automation Congress, Jinan, China, 20–22 October 2017; pp. 2099–2103. [CrossRef]
17. Zhao, Z.; He, X.; Ahn, C.K. Boundary Disturbance Observer-Based Control of a Vibrating Single-Link Flexible Manipulator. In *IEEE Transactions on Systems, Man, and Cybernetics: Systems*; IEEE: Piscataway Township, NJ, USA, 2021; Volume 51, pp. 2382–2390. [CrossRef]
18. Zhao, Z.; Shi, J.; Xiao, Y.; He, X.; He, W. Vibration control of flexible string systems with nonlinear input. In Proceedings of the 33rd Youth Academic Annual Conference of Chinese Association of Automation, Nanjing, China, 18–20 May 2018; pp. 230–234. [CrossRef]
19. Znamenskaya, L.N. *Control of Elastic Vibrations*; FIZMATLIT: Moscow, Russia, 2004.
20. Korzyuk, V.I.; Kozlovskaya, I.S. Two-point boundary problem for the equation of vibration of a string with a given velocity at a certain moment of time. I. *Proc. Inst. Math. Natl. Acad. Sci. Belarus* **2010**, *18*, 22–35.
21. Korzyuk, V.I.; Kozlovskaya, I.S. Two-point boundary problem for the equation of string vibration with a given velocity at a certain moment of time. II. *Proc. Inst. Math. Natl. Acad. Sci. Belarus* **2011**, *19*, 62–70.
22. Muravey, L.A.; Petrov, V.M.; Romanenkov, A.M. The Problem of Damping the Transverse Oscillations on a Longitudinally Moving String. *Mordovia Univ. Bull.* **2018**, *28*, 472–485. [CrossRef]
23. Tikhonov, A.N.; Samarskii, A.A. *Equations of Mathematical Physics*; Dover Publications: Mineola, NY, USA, 2011.
24. Zubov, V.I. *Lectures on Control Theory*; Nauka: Moscow, Russia, 1975.

axioms

MDPI

Article

Waves in a Hyperbolic Predator–Prey System

Andrey Morgulis [1,2]

[1] I.I. Vorovich Institute for Mathematic, Mechanics and Computer Science, Southern Federal University, 344006 Rostov-na-Donu, Russia; abmorgulis@sfedu.ru or morgulisandrey@gmail.com

[2] Southern Mathematical Institute of VSC RAS, 362027 Vladikavkaz, Russia

Abstract: We address a hyperbolic predator–prey model, which we formulate with the use of the Cattaneo model for chemosensitive movement. We put a special focus on the case when the Cattaneo equation for the flux of species takes the form of conservation law—that is, we assume a special relation between the diffusivity and sensitivity coefficients. Regarding this relation, there are pieces arguing for its relevance to some real-life populations, e.g., the copepods (*Harpacticoida*), in the biological literature (see the reference list). Thanks to the mentioned conservatism, we get exact solutions describing the travelling shock waves in some limited cases. Next, we employ the numerical analysis for continuing these waves to a wider parametric domain. As a result, we discover smooth solitary waves, which turn out to be quite sustainable with small and moderate initial perturbations. Nevertheless, the perturbations cause shedding of the predators from the main core of the wave, which can be treated as a settling mechanism. Besides, the localized perturbations make waves, colliding with the main core and demonstrating peculiar quasi-soliton phenomena sometimes resembling the leapfrog playing. An interesting side result is the onset of the migration waves due to the explosion of overpopulated cores.

Keywords: Patlak–Keller–Segel systems; the Cattaneo model of chemosensitive movement; hyperbolic models; shock waves; conservation laws

MSC: 92C17; 35L40; 92E20

Citation: Morgulis, A. Waves in a Hyperbolic Predator–Prey System. *Axioms* **2022**, *11*, 187. https://doi.org/10.3390/axioms11050187

Academic Editor: Hans J. Haubold

Received: 21 March 2022
Accepted: 14 April 2022
Published: 20 April 2022

Publisher's Note: MDPI stays neutral with regard to jurisdictional claims in published maps and institutional affiliations.

Copyright: © 2022 by the author. Licensee MDPI, Basel, Switzerland. This article is an open access article distributed under the terms and conditions of the Creative Commons Attribution (CC BY) license (https://creativecommons.org/licenses/by/4.0/).

1. Introduction

The Patlak–Keller–Segel (PKS) law provides a simple macroscopic model for a perceptual motion (taxis) of the particles ensemble in response to the spatial gradient of a stimulus (or signal) field. The commonly recognized formulation of the PKS flux reads as: $\chi p \nabla s$, where the notations of p, s, and χ stand for the density of the medium that moves in response to the signal, the intensity of the signal itself and the sensitivity coefficient, correspondingly. A multitude of parabolic PKS systems resulting from the summation of the diffusive and PKS fluxes stand as the subject of intensive research for last five decades. A considerable number of reviews expose the advances in this area, e.g., [1–3].

However, usage of the parabolic framework is not a unique way of treating the PKS law. For example, Dolak and Hillen [4] proposed a different formulation known as the Cattaneo model for chemosensitive movement. In contrast with the parabolic models, this one takes into account the inertia of the response and becomes hyperbolic. At that, the flux has a contribution from the local time derivative of itself in addition to those mentioned above. From the articles [5,6], it follows that the common parabolic model, the Cattaneomodel and several other hyperbolic models represent the approximations of the kinetics equations under different hypotheses. There is a concise review by R. Eftimie [7] of this subject that covers deriving the models and the issues of exact solutions, stability and bifurcations, etc. It makes clear that the hyperbolic models are not less natural than their parabolic counterparts, e.g., while modelling the aggregation processes in the active media. Nevertheless, the former receive much less attention than the latter. The mentioned review by R. Eftimie reports the deficiency of results for understanding the generic pattern emerging from the dynamics of the local hyperbolic models even in the case of one spatial dimension (see Table 9.1, Chapter 9) despite a considerable piece of work exposed therein.

Anyway, the hyperbolic models are the subject of continuing research. Among the recent results, those published in [8,9] have much to do with the present study. These articles address propagating the travelling shock waves and the gradual formation of the shocks from the initially smooth solutions for an infinite time extent. The analysis covers the rigorous proof of both features for a non-local hyperbolic model of a media, the density of which governs its velocity via the action of a given first-order pseudo-differential operator. Interestingly, the travelling shock waves coexist with the smooth ones. The latter are the widely studied features in the parabolic case, see, e.g., [10–12] and the references therein. The studies of their hyperbolic counterparts likely traced back to K. Hadeler [13–16].

In the present paper, we address similar issues but in a different context. We consider the Cattaneo model for a predator–prey community with the Lotka–Volterra kinetics. Regarding the predator flux, we assume that the diffusivity coefficient, $\mu = \mu(p,s)$, and the sensitivity coefficient, $\chi = \chi(p,s)$, are given functions in the species densities, p and s. In addition, we assume that 1-form $-\mu(p,s)dp + p\chi(p,s)ds$ is exact. Despite being restrictive, this assumption relies on certain biological grounds [17,18]. Formally, it entails the conservatism of the flux equation. Thanks to this circumstance, we take advantage of considering the travelling shock waves and arrive at very simple exact solutions for the limit case of sedentary prey and a highly-inertial predator. In the aforementioned articles, K. Hadeler addressed the travelling waves in a semi-linear version of the Cattaneo model. However, the model dealt with here is not semi- but quasi-linear. Besides, the conservatism allows for the elimination of the flux from the governing equations using the ansatz by K. Hadeler; see the reference above again. This reduction is helpful for calculating the numerical solution. The numerical continuation of the travelling shock waves to a wider parametric area discovered their smooth counterparts. These are the soliton-like waves. The study of their perturbations and collisions have discovered a rather peculiar interplay that resembles the quasi-soliton interactions reported in [19–21]. An interesting side result is the explosive migration waves emergent from the overpopulated cores.

Thus, there are several key findings in the present article, namely: (i) identifying a class of the Cattaneo models for the chemosensitive movement that allows the formulation of the governing equations as the conservation laws on one hand and, on the other hand, includes a biologically justified model; (ii) discovering the exact solutions describing the travelling shock waves in the limit of sedentary prey and highly-inertial predators; (iii) discovering the smooth counterparts of the shocks in the general case and their interactions such as leapfrog playing; (iv) observing the migration waves due to exploding the overpopulated kernels; (v) formation of the layers of high concentrations of the species nearby the fronts of waves emerging from the collisions of the solitary waves or upon the massive escape from the overpopulated areas.

The paper is organized as follows. In Section 2, we formulate the model and put it into the dimensionless form. In Section 3, we consider the case of the sedentary prey, in which the system becomes purely hyperbolic. In Section 4, we discuss the general issues regarding the shock waves, particularly the travelling ones. In Section 5, we distinguish a special case that allows a very simple exact solution. In Section 6, we discuss the general setting of the numerical experiments, and present their results. In Section 7, we discuss the results of our study and their applications. In the Appendix A, we have gathered the precise formulations of the data used for performing the numerical experiments.

2. The Governing Equations and Scaling

The governing equations read as:

$$p_t + q_x = F(p,s) \tag{1}$$
$$\tau q_t + \nu q = \chi(p,s) p s_x - \mu(p,s) p_x, \tag{2}$$
$$s_t = G(p,s) + D_s s_{xx}. \tag{3}$$

In this system, the dependent variables, x and t, stand for the spatial and temporal coordinates, correspondingly, $x \in \mathbb{R}$, $t > 0$. The dependent variables are $p = p(x,t)$, $q = q(x,t)$ and $s = s(x,t)$. The first and the last one play the parts of the densities of the species, say, the predators and the prey, correspondingly. The first two equations constitute the Cattaneo model for the prey-sensitive movement of the predators, so that the remaining dependent variable, q, stands for the predators' flux. The prey spreads itself purely by diffusion, and the notation of D_s stands for its diffusivity. In what follows, $D_s \equiv$ const by assumption. We also assume that the predators' diffusivity, $\mu(p,s)$, and the sensitivity, $\chi(p,s)$, are specified, and

$$p\chi(p,s) \to 0, \; p \to +0, \quad \forall s \geq 0. \tag{4}$$

We assume that the reaction terms, G and F, are prescribed, but postpone further detailing them to Section 5. The mechanical analogy suggests treating the second term on the left hand side of the second equation as a contribution of the resistance of the environment to the predator's motion. So, we will be calling the correspondent coefficient, ν, as the resistivity.

Let the notations T, X, P, S, Q, D_p, C, J_p, J_s stand for characteristic scales for the time, length, predator's density, prey's density, diffusivity, sensitivity, predator's and prey's sources densities correspondingly. The resistivity coefficient, ν, is naturally dimensionless. Since the values of D_p, D_s and C depend on the concrete species, it is natural to consider them as anyhow given. In contrast, the values of X, T, S, P, J_p and J_s are free to choose. Given this, let us set

$$T = \tau, \; X = \sqrt{\tau D_p}, \; Q = P\sqrt{\frac{D_p}{\tau}}, \; J_p = J_s = \tau^{-1}, \tag{5}$$

and postpone defining the values of P and S. We also set

$$\bar{\mu}(\bar{p},\bar{s}) = D_p^{-1}\mu(P\bar{p}, S\bar{s}), \quad \bar{\chi}(\bar{p},\bar{s}) = C^{-1}\chi(P\bar{p}, S\bar{s}). \tag{6}$$

In what follows, every variable employed is dimensionless by default.

Upon the above scaling, the dimensionless form of the governing equations reads:

$$p_t + q_x = F(p,s), \tag{7}$$
$$q_t + \nu q = \kappa\chi(p,s) p s_x - \mu(p,s) p_x, \tag{8}$$
$$s_t = G(p,s) + \delta s_{xx}, \tag{9}$$
$$\kappa = CS/D_p, \qquad \delta = D_s/D_p. \tag{10}$$

For the forthcoming analysis, it is important to distinguish the case when the right-hand side in the flux equation is integrable in the sense that:

$$\kappa\chi(p,s) p s_x - \mu(p,s) p_x = (\varphi(p,s))_x. \tag{11}$$

For such an integrability, it is necessary and sufficient to link the diffusivity to the sensitivity as follows:

$$\mu_s(p,s) = -\kappa(p\chi)_p. \tag{12}$$

3. Sedentary Prey and Hyperbolicity

Throughout this section, we consider the system (7)–(9) with $\delta = 0$. Then it becomes a first order quasi-linear PDE system, which we put into the form:

$$z_t + A(z)z_x = b(z), \quad \text{where} \tag{13}$$

$$z = \begin{pmatrix} p \\ q \\ s \end{pmatrix}, \; b : \begin{pmatrix} p \\ q \\ s \end{pmatrix} \mapsto \begin{pmatrix} F(p,s) \\ -\nu q \\ G(p,s) \end{pmatrix}, \; A : \begin{pmatrix} p \\ q \\ s \end{pmatrix} \mapsto \begin{pmatrix} 0 & 1 & 0 \\ \mu(p,s) & 0 & -\kappa p \chi(p,s) \\ 0 & 0 & 0 \end{pmatrix}. \tag{14}$$

The eigenvalues of matrix A are $\pm\sqrt{\mu(p,s)}, 0$. They are real and distinct one from another as long as $\mu(p,s) > 0$. It is always true by assumption. Hence the system of Equations (13) and (14) is strictly hyperbolic. The above triple of eigenvalues determines the triple of *characteristic speeds*—that is, every characteristic of system (13) allows a parametrization by the mapping $t \mapsto (t, X(t))$ that satisfies an equation $\dot{X}(t) = \lambda(p(X(t),t), s(X(t),t))$, where $\lambda \in \{0, \pm\sqrt{\mu(p,s)}\}$.

A question to ask about a first-order hyperbolic system is whether or not it allows diagonalizing by a pointwise transform $\rho = \mathcal{R}(z)$. Such an ansatz generally does not exist for a system that includes more than two equations. The diagonalizing is feasible, provided that the system matrix, $A = A(z)$, satisfies an integrability criterion, which allows a straight algebraic formulation. Another formulation requires satisfying the Frobenius condition $d\omega_i \wedge \omega_i = 0$ with every 1-form $\omega_i = \ell_i \cdot dz$ generated by the vectors of the dual eigenbasis $\{\ell_1, \ell_2, ...\}$ of the matrix A (this is the eigenbasis of the transposed matrix). Then for every i there exists a factor $\alpha_i = \alpha_i(z)$ such that $\alpha_i \omega_i = d\rho_i$, and $\rho_i = \mathcal{R}_i$. The last criterion is handy to use as long as there is a handy dual eigenbasis, as in the case of matrix (14), for example. Then a routine calculation reduces the diagonalization criterion to the following condition:

$$\mu_s(p,s) = \kappa p \chi(p,s) 5 \left(\ln \frac{\mu(p,s)}{p^2 \chi^2(p,s)} \right)_p. \tag{15}$$

Under condition (12) the obtained criterion simplifies to:

$$\mu(p,s) = pc(s)\chi(p,s), \tag{16}$$

where c stands for an arbitrary function. Thus, assuming the diagonalization entails restrictions that are too artificial even in the simplified form. That is why we will not be considering this option in this study anymore.

Let the condition (12) hold throughout all subsequent considerations. Then there exists a single-valued function $\varphi = \varphi(p,s)$ such that:

$$d\varphi = \mu(p,s)dp - \kappa p \chi(p,s) ds. \tag{17}$$

Hence, the system (13) and (14) consists of conservation laws, namely:

$$p_t + q_x = F(p,s), \quad q_t + \varphi_x = -\nu q, \quad s_t = G(p,s). \tag{18}$$

This feature makes feasible the generalized solutions, e.g., the shock waves.

Consider now a shock wave with discontinuities at some curve $x = X(t)$. Then the velocity of the shock, $\dot{X}$, has to satisfy the Rankine–Hugoniot conditions entailed by the conservations laws (18). They read as:

$$\dot{X}[p] = [q], \quad \dot{X}[q] = [\varphi], \quad \dot{X}[s] = 0. \tag{19}$$

Here the notation of a dependent variable put in the square brackets stands for the jump of this variable across the discontinuity—that is, the difference between the limit values evaluated for $(x,t) \to (X(t)+0,t)$ and $(x,t) \to (X(t)-0,t)$. At that,

$$[\varphi] = \int_{(p^-,s^-)}^{(p^+,s^+)} \mu(p,s)dp - \kappa p\chi(p,s)ds, \tag{20}$$

where the superscripts $+$ or $-$ stand for the unilateral limit values at the left and right shores of the discontinuity (when the observer looks forward alongside the time axis). There are two possibilities. The first is the standing wave—that is,

$$\dot{X} = 0,\ [q] = [p] = 0, \tag{21}$$

where the value of $[s]$ remains undetermined. The second is the travelling wave—that is,

$$\dot{X} \neq 0,\ [s] = 0,\ [q]^2 = [p][\varphi] = [p] \int_{(p^-,s_*)}^{(p^+,s_*)} \mu(s,p)dp = [p]^2\mu(p_*,s_*), \tag{22}$$

where $s_* = s(X(t),t)$ is the value of the prey density directly on the shock (note that function s remains continuous across the shock), and $p_* \in (p^-,p^+)$ is an unknown quantity that for each t satisfies the equation:

$$\int_{(p^-,s_*)}^{(p^+,s_*)} \mu(p,s)dp = [p]\mu(p_*,s_*). \tag{23}$$

The speed at which such a wave propagates can take two opposite values, namely:

$$\dot{X} = [q]/[p] = [\varphi]/[q] = \pm\sqrt{\mu(p_*,s_*)}. \tag{24}$$

Generically, the shock wave speeds determined in (24) differ from the characteristic ones, which are discontinuous across the shocks. It depends on the inequalities between the velocity of a specific shock wave and the limit values of the characteristic velocities whether or not this shock propagates. These inequalities are known as Lax conditions. Checking them yields a conclusion that shock waves can propagate provided that:

$$\big(\mu(p^+,s_*) - \mu(p_*,s_*)\big)\big(\mu(p^-,s_*) - \mu(p_*,s_*)\big) < 0. \tag{25}$$

Moreover, exactly one of the two possible shock waves can propagate, and the propagation speed is equal to $\sqrt{\mu(p_*,s_*)}$ provided that $\mu(p^-,s_*) > \mu(p_*,s_*)$ or $-\sqrt{\mu(p_*,s_*)}$ otherwise.

A peculiar degeneration occurs when a predator's diffusivity is independent of its density—that is, $\mu = \mu(s)$. Given this, the expressions for the shock wave speeds read as either:

$$\dot{X} = \pm\sqrt{\mu(s_*)} \tag{26}$$

or $\dot{X} = 0$. It is worth recalling that s_* stands for the trace of the prey's density, s, right on the shock. It is defined well due to the continuity of the function s across the shock. Hence, every possible wave's speed coincides with a characteristic one—that is, *the shocks spread along the characteristics.*

Matching the predator-independent diffusivity to the integrability condition, (12) and assumption (4) make the sensitivity, χ, predator-independent too. Given this, the integrability condition simplifies as follows:

$$\mu = \mu(s),\ \mu_s(s) = -\kappa\chi(s). \tag{27}$$

Modulo the scaling, relation (27) is equivalent to that deduced by Tyutyunov et al. from biological rationales in the aforementioned articles. Also, Tyutyunov et al. addressed some issues of stability and pattern formation for the corresponding PKS systems in the parabolic form [22]. Throughout the rest of the present article, we will be studying Cattaneo's systems that arise from relation (27).

4. The Travelling Shock Waves for the Predator-Independent Diffusivity

Let d/dt stand for the total derivative along a characteristic. From system (18) it follows that:

$$\frac{dq}{dt} + \lambda\frac{dp}{dt} = \kappa\chi(s)ps_x - \nu q + cF(p,s), \tag{28}$$

where the notation of $\lambda = \pm\sqrt{\mu(s)}$ stands for the corresponding characteristic speed. If this characteristic supports a shock, then the same equations hold on the shores of the shock with the limit values of every quantity involved. Subtracting them and eliminating the variable q with the use of the Rankine–Hugoniot conditions (19) lead to the following equation on the shock

$$\frac{d(\lambda_*[p])}{dt} + \lambda_*\frac{d[p]}{dt} = \kappa\chi_*[ps_x] - \nu\lambda_*[p] + c[F],\quad q = \lambda_*[p], \tag{29}$$

where subscript $_*$ indicates the quantities, which depend on the values of s_* only. Furthermore, from the continuity of the prey's density, s, it follows that the gap of ∇s across the shock is normal to it everywhere—that is,

$$[s_t] + \lambda_*[s_x] = 0. \tag{30}$$

So, we are discussing the travelling shock waves. By definition, such a wave propagates at a constant speed c, and the corresponding solution depends on only one variable,

$$\xi = t - x/c. \tag{31}$$

Then the characteristics at which the shocks occur are the parallel lines determined by equations $\xi = \xi_*$ for the values of ξ_* varying over some set, Σ, which we assume to be finite. This set gathers all the discontinuities of the solution in variable ξ. On the complement of the singular set, the system (18) reduces to an ODE system. Namely,

$$(c^2 - \mu(s))p_\xi = c^2(F(p,s) - \nu(p-r)) - \kappa\chi(s)pG(p,s),\ r_\xi = F(p,s),\ s_\xi = G(p,s), \tag{32}$$

where $r = p - q/c$, $\xi \notin \Sigma$. The conditions for matching the solutions are defined on the adjacent intervals separated by a discontinuity following from Equations (29) and (30). The latter holds trivially for every function in only variable ξ, while the former turns to a system of functional equations as follows:

$$c^2([F] - \nu[p]) - \kappa\chi_*[pG] = 0,\quad [s] = 0,\ [r] = 0. \tag{33}$$

The last equation in this set is equivalent to $[q] = c[p]$. The following constraint is for the wave speed, c, to obey:

$$\exists c : \forall \xi_* \in \Sigma\quad \lambda_* = \pm\sqrt{\mu(s_*)} = c,\ s_* = s(\xi_*). \tag{34}$$

Alluding to the values of the unknown, s, taken on Σ is correct by the continuity that is consistent with the matching condition (33). The same is true regarding another unknown, r.

The first equation in system (32) degenerates when the independent variable, ξ, is approaching a discontinuity. Let us restrict ourselves within the solutions obeying the following condition:

$$p^{\pm}_{\xi} = o\Big(\big(\xi - \xi_*\big)^{-1}\Big), \quad \xi \to \xi_*, \tag{35}$$

where the superscripts, $^+$ and $^-$, are to distinguish the solutions settled on the right and left intervals adjacent to the discontinuity point, ξ_*. This estimate entails one more condition for matching the solutions at the discontinuity. Namely, the following equations are to obey:

$$c^2(F(p^{\pm}_*, s_*) - \nu(p^{\pm}_* - r_*)) - \kappa\chi_* p^{\pm}_* G(p^{\pm}_*, s_*) = 0, \tag{36}$$

where the notations of $p^{\pm}_*$, s_* and r_* stand for the unilateral limits of variable p and the bilateral limits of variables s and r correspondingly. It is worth noting that subtracting the equalities (36) gives the first of matching condition (33).

At first, let us address the waves which allow a single shock only. Then there must be only one discontinuity on the ξ-axis, so let us place the origin there, and put $\xi_* = 0$. Consider the equation:

$$\mu(z)F(x,z) - \nu(x-y) - \kappa\chi(z)xG(x,z) = 0. \tag{37}$$

By conditions (36), a nonzero jump across the singularity in variable p due to a travelling wave presumes that Equation (37) has several distinct solutions (x_i, y, z) with the same y, z, e.g., $z = s_*, y = r_*, x_i = p^+_*$ and $x_j = p^-_*$ for some $i \neq j$.

Assume there exists a nonempty set filled with solutions (x, y, z) to Equation (37) such that $x \geq 0, z \geq 0$, and let the projection of this set onto the yz-plane alongside the x-axis cover some domain with multiplicity $N \geq 2$. For every point (z, y) in this domain, let $P_i = P_i(y,z), i = 1, 2, \ldots, N$ be the coordinates of projection of its pre-image on the x-axis alongside the yz-plane. Then constructing the travelling shock waves can go the following way: set

$$s_* = y,\ r_* = z,\ p^+_* = P_i(y,z),\ p^-_* = P_j(y,z),\ i \neq j, \tag{38}$$

and then try to find the solutions $p^{\pm} = p^{\pm}(\xi), r^{\pm} = r^{+}(\xi), s^{\pm} = s^{\pm}(\xi)$ to Equation (32), that are defined and bounded for $\pm\xi > 0$ and match the data listed above when $\xi = 0$. Note that the values of y, z are free to manipulate. One can try to obtain some pairs of conjugated waves by transposing i and j. Since we have been assuming the projection to be the N-leaf covering mapping, there are $N(N-1)/2$ conjugated pairs of the datasets. In the next subsection, we will be considering a case of a very simple implementation of the approach outlined above.

5. The Inertial Limit for the Sedentary Prey

At this point, we need more details regarding the reaction terms, $F = F(p,s)$ and $G = G(p,s)$. Henceforth, we will be assuming the following:

$$F = pf(p,s),\ f|_{s=0} < 0, \qquad G = sg(p,s),\ g(0,0) > 0,\ g_p|_{p=0} = -1,\ g_s|_{s=0} = -1. \tag{39}$$

The last two assumptions are equivalent to inequalities $g_p|_{p=0} < 0$, $g_s|_{s=0} < 0$ modulo the scaling of variables p, s since the characteristic scales of these variables, P and S, are indefinite still. For example, the Lotka–Volterra kinetics while being normalized in accord with (39) reads as

$$F = p(\gamma s - \beta),\ G = s(\alpha - s - p), \quad \alpha, \beta, \gamma = \text{const} > 0. \tag{40}$$

Thus, parameter α plays the part of the carrying capacity given the scaling adopted here.

Consider the Cauchy problem

$$S_\tau = G(0,S), \quad S|_{\tau=0} = a. \tag{41}$$

Assume there exists a number $a_0 > 0$ such that for every $a \in [0, a_0]$ problem (41) has a solution $S = S(\tau, a)$ defined on $\mathbb{R}$, and such that the mapping $a \to S(\tau, a)$ sends the segment $[0, a_0]$ to itself for every $\tau \in \mathbb{R}$. Consider also the following system of functional equations:

$$g(p,s) = 0, \quad f(p,s) = 0, \tag{42}$$

and assume there exists a positive solution $s = s_e > 0, p = p_e > 0$. The last assumption entails the existence of a strictly positive equilibrium for general system (7)–(9). By equilibrium, we mean a particular solution specified as follows: $q = 0, p = \text{const}, s = \text{const}$, so that both species distribute themselves homogeneously. A nonnegative equilibrium not being strictly positive can make sense too. For example, the Lotka–Volterra kinetics (40) allows an equilibrium with $p = 0, s = \alpha$ for every parameter setting, and there exists the strictly positive equilibrium $s_e = \beta/\gamma$, $p_e = \alpha - s_e$ provided that $\beta < \alpha\gamma$.

Throughout the rest of this section, we will be considering the inertial limit and sedentary prey. So, we put $\nu = \delta = 0$. We also assume all the listed assertions on the kinetics to be true. The Lotka–Volterra kinetics defined in (40) meet such an assumption for $\beta < \alpha\gamma$; at that, the ODE involved in the Cauchy problem (41) reads as $S_\tau = S(\alpha - S)$.

Given the assumption made, there are at least two solutions, $x = P_i(y,z)$, $i = 1,2$ to Equation (37) defined in some vicinity of every point (y, s_e). At that, $P_1 \equiv 0$, while function P_2 (in fact, in one variable, z) is that defined implicitly by equations $\mu(z)f(x,z) = \kappa\chi(z)g(x,z) = 0$, $P_i(s_e) = p_e$. However, we do not need much manipulating with the values y, z here, since the choice is evident; namely, $x_1 = 0$, $x_2 = p_e$, $z = s_e$, and the values of y are arbitrary. Then there are two pairs of the travelling shock waves, one pair propagates at speed $c = \sqrt{\mu(s_e)}$, and the other one propagates at the opposite speed. The formulae for both are the same, and they read as:

$$p = 0,\ r = y,\ s = S(\xi, s_e),\ \xi \in (-\infty, 0), \qquad p = p_e,\ r = y,\ s = s_e,\ \xi \in (0, \infty), \tag{43}$$

$$p = 0,\ r = y,\ s = S(\xi, s_e),\ \xi \in (0, \infty), \qquad p = p_e,\ r = y,\ s = s_e,\ \xi \in (-\infty, 0), \tag{44}$$

where y is an arbitrary constant.

Waves (43) and (44) are conjugated in the sense explained in the previous section but not mirroring one by the other one. Although both species in both waves take the same equilibrium values within the areas settled by predators, outside them, prey spreads itself differently. Indeed, the solutions to the Cauchy problem (41) generically behave differently for $\pm\tau > 0$ (except for the equilibria).

In Figure 1, the left and middle frames show profiles of waves (43) and (44) correspondingly. For both waves, the front of the predator's invasion moves towards the smaller concentration of prey. This feature is not as paradoxical as it can seem given the locality of the predator–prey interactions.

An overlay of waves (44) and (43) gives examples of finite predators' mass localized within a patch that moves uniformly. Such a wave propagates with two shocks at the speed $c = \sqrt{\mu(s_e)}$ or at the opposite speed. The formulae read as:

$$\begin{gathered} p = 0,\ r = y,\ s = S(\xi - \xi_0, s_e),\ \xi < \xi_0, \\ s = s_e,\ p = p_e,\ r = y,\ \xi \in (\xi_0, \xi_1), \xi_1 - \xi_0 = M/p_e \\ p = 0,\ r = y, \quad s = S(\xi - \xi_1, s_e),\ \xi > \xi_1, \end{gathered} \tag{45}$$

where the values of y and $M > 0$ are arbitrary constants. At that, the values of M stand as the mass of the patch. It is worth noting that the shapes of the patchy waves differ depending on the sign of the wave speed, c. The prey are dying out to the left (right) of the predators patch for negative (positive) c, so that the patch moves towards the smaller concentration of prey (see the right panel in Figure 1). In Section 7, we return to these waves to discuss their relevance to the population dynamics.

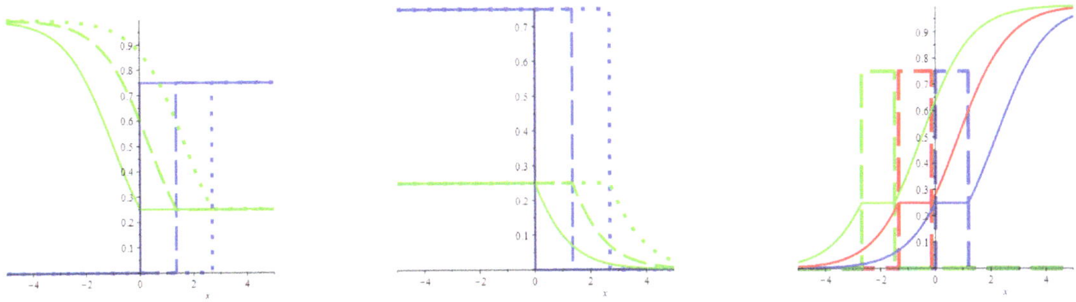

Figure 1. The figure shows the propagation of waves (43)–(45) calculated for the Lotka–Volterra kinetics (40) (from the left to the right). Each frame shows three instantaneous profiles for both the predator and the prey densities. In the left and middle frames, the blue (green) coloured lines are for the former (latter), while the solid, dashed and dotted lines picture the profiles taken at $t = 0, t = 3/2$ and $t = 3$. The right frame addresses the wave that transports a patch filled with the unit mass of predators. At that, the dashed (solid) lines mark out the densities of the predators (prey). The colours green, red and blue are for the shots taken at $t = 0$, $t = 3/2$ and $t = 3$ (so that the wave speed is negative). All three panels correspond to the Lotka–Volterra kinetics (40), where $\alpha = 1$, $\beta = 0.2$, $\gamma = 0.8$. At that, the diffusivity function reads as $\mu = (1+s)^{-1}$, so that $\kappa = 1$, and the equality (27) determines the sensitivity, χ. These definitions give the wave speed, $c = \sqrt{\mu(s_e)} \approx 0.9$. Finally, recall that the figure regards the diffusionless and inertial limit, and $\delta = \nu = 0$, hence.

6. Numerical Experiments

In this section, we present the numerical solutions to system (7)–(9) that we formulate with the use of the Lotka–Volterra kinetics (40) and predators' diffusivity $\mu = (1+\kappa s)^{-1}$. The diffusivity determines the sensitivity, $\chi(s)$, by the simplified integrability condition (27). For the numerical implementation, we eliminate the flux, q, from this system with the use of ansatz by K. Hadeler (see references provided above). As a result, we arrive at the following equations:

$$p_{tt} + \nu(p_t - F(p,s)) = (\mu(s)p)_{xx} + (F(p,s))_t, \quad (46)$$

$$s_t = G(p,s) + \delta s_{xx}. \quad (47)$$

Substituting this second-order system for the original one (which is of order 1) requires ad hoc initial conditions. These are:

$$p|_{t=0} = p_0, \quad s|_{t=0} = s_0, \quad p_t|_{t=0} = F(p_0, q_0) - q_{0x}, \quad (48)$$

where the notations of p_0, q_0, s_0 stand for the functions which determine the initial values of the dependent variables of the original system (7)–(9). The numerical implementation also requires restricting the solution within a finite spatial domain and formulating suitable boundary conditions. So, we set

$$p_x|_{x=\pm L} = 0, \quad s_x|_{x=\pm L} = 0. \quad (49)$$

It turns out that the numerical solving of the initial-boundary problems (46)–(49) is quite feasible with the *Maple* built-in PDE solver.

Implementation of the numerical solution should cover a spatial interval wide enough to put the artificial boundaries far out of the domain in which the phenomena of interest occurs. We had controlled all the results below by widening the spatial area and found them reproducing themselves sustainably for values of L higher than 8. In particular, $L = 10$ for all the figures presented in this section. In a similar way, we checked the influence of the mesh refining and varying the level of smoothing used for preparing the initial data and found the results of this inspection quite satisfactory.

The first set of numerical experiments is for answering the question of whether the shock waves persist for the positive values of the resistivity, ν and the prey diffusivity, δ. To this end, we have been taking the profiles of the species' densities, s and p from the waves (43)–(45) and putting them as the initial profiles, s_0 and p_0, which enter the boundary conditions (48). At that, we have been smoothing the shocks slightly. Further, we have been putting $q_0 = c_* p_0$, where $c_* = \sqrt{\mu(s_e)}$ is the shock wave speed. This choice is consistent with the definition of the shock waves. Appendix A provides the concrete details of setting initial conditions, the control parameters values, etc. The answer to the question formulated above is fully affirmative. The shocks become a bit smoother, but keep propagating at an almost constant speed that is nearly equal to the value of c_*. Figure 2 shows the typical behavior of the slightly smoothed counterpart of the patchy shock wave. This figure tells us that the smoothed patch spreads as a kind of soliton, which is shaped rather sharply for the small resistivity and prey diffusivity. An increase in the resistivity produces scattering of the predators behind the rear front of the wave.

The next piece of computing addresses the interactions of the observed solitary waves with some perturbations applied initially. These are:

(a) a small displacement of the species density profiles one relative to the other one with no deformation;
(b) a small deformation of the species density profiles;
(c) a small droplet of predators localized behind the main core;
(d) a small droplet of predators localized ahead of the main core.

In cases (a) and (b), the smallness means that the magnitudes of mutual displacements (deformations) are approximately ten percent of the magnitude of the main patch. The smallness of the droplet means that its mass is ten times smaller than the mass of the main patch, while they both are localized in the intervals of nearly equal lengths. In Appendix A, there is the exact formulation of the initial conditions and the concrete values of the control parameters.

In cases (a) and (b), the effects of the initial perturbations manifest themselves mostly by the predators scattering, which goes almost the same way as shown in the bottom row of frames in Figure 2, with no any qualitative distinctions. So, we do not illustrate these cases. Case (c) is similar to the above, but shedding the predators is rather intensive. We illustrate this case in the top row of frames in Figure 3. Case (d) demonstrates a very peculiar interplay of solitary waves. Namely, the main patch and the droplet attract one to the other one until they clash. Then they play leapfrog: droplet climbs over the patch and rolls down to the other side of it. The patch drops some mass due to the scattering but keeps moving at almost the same speed. The droplet keeps moving too but in the opposite direction while getting bigger and sharper. We illustrate this case in the bottom row of frames in Figure 3.

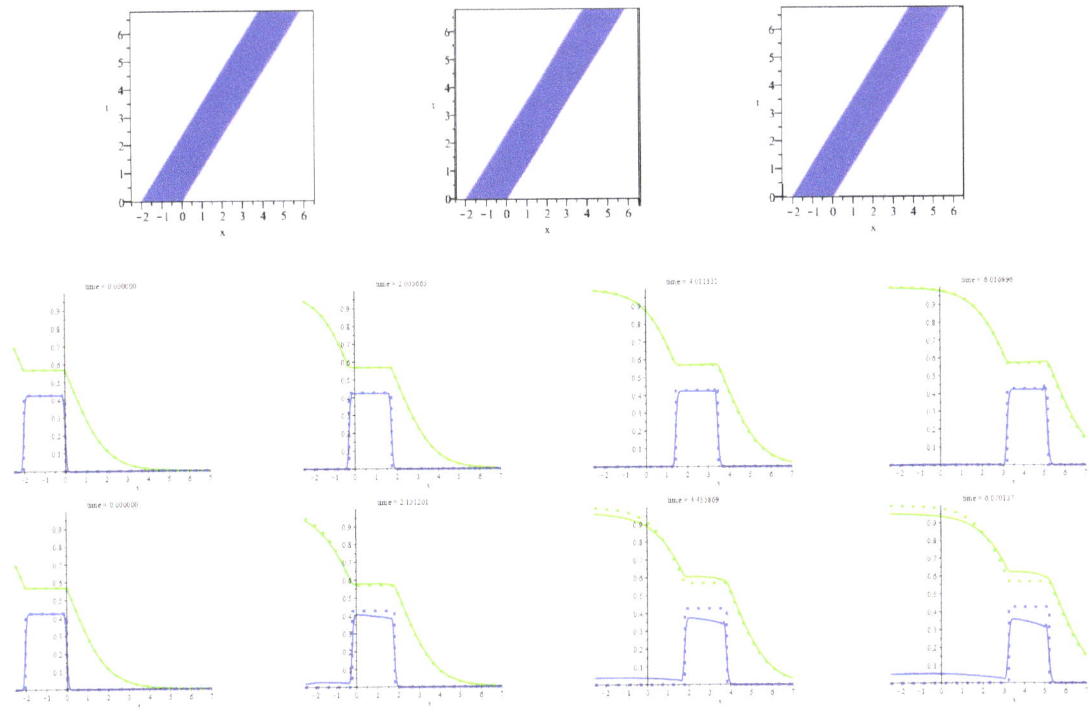

Figure 2. The figure illustrates propagating the solitary waves that are the slightly smoothed counterparts of the shock patchy wave (45) when the resistivity, ν, and prey diffusivity, δ, take some small values. In Appendix A, there is the accurate exposition of all the parameter values and initial data that we have used for producing all the pictures displayed herein. The upper row of frames shows three distributions of the predators over x, t-plane which arise from propagating the shock and smoothed patchy waves. The left upper frame displays the former while the central and right frames display the latter for two different values of the resistivity. Both values are small, but the one corresponding to the central frame is substantially smaller. The saturation of blue is in use for indicating the predators' density. The central and lower rows animate the propagation of the smoothed patchy waves shown in the middle and right frames in the upper row. For comparison, the shock patchy wave (that we have been showing in the left upper frame) is animated in the same frames synchronously. For capturing the former (latter), the solid (dotted) lines are in use, and the coloring of them distinguishes the species. Namely, for both waves, the profile of the predators' (prey) density is blue (green) colored.

The results presented above confirm the stability of the travelling patchy waves. Figure 4 demonstrates the results of a more extreme crash-test. Appendix A provides the detailed description of the initial states and other settings used for this piece of computing. It is worth recalling that every travelling patchy wave has a counterpart that propagates at the opposite speed. Gluing this pair provides the initial state for computing the next five patterns pictured in Figure 4. Overall, they illustrate the collision of two patchy waves. The remarkable feature is that the collision gives rise to an expansion wave, at the fronts of which peaks are arising and sharpening in the course of its propagation. The mirror symmetry of the patterns is due to the mirror symmetry of the problem and initial data.

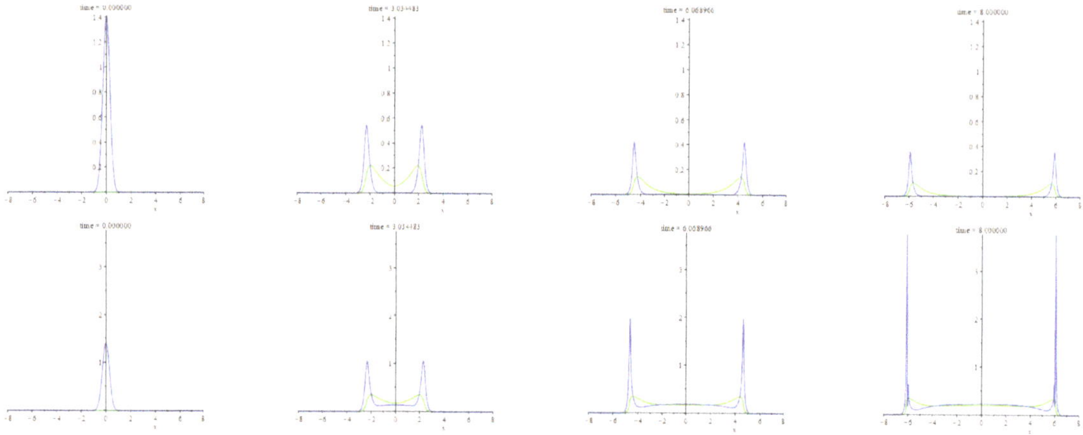

Figure 6. The figure illustrates propagating two waves due to the explosion of a smoothly localized unit mass of predators amid the uniform distribution of prey, the density of which is equal to the carrying capacity. The blue lines are for the instantaneous graphs of the predators' density. The green lines are for the deviation of the prey density from the carrying capacity. The detailed description of the initial states and other settings used for computing this figure is in Appendix A.

7. Discussion

We have addressed a Cattaneo-type dynamics of the predator–prey system with the Lotka–Volterra kinetics term in one spatial dimension. By assumption, only the predators are capable of the perceptual motions, and the flux of them generally reads as $p\chi(p,s)s_x - \mu(p,s)p_x$, where $\mu(p,s)$ and $\chi(p,s)$ are some prescribed functions. To start with, we have considered the sedentary prey. In this approximation, the governing equations turn to form a strictly hyperbolic system, for which we have formulated the criterion for reducing to Riemann's invariants in terms of sensitivity, $\chi(p,s)$, and diffusivity, $\mu(p,s)$, explicitly. It has turned out, however, that the class of systems obeying this criterion looks rather artificial. At the same time, reducing the governing equations to the conservation laws happened to be less restrictive. In particular, such a reduction is possible provided that $\mu = \mu(s)$, $\kappa\chi(s) = -d\mu/ds$. Since there are biological rationales for this structural relation (see the reference above), we have accepted it.

For the systems of conservation laws, the shock waves are natural, and we have derived a system of conditions on the shocks. In the inertial limit—that is, for $\nu = 0$—we have discovered a very simple exact solutions that describe the travelling shock waves. The wave pattern represents a semi-infinite or even finite patch of predators that propagates at a constant speed. There are no predators outside the patch while both species coexist in the equilibrium state inside, see Figure 1. The wave speed is equal to $\pm\sqrt{\mu(s_e)}$, where s_e is the prey density at the equilibrium state.

The waves carrying semi-infinite patches describes the transitions between two equilibria states. The community goes either from the extinction of both species to the coexistence at the equilibrium or from the coexistence at the equilibrium to the extinction of the predators and restoring the prey up to the carrying capacity. In this sense they resemble the KPP-Fisher waves. Here, however, the extinction of predators means that the wave either has not brought them to the areal extent under consideration yet or has moved them out already. The interesting feature is that the front of the predators' invasion always moves towards the smaller concentration of prey. This is not as paradoxical as it can seem given the locality of the predator–prey interactions. Thus, two patterns of behavior are emergent from propagating these waves. The first is restoring the resource up to the carrying capacity after the migrating predators' withdrawal. The second is the transition from the prey dying out to the mutual equilibrium with the spreading predators. The travelling shock waves carrying a finite mass of the predators combine both features. Indeed, the profile of such a wave involves the transitions from the mutual extinction via the coexistence to the extinction of predators. Besides, the patch of predators moves towards the smaller concentration of prey again. So, the corresponding pattern of behavior looks like preventing the prey from dying out and even restoring the prey population due to migrating a compact core of the predators.

We have extended the travelling shock waves to the positive values of the prey diffusivity, δ, and the resistivity, ν, numerically. It turns out that they withstand such an extension, at least while the values of δ and ν remain small enough. The shocks smooth themselves, but the speed at which they propagate is nearly equal to the value of $c = \pm\sqrt{\mu(s_e)}$.

The shock waves carrying a finite mass of predators transform themselves into the smooth soliton-like waves, to which we have paid particular attention, see Figure 2. An interesting feature is shedding the predators behind the rear front of the wave due to an increase in the resistivity. As a result, the core of migration leaves behind itself a populated areal extent, which remains settled even when the migration core moves far ahead. Thus, propagating the patchy wave can play a part in settling the predators due to migration.

A relatively small predator's droplet that occurs instantly ahead of or behind the main core of the soliton-like wave enhances the above scattering but does not cause any other noticeable changes (see Figure 3). The main core never absorbs the droplet in the case of collision, but they both keep moving after a peculiar interaction resembling the leapfrog play. Quite a different pattern of behavior arises from colliding the cores, which have gathered the equal masses of the predators (see Figures 4–6). We have examined colliding for several pairs of cores, which belong to the following classes: (i) two identical smoothed travelling patchy waves which propagate at opposite speeds; (ii) a compact group of predators that occurs suddenly ahead of the predators' patch of a travelling wave and get moving towards it; (iii) a finite mass of predators that suddenly have landed on a compact part of a greater areal extent where the resource density has attained the value of the carrying capacity. Merging the cores immediately leads to local overpopulation and a lack of prey, which, in turn, gives rise to an explosive migration. The explosion wave spreads the predators uniformly across a widening area while boundary layers of high density occur near the wave fronts. Deep in this area, the species coexist at equilibrium if feasible, or the predators become extinct, and the prey density goes back to the carrying capacity. Various numerical experiments have been reproducing this pattern of behavior sustainably though not literally, of course. Thus, we conclude that the explosive waves that we have been discussing deliver a mechanism for overcoming the local overpopulation and lack of resources.

Funding: This research received no external funding.

Institutional Review Board Statement: Not applicable.

Informed Consent Statement: Not applicable.

Data Availability Statement: Not applicable.

Acknowledgments: Author acknowledges the support from the Southern Federal university.

Conflicts of Interest: The author declares no conflict of interest.

Appendix A. Initial Data

Implementing the numerical experiments, the outcomes of which we have been discussing above, numerically solved several initial-boundary value problems, each of which comprises Equations (46) and (47) with the Lotka–Volterra kinetics (40), boundary conditions (49), and initial conditions (48). These problems depend on the control parameters, $\alpha, \beta, \gamma, \delta, \kappa, \nu, L$. We have also been using parameter ε while smoothing some initial conditions. Table A1 displays the values of these parameters and the subsections below explain setting the initial conditions for the concrete computations. We also recall that, throughout all the computations, the predators' diffusivity reads as:

$$\mu(s) = \frac{1}{1+\kappa s}. \tag{A1}$$

At that, the relation (27) determines the predators sensitivity, χ.

Table A1. The correspondence between the Figures presented above and the values of the parameters that enter Equations (46) and (47) and initial conditions (48).

Figure	α	β	γ	δ	ε	κ	ν	L
Figure 2, top row-left	1	0.4	0.7	0.00	n/a	0.6	0	n/a
Figure 2, top row-middle	1	0.4	0.7	0.01	0.05	0.6	0.005	10
Figure 2, top row-right	1	0.4	0.7	0.01	0.05	0.6	0.1	10
Figure 2, middle row	1	0.4	0.7	0.01	0.05	0.6	0.005	10
Figure 2, bottom row	1	0.4	0.7	0.01	0.05	0.6	0.1	10
Figure 3, top row	1	0.4	0.7	0.01	0.05	0.6	0.1	10
Figure 3, bottom row	1	0.4	0.7	0.01	0.05	1	0.05	10
Figure 4	1	0.4	0.7	0.01	0.05	0.6	0.1	10
Figure 5, top row	1	0.4	0.7	0.01	0.05	0.6	0.25	10
Figure 5, bottom row	1	0.4	0.7	0.01	0.05	0.6	1.5	10
Figure 6, top row	1	0.5	0.4	0.01	0.05	1	0.2	10
Figure 6, bottom row	1	0.4	0.7	0.01	0.05	1	0.2	10

Appendix A.1. Data for Figure 2

The initial conditions are:

$$2p_0 = p_e \mathrm{erf}\left(\tfrac{x-x_0}{2\varepsilon c_*}\right) - p_e \mathrm{erf}\left(\tfrac{x-x_1}{2\varepsilon c_*}\right), \quad p_e = \alpha - \tfrac{\beta}{\gamma}, \quad q_0 = c_* p_0 \tag{A2}$$

and s_0 is given by equality (45), where $\xi = x/c_*$, $\xi_0 = x_0/c_*$, $x_0 = -1/p_e$, $x_1 = 0$, $\xi_1 = x_1/c_*$, $c_* = \sqrt{\mu(s_e)}$ and $s_e = \frac{\beta}{\gamma}$.

Appendix A.2. Data for Figure 3

As we have been saying on page 10, this Figure corresponds to the initial data that read as:

$$p_0 = p_w + P p_d, \quad q_0 = c_* p_w + Q p_d, \quad s_0 = s_w, \tag{A3}$$

where the notations of p_w and s_w stand for the same functions as those defined in Appendix A.1.

$$\begin{aligned} P = 0.1,\ Q = 0,\ p_d &= \exp(-3(x+5)^2) \text{ for the top row,} \\ P = 0.1,\ Q = -0.1,\ p_d &= \exp(-3(x-3)^2) \text{ for the bottom row.} \end{aligned}$$

Appendix A.3. Data for Figure 4

In this Figure, the all of images correspond to the initial data that read as:

$$2p_0(x) = \begin{cases} p_w(x), & x > 0, \\ p_w(-x), & x < 0, \end{cases}, \quad p_w = p_e \mathrm{erf}\left(\frac{x-x_0}{2\varepsilon c_*}\right) - p_e \mathrm{erf}\left(\frac{x-x_1}{2\varepsilon c_*}\right). \tag{A4}$$

$$q_0 = -c_* \mathrm{erf}(3x) p_0(x) \tag{A5}$$

$$s_0 = \begin{cases} s_w(x), & x > 0, \\ s_w(-x), & x < 0, \end{cases}, \tag{A6}$$

where the notation of s_w stands for the same function as that defined in Appendix A.1 with $\xi = x/c_*$, $\xi_0 = x_0/c_*$, $\xi_1 = x_1/c_*$, $c_* = \sqrt{\mu(s_e)}$ and $s_e = \frac{\beta}{\gamma}$. Here $x_0 = -5 - 1/p_e$, $x_1 = -5$.

Appendix A.4. Data for Figure 5

In this Figure, both rows of images correspond to the initial data that read as:

$$p_0 = p_w + p_d, \quad q_0 = c_*(p_w - p_d), \quad s_0 = s_w + s_d, \tag{A7}$$

where p_w and s_w are the same as those defined in Appendix A.1, and

$$p_d = s_d = \frac{\sqrt{3}\exp\left(-\frac{3(x-3)^2}{2}\right)}{\sqrt{2\pi}}, \quad c_* = \sqrt{\mu(s_e)}, \quad s_e = \frac{\beta}{\gamma}.$$

Appendix A.5. Data for Figure 6

In this Figure, both rows of images correspond to the initial data that read as:

$$p_0 = \frac{5e^{-\frac{25x^2}{4}}}{2\sqrt{\pi}}, \quad q_0 = 0,\ s_0 = \alpha. \tag{A8}$$

References

1. Horstmann, D. From 1970 until present: The Keller-Segel model in chemotaxis and its consequences. II, Jahresber. *Deutsch. Math.-Verein.* **2004**, *106*, 51–69.
2. Hillen, T.; Painter, K.J. A user's guide to PDE models for chemotaxis. *J. Math. Biol.* **2009**, *58*, 183. [CrossRef]
3. Bellomo, N.; Bellouquid, A.; Tao, Y.; Winkler, M. Toward a mathematical theory of Keller-Segel models of pattern formation in biological tissues. *Math. Models Methods Appl. Sci.* **2015**, *25*, 1663–1763. [CrossRef]
4. Dolak, Y.; Hillen, T. Cattaneo models for chemosensitive movement: Numerical solution and pattern formation. *J. Math. Biol.* **2003**, *46*, 461–478 [CrossRef]
5. Filbet, F.; Laurencot, P.; Perthame, B. Derivation of hyperbolic models for chemosensitive movement. *J. Math. Biol.* **2005**, *50*, 189–207. [CrossRef]
6. Outada, N.; Vauchelet, N.; Akrid, T.; Khaladi, M. From kinetics theory of multicellular systems to hyperbolic tissue equations: Asymptotic limits and computing. *Math. Model. Methods Appl. Sci.* **2016**, *26*, 2709–2734. [CrossRef]
7. Eftimie, R. *Hyperbolic and Kinetic Models for Self-Organised Biological Aggregations. A Modelling and Pattern Formation Approach*; Springer: Cham, Switzerland, 2018. [CrossRef]
8. Fu, X.; Griette, Q.; Magal, P. A cell-cell repulsion model on a hyperbolic Keller-Segel equation. *J. Math. Biol.* **2020**, *80*, 2257–2300. [CrossRef]
9. Fu, X.; Griette, Q.; Magal, P. Sharp discontinuous traveling waves in a hyperbolic Keller-Segel equation. *Math. Model. Methods Appl. Sci.* **2021**, *31*, 861–905. [CrossRef]
10. Berezovskaya, F.S.; Karev, G.P. Bifurcations of travelling waves in population taxis models. *Phys.-Uspekhi* **1999**, *42*, 917. [CrossRef]
11. Berezovskaya, F.S.; Karev, G.P. Parametric portraits of travelling waves of population models with polynomial growth and auto-taxis rates. *Nonlinear Anal. Real World Appl.* **2000**, *1*, 123–136. [CrossRef]
12. Horstmann, D.; Stevens, A. A constructive approach to traveling waves in chemotaxis. *J. Nonlinear Sci.* **2004**, *14*, 1–25. [CrossRef]
13. Hadeler, K.P. Hyperbolic travelling fronts. *Proc. Edinb. Math. Soc.* **1988**, *31*, 89–97. [CrossRef]
14. Hadeler, K.P. Travelling fronts for correlated random walks. *Canad. Appl. Math. Wuart.* **1994**, *2*, 27–43. [CrossRef]
15. Hadeler, K.P. Reaction transport equations in biological modeling. *Math. Comput. Model.* **2000**, *31*, 75–81. [CrossRef]
16. Hadeler, K.P.; Lewis, M.A. Spatial dynamics of the diffusive logistic equation with a sedentary compartment. *Can. Appl. Math. Quart* **2002**, *10*, 473–499.
17. Tyutyunov, Y.V.; Zagrebneva, A.D.; Surkov, F.A.; Azovsky, A.I. Microscale patchiness of the distribution of copepods (Harpacticoida) as a result of trophotaxis. *Biophysics* **2009**, *54*, 355–360. http://dx.doi.org/10.1134/S000635090903018X. [CrossRef]
18. Tyutyunov, Y.V.; Zagrebneva, A.D.; Surkov, F.A.; Azovsky, A.I. Derivation of density flux equation for intermittently migrating population. *Oceanology* **2010**, *50*, 67–76. http://dx.doi.org/10.1134/S000143701001008X.
19. Tsyganov, M.A.; Brindley, J.; Holden, A.V.; Biktashev, V.N. Quasisoliton interaction of pursuitevasion waves in a predator–prey system. *Phys. Rev. Lett.* **2003**, *91*, 218102. [CrossRef]
20. Tsyganov, M.A.; Brindley, J.; Holden, A.V.; Biktashev, V.N. Soliton-like phenomena in one-dimensional cross-diffusion systems: A predator-prey pursuit and evasion example. *Phys. Nonlinear Phenom.* **2004**, *197*, 18–33. [CrossRef]
21. Tsyganov, M.A.; Biktashev, V.N. Half-soliton interaction of population taxis waves in predator–prey systems with pursuit and evasion. *Phys. Rev.* **2004**, *70*, 031901. [CrossRef]
22. Tyutyunov, Y.; Titova, L.; Senina, I. Prey-taxis destabilizes homogeneous stationary state in spatial Gause-Kolmogorov-type model for predator–prey system. *Ecol. Complex.* **2017**, *31*, 170–180. [CrossRef]

axioms

Article

A Symplectic Algorithm for Constrained Hamiltonian Systems

Jingli Fu [1], Lijun Zhang [2,3,*], Shan Cao [4], Chun Xiang [5] and Weijia Zao [6]

1 College of Information and Control Engineering, Shandong Vocational University of Foreign Affairs, Weihai 264504, China; sqfujingli@163.com
2 Department of Mathematics, Shandong University of Science and Technology, Qingdao 266590, China
3 International Institute for Symmetry Analysis and Mathematical Modelling, Department of Mathematical Sciences, North-West University, Mafikeng Campus, Mmabatho X2046, South Africa
4 Institute of Mathematical Physics, Zhejiang Sci-Tech University, Hangzhou 310018, China; qsmx522@126.com
5 College of Mechanical and Automotive Engineering, Zhejiang University of Water Resources and Electric Power, Hangzhou 310018, China; xiangchun@zjweu.edu.cn
6 Department of Mathematics, Qingdao University, Qingdao 266071, China; zhweijia@sohu.com
* Correspondence: li-jun0608@163.com

Abstract: In this paper, a symplectic algorithm is utilized to investigate constrained Hamiltonian systems. However, the symplectic method cannot be applied directly to the constrained Hamiltonian equations due to the non-canonicity. We firstly discuss the canonicalization method of the constrained Hamiltonian systems. The symplectic method is used to constrain Hamiltonian systems on the basis of the canonicalization, and then the numerical simulation of the system is carried out. An example is presented to illustrate the application of the results. By using the symplectic method of constrained Hamiltonian systems, one can solve the singular dynamic problems of nonconservative constrained mechanical systems, nonholonomic constrained mechanical systems as well as physical problems in quantum dynamics, and also available in many electromechanical coupled systems.

Keywords: constrained Hamiltonian system; canonicalization; symplectic method; numerical simulation

MSC: 37J60; 37J10; 37K05; 37K50

Citation: Fu, J.; Zhang, L.; Cao, S.; Xiang, C.; Zao, W. A Symplectic Algorithm for Constrained Hamiltonian Systems. *Axioms* **2022**, *11*, 217. https://doi.org/10.3390/axioms11050217

Academic Editor: Hans J. Haubold

Received: 18 April 2022
Accepted: 30 April 2022
Published: 7 May 2022

Publisher's Note: MDPI stays neutral with regard to jurisdictional claims in published maps and institutional affiliations.

Copyright: © 2022 by the authors. Licensee MDPI, Basel, Switzerland. This article is an open access article distributed under the terms and conditions of the Creative Commons Attribution (CC BY) license (https://creativecommons.org/licenses/by/4.0/).

1. Introduction

In 1993, symplectic algorithms for constrained Hamiltonian systems have been proposed [1]. We know that the displacements q and momenta p of an object moving freely are given by a Hamilton canonical equation in the form [2]

$$\dot{q} = \nabla_p H(p,\, q),\ \dot{p} = -\nabla_q(p,q) \tag{1}$$

where $p,\ q \in R^n$, $H:\ R^n \times R^n \to\ R^n$ is called the Hamiltonian function. A natural question is what happens when (1) is constrained by algebraic equations on q and/or p. That is, there are Hamiltonian constraints of the form $g(q) = 0$, and it leads to the constraints of Hamiltonian equations as [3,4]

$$\dot{q} = \nabla_p H(p,\, q),\ \dot{p} = -\nabla_q H(p,q) - \lambda G(q)^t, \tag{2}$$

where $g:\ R^n \to\ R^n$, $G(q) = g_q(q) \in R^{n\times m}$ and $\lambda \in R^m$. Equation (2) is called a constrained Hamiltonian system, which is not only a relatively loose concept but also a general constrained mechanical system. The flow of a Hamiltonian system like (1) possesses an important symplectic geometric structure. It has been observed in numerical experiments that symplectic methods with fixed step-size possess better long-term stability properties. Leimkuhler and Skeel [5] investigated symplectic numerical integrators of constrained Hamiltonian systems in molecular dynamics. By composition methods, Reich [6] studied symplectic integration of the constrained Hamiltonian systems. The method they proposed can reduce Hamiltonian differential-algebraic equations to ordinary differential equations in Euclidean space.

When studying symmetry properties of classical and quantum constrained systems, Li [7,8] found that via Legendre transformation, a singular Lagrangian system can be transformed into the phase space determined by generalized momenta and generalized coordinates. Since there are inherent constraints between generalized momenta and generalized coordinates, it is named a constrained Hamiltonian system. A lot of important physical systems belong to this system, such as quantum electrodynamics, quantum flavor dynamics, and so on. Even many electromechanical coupled systems belong to constrained Hamiltonian systems. For a Lagrangian system, if the value of determinant $\det\left(\frac{\partial^2 L}{\partial q_s \partial q_k}\right)$ vanishes, then it is named as a singular Lagrange system. The Lagrangian function of supersymmetry, supergravity, and string theory are all singular. Therefore, the fundamental theory of constrained Hamiltonian systems acts an important role in modern quantum field theory [9].

In the late 1980s, Feng et al. established the so-called symplectic algorithms to study the equations in Hamiltonian form and showed that these methods are more superior over a long time by combining theoretical analysis and computer experimentation [10,11]. The symplectic method has been widely recognized as a suitable numerical integrator with global conservation properties for canonical Hamiltonian systems. It has been well applied in testing particle simulation and some physical experiments in plasma physics, and thus derived a series of results, for instance, a variational multi-symplectic particle-in-cell algorithm of the Vlasov-Maxwell system [12], the practical symplectic partitioned Runge-Kutta and Runge-Kutta-Nystrom methods [13], the symplectic integrations of Hamiltonian systems [14], symplectic integrators of the Ablowitz–Ladik discrete nonlinear Schrödinger equation [15], etc. The standard symplectic scheme normally works for a canonical structure of the dynamical system. However, the symplectic simulation for the constrained Hamiltonian systems is beset with difficulties since the constrained Hamiltonian systems are usually non-canonical.

In this paper, we will present a general procedure for constructing the canonical coordinates of constrained Hamiltonian systems. By defining a variable transformation and calculations, the canonical variables for constrained Hamiltonian systems can be derived, and thus the constrained Hamiltonian systems are canonicalized. Once the canonical coordinates of constrained Hamiltonian systems are derived, one can employ the standard canonical symplectic methods to study the constrained Hamiltonian systems. The method we proposed is of importance in the study of constrained Hamiltonian systems. We believe that the symplectic method of constrained Hamiltonian systems given in this paper can be used in the study of quantum dynamics, electromechanical coupled systems, and strange constrained dynamics as well.

To verify the effect of the canonicalization and illustrate the advantage of the canonical symplectic simulation, a numerical example of the constrained Hamiltonian system is presented. Clearly, the numerical results derived by the canonical symplectic method are more accurate in the long-term simulation since they can maintain conservation properties.

2. Canonicalization of Constrained Hamiltonian Systems

Assume that a mechanical system is determined by the generalized coordinates $q_i(i = 1, 2, \ldots, n)$, and the Lagrangian function $L = L(t, q_i, \dot{q}_i)$ satisfies $\det\left(\frac{\partial^2 L}{\partial q_s \partial q_k}\right) = 0$ When the generalized momenta and Hamiltonian of the system are constructed, there are inherent constraints between the canonical variables in the phase space

$$\phi_j(t, q_i, p_i) = 0 \quad (j = 1, 2, \ldots, n - r, i = j = 1, 2, \ldots, n) \tag{3}$$

this is the constraint equation that should be obtained between the generalized coordinates and the generalized momenta of the constrained Hamiltonian system.

Then the motion equations of a singular system can be written as [11]

$$\dot{q}_i = \frac{\partial H_c}{\partial p_i} + \lambda_j \frac{\partial \varphi_j}{\partial p_i}, \qquad \dot{p}_i = -\frac{\partial H_c}{\partial q_i} - \lambda_j \frac{\partial \varphi_j}{\partial q_i} \qquad (i = 1, 2, \ldots, n) \tag{4}$$

where H_c is the Hamiltonian of the system and λ_j is the Lagrange multiplier. The multiplier in Formula (4) can be given by Equations (3) and (4).

The motion Equation (4) of the constrained Hamiltonian system can be rewritten as

$$\begin{pmatrix} \dot{p}_1 \\ \vdots \\ \dot{p}_i \\ \dot{q}_1 \\ \vdots \\ \dot{q}_n \end{pmatrix} = \begin{pmatrix} 0_n & S_n \\ T_n & 0_n \end{pmatrix} \begin{pmatrix} \frac{\partial H_c}{\partial p_1} \\ \vdots \\ \frac{\partial H_c}{\partial p_i} \\ \frac{\partial H_c}{\partial q_1} \\ \vdots \\ \frac{\partial H_c}{\partial q_i} \end{pmatrix} = M_{2n\times 2n} \begin{pmatrix} \frac{\partial H_c}{\partial p_1} \\ \vdots \\ \frac{\partial H_c}{\partial p_i} \\ \frac{\partial H_c}{\partial q_1} \\ \vdots \\ \frac{\partial H_c}{\partial q_i} \end{pmatrix}, \tag{5}$$

where

$$S_n = \begin{pmatrix} -1 - \lambda_j \frac{\partial \varphi_j}{\partial H_c} & \cdots & 0 \\ \vdots & \ddots & \vdots \\ 0 & \cdots & -1 - \lambda_j \frac{\partial \varphi_j}{\partial H_c} \end{pmatrix}_{n\times n}, T_n = \begin{pmatrix} 1 + \lambda_j \frac{\partial \varphi_j}{\partial H_c} & \cdots & 0 \\ \vdots & \ddots & \vdots \\ 0 & \cdots & 1 + \lambda_j \frac{\partial \varphi_j}{\partial H_c} \end{pmatrix}_{n\times n} \tag{6}$$

and $M_{2n\times 2n}$ is an anti-symmetric matrix.

Let $v = (p, q)^T$, where $p = (p_1, p_2, \ldots, p_n)$, $q = (q_1, q_2, \ldots, q_n)$, then Equation (5) can be rewritten as

$$\dot{v} = K(v)^{-1} \nabla H_c(v), \tag{7}$$

where

$$K(v) = \begin{pmatrix} 0_n & T_n^{-1} \\ S_n^{-1} & 0_n \end{pmatrix}. \tag{8}$$

It is easy to see that Equations (5) and (7) are non-canonical Hamiltonian systems.

To rewrite the non-canonical Hamiltonian system in canonical form, we let $Z = \Psi(v)$ be the corresponding canonical variables which is a transformation from R^{2n} to R^{2n}. $Z = (\widetilde{p}, \widetilde{q})^T$ are new variables after canonicalization. By the chain rule, the canonicalization of Equation (7) can be written as [11]

$$\dot{Z} = \left(\frac{\partial \Psi}{\partial v}\right) K(v)^{-1} \left(\frac{\partial \Psi}{\partial v}\right)^T \nabla \widetilde{H}(Z), \tag{9}$$

where $\widetilde{H}(Z) = H_c(v)$. If we let $\left(\frac{\partial \Psi}{\partial v}\right) K(v)^{-1} \left(\frac{\partial \Psi}{\partial v}\right)^T = J^{-1}$, i.e.,

$$K(v) = \left(\frac{\partial \Psi}{\partial v}\right)^T J \left(\frac{\partial \Psi}{\partial v}\right) \tag{10}$$

Note that $K(v)$ is a given matrix and $v = (p, q)^T$ is the original variable, so we can get $\Psi(v)$ through this transformation, which is a set of canonical new generalized momenta $\widetilde{p} = (\widetilde{p}_1, \widetilde{p}_2, \ldots, \widetilde{p}_n)$ and generalized coordinates $\widetilde{q} = (\widetilde{q}_1, \widetilde{q}_2, \ldots, \widetilde{q}_n)$. Now, we have transformed the non-canonical Hamiltonian system into a canonical Hamiltonian system.

By substituting the new variables into the original Hamiltonian of the constrained system, it becomes canonical. Based on the canonical Hamiltonian equations, one can examine their properties and hence some useful algorithms can be applied to examine the numerical solutions and numerical simulation of the constrained Hamilton systems.

The results of the original system can be obtained by replacing the new variables with the old ones.

3. Symplectic Method for Constrained Hamiltonian Systems

The constrained Hamiltonian systems are transformed in the canonical form (9):

$$\frac{dZ}{dt} = J^{-1}\widetilde{\nabla}H(Z), \tag{11}$$

that is, the canonical Hamiltonian system is

$$\tfrac{dZ}{dt} = J^{-1}\widetilde{\nabla}H(Z), \quad J = \begin{pmatrix} 0 & I_n \\ -I_n & 0 \end{pmatrix}, \quad Z \in R^{2n} \tag{12}$$

We now show that the properties, conclusions, and calculation methods of canonical Hamiltonian systems can be extended to constrained Hamiltonian systems. We give the symplectic method for constrained Hamiltonian systems as follows.

A transformation of the constrained Hamiltonian system

$$\Psi : R^{2n} \to R^{2n}, \quad v = \begin{pmatrix} p \\ q \end{pmatrix} \to \widetilde{Z} = \begin{pmatrix} \widetilde{p} \\ \widetilde{q} \end{pmatrix} \tag{13}$$

is called the symplectic transformation for a system if its Jacobian is a symplectic matrix

$$\left(\frac{d\widetilde{Z}}{dZ}\right)^T J \left(\frac{d\widetilde{Z}}{dZ}\right) = J \Leftrightarrow \sum_{k=1}^{n} d\widetilde{p}_k \wedge d\widetilde{q}_k = \sum_{k=1}^{n} dp_k \wedge dq_k. \tag{14}$$

For the canonical Hamiltonian system (9), if

$$\widetilde{p} = p - \tau\tfrac{\partial H}{\partial q}(\widetilde{p}, q), \quad \widetilde{q} = q + \tau\tfrac{\partial H}{\partial p}(\widetilde{p}, q), \tag{15}$$

then it is a first-order symplectic scheme. When $H(p,q) = U(p) + V(q)$, Equation (15) becomes

$$\widetilde{p} = p - \tau\tfrac{\partial V}{\partial q}(q), \quad \widetilde{q} = q + \tau\tfrac{\partial U}{\partial p}(\widetilde{p}), \tag{16}$$

which is an explicit symplectic scheme. For the canonical Hamiltonian system (9), the Euler midpoint rule is

$$\widetilde{Z} = Z + \tau J^{-1}\nabla H(\frac{\widetilde{Z} + Z}{2}), \tag{17}$$

which is a second-order symplectic scheme. A Runge-Kutta method

$$\widetilde{Z} = Z + \tau\sum_{i=1}^{m} b_i J^{-1}\nabla H(K_i), \quad K_i = Z + \tau\sum_{i=1}^{m} a_{ij} J^{-1}\nabla H(K_j), \ i = 1, \ldots, m, \tag{18}$$

is symplectic if and only if $b_i b_j - b_i a_{ij} - b_j a_{ij} = 0$. In Equations (15)–(18), τ represents the time step size.

4. Example

The Lotka-Volterra model can be expressed as a non-canonical Hamiltonian system with $n = 1$

$$\begin{pmatrix} \dot{p} \\ \dot{q} \end{pmatrix} = \begin{pmatrix} 0 & -pq \\ pq & 0 \end{pmatrix} \nabla H(p,q), \tag{19}$$

where $H(p,q) = p - 2\log p + q - \log q$.

The Hamiltonian H can be rewritten as $H = H_1 + H_2$ with $H_1 = p - 2\log p$ and $H_2 = q - \log q$. According to the canonialization method shown in Section 2, we have

$$K = \begin{pmatrix} 0 & \frac{1}{pq} \\ -\frac{1}{pq} & 0 \end{pmatrix}. \tag{20}$$

According to Equation (10), we get

$$\frac{\partial \widetilde{p}}{\partial p}\frac{\partial \widetilde{q}}{\partial p} - \frac{\partial \widetilde{p}}{\partial p}\frac{\partial \widetilde{q}}{\partial p} = 0, \quad \frac{\partial \widetilde{p}}{\partial p}\frac{\partial \widetilde{q}}{\partial q} - \frac{\partial \widetilde{q}}{\partial p}\frac{\partial \widetilde{p}}{\partial q} = \frac{1}{pq} \tag{21}$$

and

$$\widetilde{p} = \log(p), \quad \widetilde{q} = \log(q). \tag{22}$$

Hence, we have

$$p = \exp(\ \widetilde{p}), \quad q = \exp(\widetilde{q}) \tag{23}$$

and thus

$$\widetilde{H}(\widetilde{p}, \widetilde{q}) = \exp(\widetilde{p}) - 2\widetilde{p} + \exp(\widetilde{q}) - \widetilde{q}, \tag{24}$$

which is a canonical Hamiltonian system. Using the second-order explicit symplectic scheme on the basis of the canonicalization, we get the trajectory of the canonical variable $\widetilde{p}, \ \widetilde{q}$, where $\widetilde{p}(0) = \ln 2, \widetilde{q}(0) = \ln 3,$ and time step size $\tau = 0.1$ (see Figure 1).

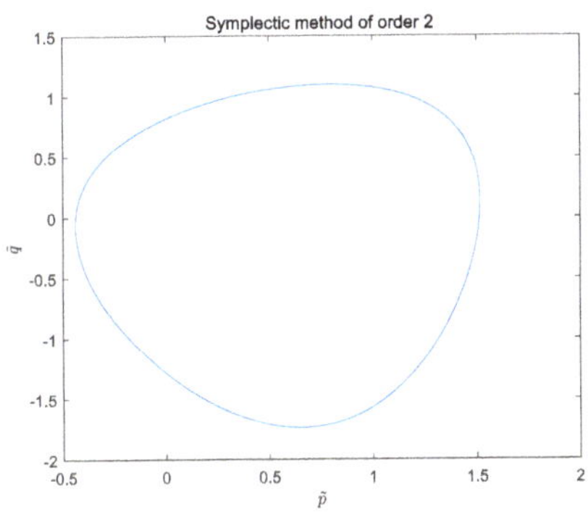

Figure 1. Trajectory of the canonical variable $\widetilde{p}, \ \widetilde{q}$.

Using Equation (23) we can obtain p, q, and $p(0) = 2$, $q(0) = 3$, and time step size $\tau = 0.1$, then using the second-order explicit symplectic scheme on the basis of p, q, we get the trajectory of the non-canonical variable p, q (see Figure 2).

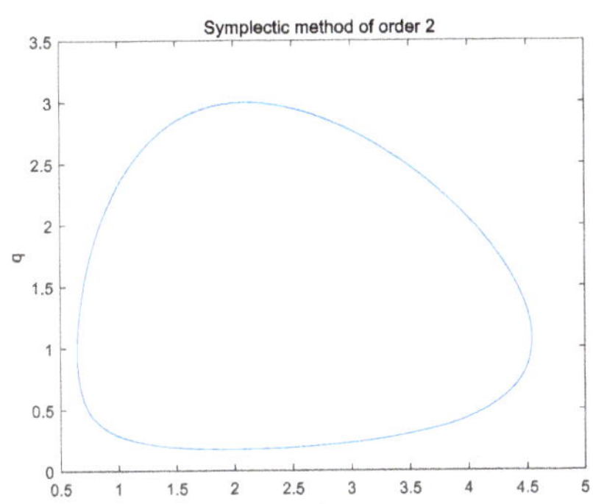

Figure 2. Trajectory of the non-canonical variable p, q.

In addition, the implicit Runge-Kutta method of order 3 is applied directly to the non-canonical Hamiltonian system directly, and then we get the trajectory of the original variables p, q, and $p(0) = 2$, $q(0) = 3$ and time step size $\tau = 0.1$ (see Figure 3).

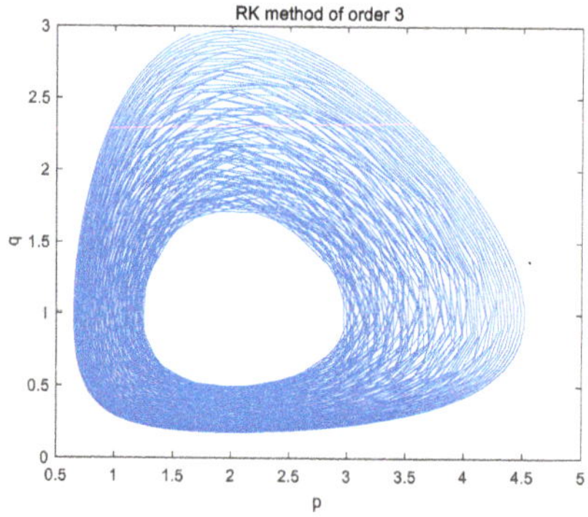

Figure 3. Trajectory of the non-canonical variable.

As can be seen from Figures 1 and 2, the trajectory diagrams of regularized variables and initial variables are kept unchanged by a symplectic algorithm. After 1,000,000 steps, the graph remains basically unchanged, which indicates that the symplectic algorithm of constrained Hamiltonian systems has the property of preserving structure. Namely, the physical properties of constrained Hamiltonian systems can be maintained by a symplectic method. One can see from Figure 3 that the graph using the third-order Runge Kutta method (or general numerical calculation method) is very unstable. This method does not have the property of preserving the structure, that is, it cannot maintain the physical properties of the constrained Hamiltonian systems. It is shown clearly from the three figures that the symplectic algorithm has better structure-preserving properties. It is of great significance to study the constrained Hamiltonian systems using the symplectic algorithm.

5. Conclusions

In this paper, we discuss the canonicalization method of the constrained Hamiltonian systems, then the symplectic method is applied to the constrained Hamiltonian systems on the basis of the canonicalization. Compared with the traditional Runge-Kutta method, they have better structural preservation properties. Consequently, the symplectic methods can be applied to more noncanonical Hamiltonian systems, which will be further investigated in our next work.

Author Contributions: Conceptualization, methodology, validation, and supervision, J.F. and L.Z.; formal analysis and investigation, C.X.; writing—original draft preparation, S.C.; writing—review and editing, L.Z.; W.Z. symplectic algorithm numerical simulation. All authors have read and agreed to the published version of the manuscript.

Funding: This research was funded by the National Nature Science Foundation of China, 11872335, 12172199, 11672270.

Institutional Review Board Statement: Not applicable.

Informed Consent Statement: Not applicable.

Data Availability Statement: Not applicable.

Conflicts of Interest: The authors declare no conflict of interest.

References

1. Reich, S. *Symplectic Integration of Constrained Hamiltonian Systems by Runge-Kutta Methods*; Technical Report 93-13; Department of Computer Science, University of British Columbia: Vancouver, BC, Canada, 1993.
2. Hildebrand, F.B. *Methods of Applied Mathematics*, 2nd ed.; Prentice-Hall: Englewood Cliffs, NJ, USA, 1965.
3. Dirac, P.A.M. *Lecture on Quantum Mechanics*; Yeshiva University Press: New York, NY, USA, 1964.
4. Arnold, V.I. *Mathematical Methods of Classical Mechanics, Graduate Texts in Mathematics, No. 60*; Springer: New York, NY, USA, 1975.
5. Leimkuhler, B.J.; Skeel, R.D. Symplectic numerical integrators in constrained Hamiltonian systems. *J. Comput. Phys.* **1994**, *112*, 117–125. [CrossRef]
6. Reich, S. Symplectic Integration of Constrained Hamiltonian Systems by Composition Methods. *SIAM J. Numer. Anal.* **1996**, *33*, 475–491. [CrossRef]
7. Li, Z.P. *Classical and Quantal Dynamics of Constrained System and Their Symmetrical Properties*; Beijing Polytechnic Univ. Press: Beijing, China, 1993.
8. Li, Z.P.; Jiang, J.H. *Symmetries in Constrained Canonical System*; Beijing Science Press: Beijing, China, 2002.
9. Holod, I.; Lin, Z. Verification of electromagnetic fluid-kinetic hybrid electron model in global Gyrokinetic particle simulation. *Phys. Plasmas* **2013**, *20*, 032309. [CrossRef]
10. Feng, K.; Qin, M. *Symplectic Geometric Algorithms for Hamiltonian Systems*; Zhejiang Science and Technology Publishing House: Hangzhou, China, 2010.
11. Feng, K.; Qin, M. Symplectic difference schemes for Hamiltonian systems. *J. Comp. Math.* **1991**, *1*, 86–96.
12. Xiao, J.; Liu, J.; Qin, H. A variational multi-symplectic particle-in-cell algorithm with smoothing functions for the Vlasov-Maxwell system. *Phys. Plasmas* **2013**, *20*, 102517. [CrossRef]
13. Blanes, S.; Moan, P.C. Practical symplectic partitioned Runge-Kutta and Runge-Kutta-Nystrom methods. *J. Comp. Appl. Math.* **2000**, *142*, 313–330. [CrossRef]
14. Channell, P.J.; Scovel, C. Symplectic integration of Hamiltonian systems. *Nonlinearity* **1990**, *3*, 231–259. [CrossRef]
15. Schober, C.M. Symplectic integrators for the Ablowitz–Ladik discrete nonlinear Schrödinger equation. *Phys. Lett. A* **1999**, *259*, 140–151. [CrossRef]

axioms

MDPI

Article

A Theoretical Dynamical Noninteracting Model for General Manipulation Systems Using Axiomatic Geometric Structures

Paolo Mercorelli

Institute of Product and Process Innovation, Leuphana University of Lueneburg, Universitaetsallee 1, D-21335 Lueneburg, Germany; mercorelli@uni.leuphana.de; Tel.: +49-4131-677-1896

Abstract: This paper presents a new theoretical approach to the study of robotics manipulators dynamics. It is based on the well-known geometric approach to system dynamics, according to which some axiomatic definitions of geometric structures concerning invariant subspaces are used. In such a framework, certain typical problems in robotics are mathematically formalised and analysed in axiomatic form. The outcomes are sufficiently general that it is possible to discuss the structural properties of robotic manipulation. A generalized theoretical linear model is used, and a thorough analysis is made. The noninteracting nature of this model is also proven through a specific theorem.

Keywords: manipulation system; geometric approach; noninteraction

MSC: 19L64; 70Q05; 14L24

Citation: Mercorelli, P. A Theoretical Dynamical Noninteracting Model for General Manipulation Systems Using Axiomatic Geometric Structures. *Axioms* **2022**, *11*, 309. https://doi.org/10.3390/axioms11070309

Academic Editor: Humberto Bustince

Received: 21 April 2022
Accepted: 13 June 2022
Published: 25 June 2022

Publisher's Note: MDPI stays neutral with regard to jurisdictional claims in published maps and institutional affiliations.

Copyright: © 2022 by the authors. Licensee MDPI, Basel, Switzerland. This article is an open access article distributed under the terms and conditions of the Creative Commons Attribution (CC BY) license (https://creativecommons.org/licenses/by/4.0/).

1. Introduction

To briefly describe the history of robotics up to the present day is not a superfluous task. It is curious to first search for the original meaning of the word. Some philologists propose that the term "robot" came from the Latin root of the word "robor-roboris", one of the meanings of which is "force". In any case, the term "robot" was introduced for the first time in 1921 by the Czech writer Karel Capek in his satirical work entitled "Rossum's Universal Robots"; in Czech, "robota" means "work". Some consider the source to be Indo-European, so it might be useful, in this endeavor, to track down various corruptions, such as "labor-laboris", and hence "work". Capek's satirical work emphasizes the difference between the machine and the human and, in particular, the substantial difference consists in the fact that robots never get tired.

After World War II, the need to manipulate radioactive material generated the need to build the first mechanical manipulators that are remotely controlled. They were made in the laboratories of Argonne and Oack Ridge (USA), and were of the master–slave type, which are manipulators consisting of a "slave part" driven by the human operator whose movements were duplicated on the slave part through a series of mechanical linkages. General Electric, together with General Mills, called these teleoperators. However, the teleoperators were certainly not the only expression of robotics in the years following World War II; the CNC-machines (Computerized Numerical Control machines), initially used for the lamination of some parts of the aircraft, joined these. In fact, the numerical control machines had a considerable weight in history of robotics. Their great merit was to fully replace the human operator in the teleoperators. In 1954, for the first time, George Devol replaced in teleoperators the human operator with a programmable controller similar to that present in the numerical control machines, giving rise to the first real robot: "real" to emphasize the fact that a machine during the execution of tasks does not depend in any way on a human. Devol's patent rights were bought by Eledemberg Joseph, a student at Columbia University, who in 1956 built Unimation. This company in 1961 installed in the plants of General Motors the first robot that, because of its programmability, was able to perform a wide range of operations, multiplying the flexibility of the chain of assembly.

Mechanically, the robot was with an open kinematic chain consisting of many degrees of freedom and consequently its control was not an easy task. To improve the pilotability of the robots and strengthen their capacity, in 1962, Ernst inserted the first force sensors in the structure of the mechanical robots.

The sensing robots went on in different forms: Tomovic and Bono developed a pressure sensor for taking robotics; McCarthy developed a vision system binary, etc. The entire activity research concretized in the first manipulator industrial computer control of the Cincinnati Milacron, baptized "The Tomorrow Tool *T*3 (1973). In addition, in the 1970s, the Unimation begins to produce the PUMA (Programmable Universal Machine for Assembly), which represents one of the cornerstones in the history of robotics.

During the 1980s, the research, aimed at improving performance of industrial robots, was developed and the first techniques to control the position and the stiffness of robots in feedback. In the past few years, the trend has been to build very versatile devices. An example useful for all is the Robotworld of Automatix structured into several modules, each with four degrees of freedom, which can be connected to different tools for the execution of various operations.

The potential applications of the research in the field of robotic manipulation have been and still are the reasons driving the research. For example, think of the possibilities of their use in operating in environments hostile to man (or applications in space exploration in central nuclear, removal of toxic waste), or the robotization of work notoriously difficult and/or dangerous to humans, or, finally, medical applications (robotic protheses, using robotic surgery).

Technological development has not only increased the use of robots in industrial fields, but has also made it possible for them to actually use different applications, such as medical implants instruments for non-invasive surgery. These applications require high-precision performance and often also a high execution speed. In general, therefore, studies highlighting the extent to which the potentialities and prospects of such devices can be improved are required. Robotic manipulation systems are of a great importance due to their flexibility for application in any industrial sector. Their flexibility is a result of the multifunctionality of robotic hands, which allows for their application to industrial processes in many fields and for the possibility of interactions and cooperation with other robotic structures. This is connected with the fact that manipulation skills are, together with speech, probably the most important features that distinguish humans from animals. A certain evolutionary biology believes that a certain part of human supremacy over other primates is also due to the prensility of our upper limbs that allowed the immediate application of ideas in what is usually called actions, a process otherwise definitely more complicated. In other words, one might wonder what would be without prensility of the upper limbs. This seems to be unimaginable, to have a reality different from what we have today. Due to this consideration, this can seem elementary, and leads us to think about how important it is that a machine performs a certain function, or better, a certain action. In this sense, to be able to affect the environment, a manipulator needs more versatile hands. As a result of technological development, the application of robotics is on the rise in many industrial sectors, and even in the medical field (e.g., micro-manipulation of internal tissues or laparoscopy). Due to the high mechanical efficiency and the vast possibilities of application of robots, in the past years the manipulation was followed with great interest in both academic and industrial world, refs. ([1–9]). For these advanced applications, robotic devices with high performances in terms of precision and speed are required. In order to achieve high performances, a general strategy in robotics is represented by the decoupling control technique.

1.1. Coupling and Decoupling

The decoupling of coupled systems is one of the most interesting problems in system theory and control. The decoupling control strategies allow us to simplify the control itself and also the identification procedure of the parameters of the robot. The couplings which are contained in the mathematical description of the robot model through the motor inertia, the mass inertia, the stiffness and the damping matrix within the joints should be decoupled by the control. These couplings lead to an eighth-order multivariable system for each joint. The decoupling within joints is achieved by a novel MIMO state controller motor positions and output torques and their derivatives as states. In general, in order to design the controller of robots taking into account the coupling, the system is broken down into two decoupled subsystems using modal decoupling and subsequently considered as two separate single-variable systems (SISO). Thus, the parameters of the SISO state controller can be determined for the respective subsystem and two independent controllers are to be designed [10]. A globally asymptotic stability for the entire system can be achieved with the MIMO state controller. The controller significantly expands the approaches from [11], both theoretically and practically. The decoupling control represents one of the most interesting controller structures that have been implemented on robots. Multivariable systems, in which several output variables can be dependent on several input variables at the same time, are characterized by the mutual coupling of inputs and outputs. It is therefore the task of the control design for multivariable systems to minimize the influence of the coupling so that, in the ideal case, each output variable is only influenced by a corresponding virtual manipulated variable and thus the controlled system achieves the desired smooth dynamic behavior, ref. [12]. When designing a controller for a linear multivariable system, there are basically two options:

- central controller design
- decentralized controller design.

The modal method for the controller design is used to decouple the system from the controller. While central controller design is based on the overall system, decentralized controller design uses several decoupled, lower-order subsystems instead of the high-order, coupled system. In the following, the two methods are presented and analyzed one after the other. In a centralized decoupled control design, the state controller can be designed through the complete modal synthesis, whereby the closed overall system is decoupled. In the context of a decentralized decoupled control design, if a multivariable system is decoupled, the synthesis problem is reduced to the case of single control. For this, the system is first transformed into a modal form in which it can be divided into several small, decoupled subsystems. The decoupling control finds application, not only in manipulation systems, but also in other systems. One example is represented by the mobile robots. For instance, in [13], an explicit model predictive control (MPC), in combination with sliding mode control in the context of a decoupling controller, is proposed. The decoupling control is particularly important for MPC in order to reduce the computational load. In addition, recent MPC contribution takes into account the problem of computational load in the field of tracking of different trajectories mark progress in optimal design for model predictive control based on a new improved intelligent technique and it is named the modified multitracker optimization algorithm, such as, for instance, in [14]. This modification improves the exploration behavior to prevent it from becoming trapped in a local optimum. The proposed method is applied on the robotic manipulator to track trajectories. In addition, more recently, in [15], an optimized algorithm in MPC context for autonomous vehicle is proposed. More in general, the decoupling proposed in this contribution can be integrated in with the methods proposed in [16], as well as in [17,18], in which the D-decomposition method is used in order to compute optimized controller gains that provide fine performance in different engineering applications.

1.2. Main Contribution of the Paper

The present paper presents a new approach to the study of robotics manipulators dynamics based on the well-known geometric control approach to system dynamics. Using this framework several typical problems in robotics are mathematically formalised and analysed. The outcomes are sufficiently general that it is possible to discuss the structural properties of robotic manipulation, which are obtained using a geometric approach. The geometric approach was pioneered in the 1970s, in [19–22]. The approach used for the derivation of these properties is decidedly new in this kind of literature that refers especially to [23–26]. The novelty consists in using the geometric approach (theory of invariants subspaces) analysis and then the derivation of the properties as listed above for the synthesis of control systems that guarantee and then allow to exploit these properties in any operating condition. The seminal references for this approach were [20,27]. The problem of the noninteracting force motion model is here investigated, a generalised linear model is used, and a careful analysis is performed. Contributions to the topic of manipulation using the geometric approach further progressed through the use of linear algebra. Recent contributions, such as in [28–30] have led to progress in the analysis and synthesis of geometric controllers for application to electro-mechanical systems. In [31–33], a geometric approach guarantees robustness and many practical advantages in possible real applications; see [34]. In particular, the geometric approach can be focused on the disturbance decoupling problem [35], an issue that has attracted many scientists. Furthermore, in [36–38], interesting and interpretable results are proposed. For a broad overview of the manipulation control problem, the reader is referred to [26] and the references therein. The present paper aims at analysing the structural properties of noninteraction with respect to rigid-body object motions and reachable contact forces along with possible mechanism redundancy. More recently, refs. [39–41] underline the importance of a noninteraction in the control strategy to simplify the structure of the controller. In the same way, ref. [42–44] point out the importance of the position/force control in robotic manipulation.

The present study is conducted using geometric techniques. Some axiomatic definitions of geometric structures concerning invariant subspaces are used as a possible framework in order to derive some structural properties in the considered system. This paper follows the contributions published in [31,35] and, more recently, in [32,34,45]. These studies on geometric control represent an interesting line of research in which problems such as decoupling, noninteraction and disturbance rejection are taken into account in the context of mechanical systems.

1.3. Structure of this Contribution

The present paper is structured as follows. In Section 2, the linearized dynamical model is derived. Section 3 is dedicated to the reachable internal contact forces and a fundamental theorem is demonstrated. In Section 4, the noninteraction property is presented. In Section 5, a possible reinterpretation of the theoretical results is proposed and a case study with its simulations is shown. The paper closes with a conclusion and an appendix in which the proof of the theorem that states the structural property of noninteraction is proposed.

2. Dynamic Model

For the dynamic model, $\mathbf{q} \in \mathbb{R}^q$ denotes the vector of manipulator joint positions, $\tau \in \mathbb{R}^q$ denotes the vector of joint actuator torques, $\mathbf{u} \in \mathbb{R}^d$ denotes the vector locally describing the position and the orientation of a frame attached to the object, and $\mathbf{w} \in \mathbb{R}^d$ denotes the vector of forces and torques resulting from external forces acting directly on the object. In the literature, $\mathbf{w}$ usually refers to the disturbance vector. The force/torque interaction $\mathbf{t}_i$ at the i-th contact is taken into account by using a lumped-parameter $(\mathbf{K}_i, \mathbf{B}_i)$ model of visco elastic phenomena. According to this model, the contact force vector $\mathbf{t}_i$ is as follows:

$$\mathbf{t}_i = \mathbf{K}_i({}^h\mathbf{c}_i - {}^o\mathbf{c}_i) + \mathbf{B}_i({}^h\dot{\mathbf{c}}_i - {}^o\dot{\mathbf{c}}_i), \tag{1}$$

where vectors ${}^h\mathbf{c}_i$ and ${}^o\mathbf{c}_i$ describe the postures of two contact frames, the first on the manipulator and the second on the object, where the i-th contact spring and damper are anchored. Matrices $\mathbf{K}_i$ and $\mathbf{B}_i$ are symmetric and positive definite and the dimensions of vectors involved in Equation (1) depend on the particular model used to describe the contact interaction [46]. The Jacobian $\mathbf{J}$ and grasp matrix $\mathbf{G}$ of the manipulation system, see [25,47], are defined by the linear maps relating the velocities of vectors ${}^h\mathbf{c}$ and ${}^o\mathbf{c}$ with the joint and object velocities $\dot{\mathbf{q}}$ and $\dot{\mathbf{u}}$, respectively:

$$\begin{aligned} {}^h\dot{\mathbf{c}} &= \mathbf{J}\dot{\mathbf{q}}, \\ {}^o\dot{\mathbf{c}} &= \mathbf{G}^T\dot{\mathbf{u}}. \end{aligned} \tag{2}$$

Note that $\mathbf{J}^T\mathbf{t}$ and $\mathbf{G}\mathbf{t}$ dually represent the effects of contact forces $\mathbf{t}$ on the manipulation and object dynamics whose full nonlinear models are, respectively:

$$\begin{aligned} \mathbf{M}_h\ddot{\mathbf{q}} + \mathbf{Q}_h &= -\mathbf{J}^T\mathbf{t} + \boldsymbol{\tau}, \\ \mathbf{M}_o\ddot{\mathbf{u}} + \mathbf{Q}_o &= \mathbf{G}\mathbf{t} + \mathbf{w}. \end{aligned} \tag{3}$$

Here, $\mathbf{M}_h$ and $\mathbf{M}_o$ are inertia symmetric and positive definite matrices, while $\mathbf{Q}_h$ and $\mathbf{Q}_o$ are terms including velocity-dependent and gravity forces of the manipulator and the object, respectively. To proceed with the analysis of the linearised model of the full manipulation system, consider a reference equilibrium configuration

$$\begin{aligned} &\mathbf{q} = \mathbf{q}_o, \quad \mathbf{u} = \mathbf{u}_o, \quad \dot{\mathbf{q}} = \dot{\mathbf{u}} = 0, \\ &\boldsymbol{\tau} = \boldsymbol{\tau}_o, \quad \mathbf{w} = \mathbf{w}_o \quad \mathbf{t} = \mathbf{t}_o, \end{aligned}$$

such that

$$\boldsymbol{\tau}_o = \mathbf{J}^T\mathbf{t}_o \qquad \text{and} \qquad \mathbf{w}_o = -\mathbf{G}\mathbf{t}_o.$$

The linear approximation of the manipulation system in the neighbourhood of this equilibrium is given by

$$\dot{\mathbf{x}} = \mathbf{A}\mathbf{x} + \mathbf{B}_\tau\delta\boldsymbol{\tau} + \mathbf{B}_w\delta\mathbf{w}, \tag{4}$$

where the state and input vectors are defined as the departures from the reference equilibrium configuration as follows:

$$\begin{aligned} \mathbf{x} &= \left[\delta\mathbf{q}^T, \delta\mathbf{u}^T, \delta\dot{\mathbf{q}}^T, \delta\dot{\mathbf{u}}^T\right]^T = \left[(\mathbf{q}-\mathbf{q}_o)^T \ (\mathbf{u}-\mathbf{u}_o)^T \ \dot{\mathbf{q}}^T \ \dot{\mathbf{u}}^T\right]^T, \\ \delta\boldsymbol{\tau} &= \boldsymbol{\tau} - \mathbf{J}^T\mathbf{t}_o, \\ \delta\mathbf{w} &= \mathbf{w} + \mathbf{G}\mathbf{t}_o. \end{aligned} \tag{5}$$

The dynamic, input and disturbance matrices are

$$\mathbf{A} = \begin{bmatrix} 0 & \mathbf{I} \\ \mathbf{L}_k & \mathbf{L}_b \end{bmatrix}, \quad \mathbf{B}_\tau = \begin{bmatrix} 0 \\ 0 \\ \mathbf{M}_h^{-1} \\ 0 \end{bmatrix}, \quad \mathbf{B}_w = \begin{bmatrix} 0 \\ 0 \\ 0 \\ \mathbf{M}_o^{-1} \end{bmatrix}. \tag{6}$$

To simplify the notation, we will henceforth omit the symbol δ. According to [47], by neglecting gravity, assuming a locally isotropic model of visco–elastic phenomena (where the stiffness matrix $\mathbf{K}$ is proportional to the damping matrix $\mathbf{B}$), and assuming that the local variations of the Jacobian and grasp matrices are small, blocks $\mathbf{L}_k$ and $\mathbf{L}_b$ in $\mathbf{A}$ can be simply obtained as

$$\mathbf{L}_k = -\mathbf{M}^{-1}\mathbf{P}_k, \qquad \mathbf{L}_b = -\mathbf{M}^{-1}\mathbf{P}_b, \tag{7}$$

where

$$\begin{aligned} \mathbf{M} &= \mathrm{diag}(\mathbf{M}_h, \mathbf{M}_o), \\ \mathbf{P}_k &= \begin{bmatrix} \mathbf{J}^T \\ -\mathbf{G} \end{bmatrix} \mathbf{K} \begin{bmatrix} \mathbf{J} & -\mathbf{G}^T \end{bmatrix}, \\ \mathbf{P}_b &= \begin{bmatrix} \mathbf{J}^T \\ -\mathbf{G} \end{bmatrix} \mathbf{B} \begin{bmatrix} \mathbf{J} & -\mathbf{G}^T \end{bmatrix}. \end{aligned}$$

Concerning the contact forces, we then obtain

$$\mathbf{t}' = \mathbf{t} - \mathbf{t}_o = \mathbf{K}(\mathbf{J}\delta\mathbf{q} - \mathbf{G}^T\delta\mathbf{u}) + \mathbf{B}(\mathbf{J}\delta\dot{\mathbf{q}} - \mathbf{G}^T\delta\dot{\mathbf{u}}), \tag{8}$$

and in terms of matrices, we have

$$\mathbf{t}' = \mathbf{C}_t\mathbf{x},$$

where the output matrix of the contact force is as follows:

$$\mathbf{C}_t = \begin{bmatrix} \mathbf{KJ} & -\mathbf{KG}^T & \mathbf{BJ} & -\mathbf{BG}^T \end{bmatrix}. \tag{9}$$

The properties of grasping defined as follows have a relevant influence on the dynamic behaviour of the manipulation system, refs. [25,47]. These properties are based on the existence of the null spaces of the Jacobian and grasp matrices **J** and **G** and of their transpose matrices.

Definition 1. *A grasp (or manipulation system) is considered defective if* $\ker(\mathbf{J}^T) \neq \mathbf{0}$.

Definition 2. *A grasp is considered indeterminate if* $\ker(\mathbf{G}^T) \neq \mathbf{0}$.

If a grasp is indeterminate, there exist motions of the objects under which no variations of contact forces occur. In other words, indeterminacy implies that the object is not firmly grasped.

Definition 3. *A manipulation system is considered graspable if* $\ker(\mathbf{G}) \neq \mathbf{0}$.

If a system is graspable, it is possible to exert contact forces with zero resultant forces on the object. Usually in the literature, the forces belonging to the null space of **G** are referred to as internal forces. Finally, the well-known notion of manipulator redundancy is formalised as follows.

Definition 4. *A grasp is considered redundant if* $\ker(\mathbf{J}) \neq \mathbf{0}$.

Proposition 1. *If a system is not indeterminate, i.e.,* $\ker(\mathbf{G}^T) = \mathbf{0}$, *then the minimal* **A***-invariant subspace containing* $\mathrm{im}(\mathbf{B}_\tau)$, $\min\mathcal{I}(\mathbf{A}, \mathbf{B}_\tau)$, *is externally stable.*

From now on, we will assume that the considered system is not indeterminate $\ker(\mathbf{G}^T) = \mathbf{0}$. Concerning the coordinate movements of the object, the following proposition [47] show that the subspace $\mathbf{J}\mathbf{\Gamma}_{qc}^T = \mathbf{G}^T\mathbf{\Gamma}_{uc}^T$. of rigid-body motions is reachable.

Proposition 2. *The rigid kinematics are described by the base matrix* $\mathbf{\Gamma}$ *whose columns form a basis for*

$$\ker\begin{bmatrix} \mathbf{J} & \mathbf{G}^T \end{bmatrix} = \mathrm{im}(\mathbf{\Gamma}), \tag{10}$$

where $\mathbf{\Gamma} = [\mathbf{\Gamma}_{qc}^T \ \ \mathbf{\Gamma}_{uc}^T]$.

Proposition 3. *Let the subspace of rigid-body motions be defined as the column space of* $\mathbf{T}_c$, *where*

$$\mathbf{T}_c = \begin{bmatrix} \mathbf{\Gamma}_{qc} & \mathbf{0} \\ \mathbf{\Gamma}_{uc} & \mathbf{0} \\ \mathbf{0} & \mathbf{\Gamma}_{qc} \\ \mathbf{0} & \mathbf{\Gamma}_{uc} \end{bmatrix}. \tag{11}$$

Accordingly, the following holds:

$$\mathrm{im}(\mathbf{T}_c) \subseteq \min\mathcal{I}(\mathbf{A}, \mathbf{B}_\tau).$$

3. Reachable Internal Contact Forces

Contact forces $\mathbf{t}$ are exerted on an object by the manipulation system in order to maintain the grasp, reject disturbance wrenches $\mathbf{w}$ and control the object motion. Therefore, the control of contact forces is a fundamental part of the manipulation control problem, as the better the control of forces, the finer the manipulation. In [47], the reachable subspace of contact forces as an outputs of the dynamic system given in Equation (4) was studied. The following theorem provides an explicit formula for the subspace of reachable internal forces.

Theorem 1. *Under the hypothesis stating that* $\mathbf{K}$ *is proportional to* $\mathbf{B}$,

$$\mathcal{R}_{ti,\tau} = \mathrm{im}((\mathbf{I} - \mathbf{K}\mathbf{G}^T(\mathbf{G}\mathbf{K}\mathbf{G}^T)^{-1}\mathbf{G})\ \mathbf{C}_t) = \mathrm{im}((\mathbf{I} - \mathbf{K}\mathbf{G}^T(\mathbf{G}\mathbf{K}\mathbf{G}^T)^{-1}\mathbf{G})\ \mathbf{K}\mathbf{J}).$$

Then, the output matrix is defined as follows:

$$\mathbf{e}_{ti} = \mathbf{E}_{ti}\mathbf{x}; \quad \text{with} \quad \mathbf{E}_{ti} = (\mathbf{I} - \mathbf{K}\mathbf{G}^T(\mathbf{G}\mathbf{K}\mathbf{G}^T)^{-1}\mathbf{G})\mathbf{C}_t = \left[\begin{array}{cccc} \mathbf{Q}_k & 0 & \mathbf{Q}_\beta & 0 \end{array}\right], \tag{12}$$

where

$$\mathbf{Q}_k = (\mathbf{I} - \mathbf{K}\mathbf{G}^T(\mathbf{G}\mathbf{K}\mathbf{G}^T)^{-1}\mathbf{G})\mathbf{K}\mathbf{J}. \tag{13}$$

and

$$\mathbf{Q}_\beta = (\mathbf{I} - \mathbf{B}\mathbf{G}^T(\mathbf{G}\mathbf{B}\mathbf{G}^T)^{-1}\mathbf{G})\mathbf{B}\mathbf{J}. \tag{14}$$

It should be remarked that $\mathrm{im}(\mathbf{Q}_k) = \mathrm{im}(\mathbf{Q}_\beta)$ under the hypothesis $\mathrm{im}(\mathbf{K}) = \mathrm{im}(\mathbf{B})$.

4. Noninteraction as a Structural Property

The present section aims to analyse noninteraction as a control property for a general grasping mechanism with respect to the rigid-body object motions and the reachable contact forces together with the possible mechanism redundancy. The geometric approach is used for this analysis. It should be remarked that the earliest geometric approaches to noninteracting control where proposed by Basile and Marro [19,20] and to Wonham and Morse [21,22,27]. The results of this section address the force/motion noninteracting control of general manipulation mechanisms and are based on necessary and sufficient conditions for the existence of the noninteraction control law given in [19,20]. We now proceed to analyse noninteraction as a structural property of general manipulation systems by formalizing the notion of force/motion noninteraction.

Definition 5. *A control law for the dynamic system in Equation (4) is noninteracting with respect to the regulated outputs* $\mathbf{e}_{uc}$, $\mathbf{e}_{ti}$ *and* $\mathbf{e}_{qr}$ *if there exists a partition* $\boldsymbol{\tau}_{uc}$, $\boldsymbol{\tau}_{ti}$ *and* $\boldsymbol{\tau}_{qr}$ *of the input vector* $\boldsymbol{\tau}$ *such that for an initial condition of zero, each input* $\boldsymbol{\tau}_{(\cdot)}$ *(with all the other inputs, also being zero) only affects the corresponding output* $\mathbf{e}_{(\cdot)}$.

The Fundamental Theorem

The following theorem shows that the noninteraction of the regulated outputs $\mathbf{e}_{uc}$, $\mathbf{e}_{ti}$ and $\mathbf{e}_{qr}$ for the dynamic system in Equation (4), is an intrinsic structural property of general manipulation systems. Assume that following hypothesis:

H1. *The manipulation mechanism is not indeterminate that is,* $\ker(\mathbf{G}^T) = \mathbf{0}$.

Then, the following theorem holds.

Theorem 2 (Noninteraction). *Consider the linearized manipulation system given in Equation (4). Under Hypothesis H1, there exists a noninteracting control law decoupling the following outputs:*

(a) *the rigid–body object motions* $\mathbf{e}_{uc}$,
(b) *the reachable internal forces* $\mathbf{e}_{ti}$,
(c) *the mechanism redundancy* $\mathbf{e}_{qr}$.

Proof. Under hypothesis H1, the couple $(\mathbf{A}, \mathbf{B}_\tau)$ is stabilisable (Proposition 1). Moreover, under H1 for the linearized system in Equation (4), it is a simple matter to verify that the system is detectable based on the informative output $\mathbf{y} = (\mathbf{q}^T, \mathbf{t}^T)^T$. Then, there exists an observer–based controller noninteracting with respect to the regulated outputs $\mathbf{e}_{uc}$, $\mathbf{e}_{ti}$ and $\mathbf{e}_{qr}$. Recall that following:

$$\mathbf{e}_{uc} = \mathbf{E}_{uc}\mathbf{x} = \begin{bmatrix} 0 & \mathbf{\Gamma}_{uc}^P & 0 & 0 \end{bmatrix}\mathbf{x}; \tag{15}$$

$$\mathbf{e}_{ti} = \mathbf{E}_{ti}\mathbf{x} = \begin{bmatrix} \mathbf{Q}_k & 0 & \mathbf{Q}_\beta & 0 \end{bmatrix}\mathbf{x}; \tag{16}$$

$$\mathbf{e}_{qr} = \mathbf{E}_{qr}\mathbf{x} = \begin{bmatrix} \mathbf{\Gamma}_r^P \mathbf{M}_h & 0 & 0 & 0 \end{bmatrix}\mathbf{x}. \tag{17}$$

Based on the theorem in [20], emerges that the outputs $\mathbf{e}_{uc}$, $\mathbf{e}_{ti}$ and $\mathbf{e}_{qr}$ are noninteracting if and only if

$$\begin{aligned} \mathbf{E}_{uc}\mathcal{R}_{\mathcal{K}_{uc}} &= \mathrm{im}(\mathbf{E}_{uc}), \\ \mathbf{E}_{ti}\mathcal{R}_{\mathcal{K}_{ti}} &= \mathrm{im}(\mathbf{E}_{ti}), \\ \mathbf{E}_{qr}\mathcal{R}_{\mathcal{K}_{qr}} &= \mathrm{im}(\mathbf{E}_{qr}), \end{aligned} \tag{18}$$

where

$$\begin{aligned} \mathcal{K}_{uc} &= \ker(\mathbf{E}_{ti}) \cap \ker(\mathbf{E}_{qr}), \\ \mathcal{K}_{ti} &= \ker(\mathbf{E}_{uc}) \cap \ker(\mathbf{E}_{qr}), \\ \mathcal{K}_{qr} &= \ker(\mathbf{E}_{ti}) \cap \ker(\mathbf{E}_{uc}). \end{aligned} \tag{19}$$

Here, $\mathcal{R}_{\mathcal{K}_{(\cdot)}}$ denotes the $\mathcal{K}_{(\cdot)}$-constrained controllability subspace, which is the subspace of all the points reachable through trajectories leaving the origin and belonging to $\mathcal{K}_{(\cdot)}$. We go on to prove the equalities in Equation (18). To simplify the proof, we replace the intersection subspaces in Equation (19) with suitable subspaces whose constrained controllability sets suffice for our purposes. The demonstration is provided in Appendixes A and B. □

5. Case Study

Considering theorem in [34] which states that for the linearized manipulation system under the hypothesis $\ker(\mathbf{G}^T) = \{\mathbf{0}\}$, it is possible to find a stabilizing state–feedback control law, $\tau = \mathbf{F}\mathbf{x} + \tau^*$ and an input partition $\tau^* = \mathbf{U}_{ti}\mathbf{u}_{ti} + \mathbf{U}_{uc}\mathbf{u}_{uc}$ which realize a noninteracting control of the reachable internal forces $\mathbf{t}_i$ and rigid–body object motions $\mathbf{u}_c$ as follows:

$$\begin{aligned} &(\mathbf{E}_{ti}, \mathbf{A} + \mathbf{B}_\tau\mathbf{F}, \mathbf{B}_\tau\mathbf{U}_{ti}), \\ &(\mathbf{E}_{uc}, \mathbf{A} + \mathbf{B}_\tau\mathbf{F}, \mathbf{B}_\tau\mathbf{U}_{uc}), \end{aligned} \tag{20}$$

it holds:

$$\begin{aligned} &\mathcal{R}_{ti} = min\mathcal{I}(\mathbf{A} + \mathbf{B}_\tau\mathbf{F}, \mathbf{B}_\tau\mathbf{U}_{ti}) \subseteq \ker(\mathbf{E}_{uc}), \\ &\mathbf{E}_{ti}\mathcal{R}_{ti} = \mathrm{im}(\mathbf{E}_{ti}), \end{aligned} \tag{21}$$

$$\begin{aligned} &\mathcal{R}_{uc} = min\mathcal{I}(\mathbf{A} + \mathbf{B}_\tau\mathbf{F}, \mathbf{B}_\tau\mathbf{U}_{uc}) \subseteq \ker(\mathbf{E}_{ti}), \\ &\mathbf{E}_{uc}\mathcal{R}_{uc} = \mathrm{im}(\mathbf{E}_{uc}). \end{aligned} \tag{22}$$

The partition matrices $\mathbf{U}_{uc}$ and $\mathbf{U}_{ti}$ are such that the following conditions are satisfied:

$$\begin{aligned} \mathrm{im}(\mathbf{B}_\tau\mathbf{U}_{uc}) &= \mathrm{im}(\mathbf{B}_\tau) \cap \mathcal{R}_{uc}, \\ \mathrm{im}(\mathbf{B}_\tau\mathbf{U}_{ti}) &= \mathrm{im}(\mathbf{B}_\tau) \cap \mathcal{R}_{ti}, \end{aligned} \tag{23}$$

and matrix $\mathbf{F}$ satisfies the following conditions:

$$\begin{aligned} (\mathbf{A} + \mathbf{B}_\tau\mathbf{F})\mathcal{R}_{uc} &\subseteq \mathcal{R}_{uc}, \\ (\mathbf{A} + \mathbf{B}_\tau\mathbf{F})\mathcal{R}_{ti} &\subseteq \mathcal{R}_{ti}. \end{aligned} \tag{24}$$

The decoupling controller is that sketched in Figure 1.

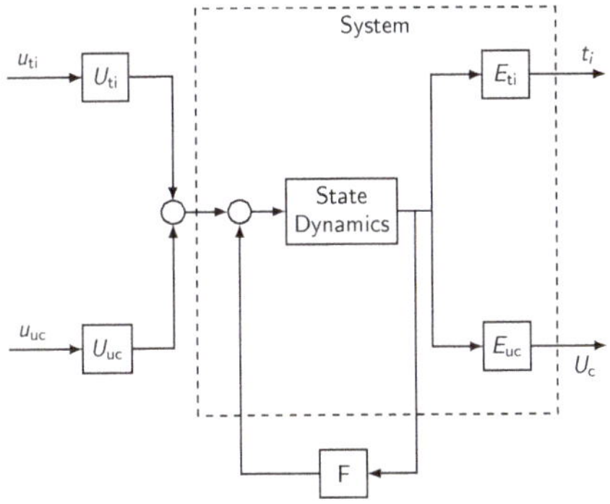

Figure 1. Force/motion decoupling controller.

In this section numerical results are reported for the simple defective gripper pictorially described in Figure 2.

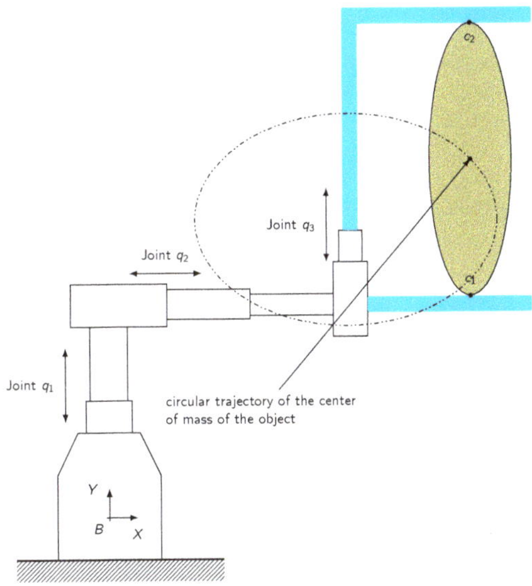

Figure 2. Planar 3–DoF's Cartesian manipulator. It exhibits a defective ($\ker(\mathbf{J}^T) = \mathbf{0}$) grasp.

It is a planar 3–DoF's Cartesian manipulator and has been chosen in order to show the effectiveness of previous results for industrial grippers. In the base frame B, the contact *centroids*, see [48], are $\mathbf{c}_1 = (2, 2)$, $\mathbf{c}_2 = (2, 3)$ and object center of mass is $c_b = (2, 2.5)$ while the transpose of the Jacobian and the grasp matrix assume the following values

$$\mathbf{J}^T = \begin{bmatrix} 0 & 1 & 0 & 0 \\ 1 & 0 & 1 & 0 \\ 0 & 0 & 0 & 1 \end{bmatrix}; \ \mathbf{G} = \begin{bmatrix} 1 & 0 & 1 & 0 \\ 0 & 1 & 0 & 1 \\ 0.5 & 0 & -0.5 & 0 \end{bmatrix}.$$

The inertia matrices of the object and manipulator along with stiffness and damping matrices at the contacts are assumed to be normalized to the identity matrix. The controlled outputs are (a) the projection $\mathbf{t}_i$ of the contact forces along the 1–dimensional subspace of reachable contact force $\mathrm{im}([0 1 0 -1]^T)$ and (b) the projection of the rigid–body motion in the 2–dimensional subspace of object motions $\mathrm{im}\begin{bmatrix} 1 & 0 \\ 0 & 1 \\ 0 & 0 \end{bmatrix}$ which, since $\mathbf{u} = [\delta\mathbf{x}\ \delta\mathbf{y}\ \delta\theta]^T$, corresponds to translations of the object.

General Procedure

The objective of the control is twofold. First, force and motion control must be decoupled, then the perfect tracking of desired trajectories $\mathbf{t}_{id}$ and $\mathbf{u}_{cd}$ can be achieved. The decoupling controller is pictorially described in Figure 1 and has been synthesized, according to Section 4, Equations (20), (23) and (24). State–feedback matrix $\mathbf{F}$ and input partition matrix $\mathbf{U} = \begin{bmatrix} \mathbf{U}_{ti} & \mathbf{U}_{uc} \end{bmatrix}$ are obtained respectively according to the following procedure:

- Item 1: Considering Equation (21), the reachable subspace of the internal contact force is calculated:
$$\mathcal{R}_{ti} = (\mathbf{E}_{ti}^T\mathbf{E}_{ti})^{-1}\mathbf{E}_{ti}^T \mathrm{im}(\mathbf{E}_{ti}). \quad (25)$$
- Item 2: Once $\mathcal{R}_{ti}$ is obtained, partition $\mathbf{U}_{ti}$ using (23), is obtained as follows:
$$\mathrm{im}(\mathbf{U}_{ti}) = (\mathbf{B}_\tau^T\mathbf{B}_\tau)^{-1}\mathbf{B}_\tau^T \mathrm{im}(\mathbf{B}_\tau) \cap \mathcal{R}_{ti}. \quad (26)$$
- Item 3: Considering Equation (21), matrix $\mathbf{F}_{ti}$ is calculated such that the following condition
is satisfied:
$$\mathcal{R}_{ti} = min\mathcal{I}(\mathbf{A} + \mathbf{B}_\tau\mathbf{F}_{ti}, \mathbf{B}_\tau\mathbf{U}_{ti}) \subseteq \ker(\mathbf{E}_{uc}). \quad (27)$$
- Item 4: Considering Equation (22), the reachable subspace of the internal coordinated movements is calculated as follows:
$$\mathcal{R}_{uc} = (\mathbf{E}_{uc}^T\mathbf{E}_{uc})^{-1}\mathbf{E}_{uc}^T \mathrm{im}(\mathbf{E}_{uc}). \quad (28)$$
- Item 5: Once $\mathcal{R}_{uc}$ is obtained, partition $\mathbf{U}_{uc}$ using (23), is obtained as follows:
$$\mathrm{im}(\mathbf{U}_{uc}) = (\mathbf{B}_\tau^T\mathbf{B}_\tau)^{-1}\mathbf{B}_\tau^T \mathrm{im}(\mathbf{B}_\tau) \cap \mathcal{R}_{uc}. \quad (29)$$
- Item 6: Considering Equation (22), matrix $\mathbf{F}_{uc}$ is calculated such that the following condition
is satisfied:
$$\mathcal{R}_{uc} = min\mathcal{I}(\mathbf{A}_{ti} + \mathbf{B}_\tau\mathbf{F}_{uc}, \mathbf{B}_\tau\mathbf{U}_{uc}) \subseteq \ker(\mathbf{E}_{ti}). \quad (30)$$
- Item 7 : The final state–feedback noninteracting matrix is the following:
$$\mathbf{F} = \mathbf{F}_{ti} + \mathbf{F}_{uc}. \quad (31)$$

End

Matrix $\mathbf{F}_{ti}$ realizes the invariance of the internal contact forces. In this context, it is possible to squeeze the object without moving it. Using matrix $\mathbf{F}_{ti}$ the following noninteracting transition matrix is obtained:
$$\mathbf{A}_{ti} = \mathbf{A} + \mathbf{B}_\tau\mathbf{F}_{ti}. \quad (32)$$
In the same way, matrix $\mathbf{A}_{ti}$, defined in Equation (32), together with matrix $\mathbf{F}_{uc}$ realize the invariance of the subspace of the object motions. In this context, it is possible to move the object without squeezing it. Thanks to matrix $\mathbf{F}_{uc}$ the following transition matrix which realizes the noninteracting control system is obtained:
$$\mathbf{A}_{dec} = \mathbf{A}_{ti} + \mathbf{B}_\tau\mathbf{F}_{uc}. \quad (33)$$
To go more in depth, Equation (31) is obtained from $\mathbf{A}_{ti} = \mathbf{A} + \mathbf{B}_\tau\mathbf{F}_{ti}$ and $\mathbf{A}_{dec} = \mathbf{A}_{ti} + \mathbf{B}_\tau\mathbf{F}_{uc}$. A combination of these two relations yields:
$$\mathbf{A}_{dec} = \mathbf{A} + \mathbf{B}_\tau(\mathbf{F}_{ti} + \mathbf{F}_{uc}), \quad (34)$$

and Equation (31) comes from Equation (34). Considering numerical data, the following matrices are obtained:

$$\mathbf{F} = \begin{bmatrix} -7 & 6.5 & -6 & -1 & -41 & 0 & -7.5 & -0.02 & -5.5 & -3 & -22 & 0 \\ 10 & -120 & 10 & -72 & 5 & 0 & 0.29 & -16 & 0.29 & 7.2 & -6.2 & 0 \\ -6.1 & 6.5 & -7.1 & -0.97 & -41 & 0 & -5.5 & -0.021 & -7.5 & -3.1 & -22 & 0 \end{bmatrix},$$

$$\mathbf{U}_{ti} = \begin{bmatrix} -0.707 \\ 0 \\ 0.707 \end{bmatrix}, \quad \mathbf{U}_{uc} = \begin{bmatrix} 0 & -0.707 \\ 1 & 0 \\ 0 & -0.707 \end{bmatrix}.$$

Considering an angular velocity of 0.1 rad/s and $\mathbf{u}_o$ of coordinates $(2.5, 1)$ as possible starting point, see Figure 2, the control task consists of maintaining the contact force to the constant value of $\mathbf{t}_o = [0; 1; 0; -1]^T$. The joint forces $\tau^* = \mathbf{U}_{ti}\mathbf{u}_{ti} + \mathbf{U}_{uc}\mathbf{u}_{uc}$ represents the control law which guarantees the perfect tracking of desired object motions with the desired internal force $\mathbf{t}_i$, see Figure 3. The required circular trajectory of the center of mass of the object is represented in Figure 2.

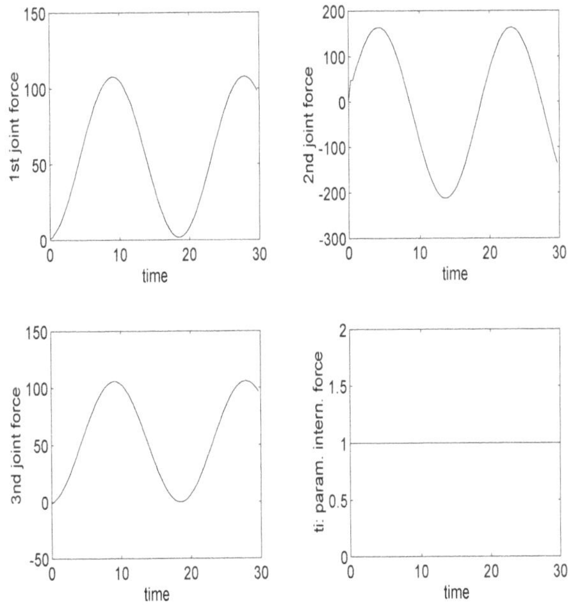

Figure 3. Internal force $\mathbf{t}_i$ perfectly tracks the constant internal force while the object center of mass perfectly tracks the unit circle as depicted in Figure 2.

It is worthwhile to remark that for a simple industrial gripper, under the reasonable hypothesis that the angular dynamics of the object can be disregarded, linearized dynamics represents the complete description of manipulation system dynamics.

6. Conclusions and Future Work

This paper considered the problem of noninteracting control in a linearized general manipulation systems. The geometric approach was used throughout the paper. The main results demonstrate that, in general, there always exists an observer-based control law that is noninteracting with respect to the aforementioned outputs. Note that the generality of our approach allows for the consideration of this force/motion noninteraction as a structural property of general manipulation systems. A possible future work can include the analysis of the robustness of the proposed theorem including also a robust design of the controller. Moreover, also a possible noninteraction realized using a feedback from the measured output and its corresponding robust control design should be taken into consideration.

Funding: This research received no external funding.

Data Availability Statement: The data used to support this research article are available upon request to the author.

Conflicts of Interest: The authors declare no conflict of interest.

Nomenclature

$\mathbf{q} \in \mathbb{R}^q$	vector of manipulator joint positions
$\tau \in \mathbb{R}^q$	vector of joint actuator torques
$\mathbf{u} \in \mathbb{R}^d$	vector locally describing the position and the orientation of a frame attached to the object
$\mathbf{w} \in \mathbb{R}^d$	vector of forces and torques resultant from external forces acting directly on the object
$\mathbf{t}_i$	force/torque interaction $\mathbf{t}_i$ at the i-th contact
$(\mathbf{K}_i, \mathbf{B}_i)$	lumped parameters of visco-elastic phenomena
$x(t)$	position of the armature
${}^h\mathbf{c}_i$	vector describing the posture of the contact frame on the manipulator
${}^o\mathbf{c}_i$	vector describing the posture of the contact frame on the object
$\mathbf{J}$	Jacobian matrix of the manipulator
$\mathbf{G}$	grasp matrix of the manipulator
$\mathbf{M}_h$ and $\mathbf{M}_o$	inertia symmetric and positive definite matrices
$\mathbf{Q}_h$ and $\mathbf{Q}_o$	terms including the velocity-dependent and gravity forces of the manipulator and object, respectively
$\mathbf{x}$	state space
$\mathbf{A}$	dynamic matrix
$\mathbf{B}_\tau$	input matrix
$\mathbf{B}_w$	disturbance matrix
$\mathcal{B}_\tau = \mathrm{im}\mathbf{B}_\tau$	image of matrix $\mathbf{B}_\tau$ (subspace spanned by the columns of matrix $\mathbf{B}_\tau$)
$\mathbf{y} = (\mathbf{q}^T, \mathbf{t}^T)^T$	informative output
$\mathbf{C}_\mathbf{t}$	output contact forces
$\mathbf{T}_c$	subspace of rigid body
$\mathbf{T}_a$	complementary subspace of rigid body
$\min\mathcal{I}(\mathbf{A}, \mathcal{B}) = \sum_{i=0}^{n-1} \mathbf{A}^i \mathrm{im}\mathbf{B}$	minimum $\mathbf{A}$-invariant subspace containing $\mathrm{im}(\mathbf{B})$ (controllable subspace)
$\max\mathcal{V}(\mathbf{A}, \mathcal{B}, \mathcal{C})$	maximum $(\mathbf{A}, \mathcal{B})$ controlled invariant subspace contained in $\mathcal{C}$
$\min\mathcal{S}(\mathbf{A}, \mathcal{C}, \mathcal{D})$	minimum $(\mathbf{A}, \mathcal{C})$ conditioned invariant subspace containing $\mathcal{D}$
$\mathcal{R}_{ti,\tau}$	subspace of reachable internal forces
$\mathbf{e}_{uc}$, $\mathbf{e}_{ti}$ and $\mathbf{e}_{qr}$	rigid–body object motions, reachable internal forces and mechanism redundancy outputs
$\mathcal{R}_{\mathcal{K}_{(\cdot)}}$: $\mathcal{K}_{(\cdot)}$-*constrained controllability subspace*	subspace of all the points reachable through trajectories leaving the origin and belonging to $\mathcal{K}_{(\cdot)}$.
$\mathrm{im}(\mathbf{S}_q)$	subspace of manipulator movements reachable from movement of the object
$\mathrm{im}(\mathbf{S}_u)$	subspace of object movements reachable from movement of the manipulator

Appendix A

This appendix outlines Theorem 2 in all its formal aspects.

Before proceeding to prove Theorem 2, certain additional notation and results are required.

Let us define the subspaces of the state spaces $\mathrm{im}(\mathbf{T}_r)$, $\mathrm{im}(\mathbf{T}_i)$, $\mathrm{im}(\mathbf{T}_h)$ and $\mathrm{im}(\mathbf{T}_d)$, where

$$\mathbf{T}_r = \begin{bmatrix} \boldsymbol{\Gamma}_r & \mathbf{0} \\ \mathbf{0} & \mathbf{0} \\ \mathbf{0} & \boldsymbol{\Gamma}_r \\ \mathbf{0} & \mathbf{0} \end{bmatrix}; \quad \mathbf{T}_i = \begin{bmatrix} \mathbf{0} & \mathbf{0} \\ \boldsymbol{\Gamma}_i & \mathbf{0} \\ \mathbf{0} & \mathbf{0} \\ \mathbf{0} & \boldsymbol{\Gamma}_i \end{bmatrix}, \quad \mathbf{T}_h = \begin{bmatrix} \boldsymbol{\Gamma}_h & \mathbf{0} \\ \mathbf{0} & \mathbf{0} \\ \mathbf{0} & \boldsymbol{\Gamma}_h \\ \mathbf{0} & \mathbf{0} \end{bmatrix}, \quad \mathbf{T}_d = \begin{bmatrix} \mathbf{0} & \mathbf{0} \\ \boldsymbol{\Gamma}_d & \mathbf{0} \\ \mathbf{0} & \mathbf{0} \\ \mathbf{0} & \boldsymbol{\Gamma}_d \end{bmatrix}. \tag{A1}$$

Here $\boldsymbol{\Gamma}_r$, $\boldsymbol{\Gamma}_i$, $\boldsymbol{\Gamma}_h$ and $\boldsymbol{\Gamma}_d$ are the basis matrices for the subspaces previously defined. In particular, $\boldsymbol{\Gamma}_r$ is a basis matrix for $\ker(\mathbf{J})$, and $\mathrm{im}(\boldsymbol{\Gamma}_i) = \mathbf{0}$ because our system is not indeterminate. Regarding the other subspaces, the following is established:

$$\begin{aligned} \boldsymbol{\Gamma}_h &= \text{b.m. of } \mathrm{im}(\mathbf{M}_h^{-1}\mathbf{J}^T) \cap \max\mathcal{I}(\mathbf{M}_h^{-1}\mathbf{J}^T\mathbf{K}\mathbf{J}, \ker(\mathbf{G}\mathbf{K}\mathbf{J})), \\ \boldsymbol{\Gamma}_d &= \text{b.m. of } \mathrm{im}(\mathbf{M}_o^{-1}\mathbf{G}) \cap \max\mathcal{I}(\mathbf{M}_o^{-1}\mathbf{G}\mathbf{K}\mathbf{G}^T, \ker(\mathbf{J}^T\mathbf{K}\mathbf{G}^T)). \end{aligned} \tag{A2}$$

Appendix A.1. Demonstration of the Noninteraction Theorem

We begin with the calculation of $\mathcal{R}_{\mathcal{K}_{(\cdot)}}$ and, in particular, with the calculation of the subspaces included in $\mathcal{R}_{\mathcal{K}_{(\cdot)}}$. In this appendix, $\ker(\mathbf{Q}_k)$ and $\ker(\mathbf{Q}_\beta)$ will be calculated. It may be useful to remark that $\ker(\mathbf{Q}_k) = \ker(\mathbf{Q}_\beta)$ under the hypothesis of proportionality outlined above.

$$\ker(\mathbf{E}_{ti}) = \ker\begin{bmatrix} \mathbf{Q}_k & \mathbf{0} & \mathbf{Q}_\beta & \mathbf{0} \end{bmatrix} \supseteq \mathrm{im}(\mathbf{L}_{ti}),$$

where

$$\mathbf{L}_{ti} = \begin{bmatrix} \boldsymbol{\Gamma}_{qr} & \mathbf{0} & \boldsymbol{\Gamma}_{qc} & \mathbf{0} & \boldsymbol{\Gamma}_{qc} & \mathbf{0} & \boldsymbol{\Gamma}_{qc} & \mathbf{0} & \boldsymbol{\Gamma}_{qc} & \mathbf{0} & .. \\ \mathbf{0} & \mathbf{0} & \boldsymbol{\Gamma}_{uc} & \mathbf{0} & -\boldsymbol{\Gamma}_{uc} & \mathbf{0} & -\mathbf{H}\boldsymbol{\Gamma}_{uc} & \mathbf{0} & -\mathbf{H}^2\boldsymbol{\Gamma}_{uc} & \mathbf{0} & .. \\ \mathbf{0} & \boldsymbol{\Gamma}_{qr} & \mathbf{0} & \boldsymbol{\Gamma}_{qc} & \mathbf{0} & \boldsymbol{\Gamma}_{qc} & \mathbf{0} & \boldsymbol{\Gamma}_{qc} & \mathbf{0} & \boldsymbol{\Gamma}_{qc} & .. \\ \mathbf{0} & \mathbf{0} & \mathbf{0} & \boldsymbol{\Gamma}_{uc} & \mathbf{0} & -\boldsymbol{\Gamma}_{uc} & \mathbf{0} & -\mathbf{H}\boldsymbol{\Gamma}_{uc} & \mathbf{0} & -\mathbf{H}^2\boldsymbol{\Gamma}_{uc} & .. \end{bmatrix}$$

and $\mathbf{H} = \mathbf{M}_o^{-1}\mathbf{G}\mathbf{B}\mathbf{G}^T$. It can be recalled that $\mathbf{B}$ is proportional to $\mathbf{K}$. In the same way

$$\ker(\mathbf{E}_{uc}) = \ker\begin{bmatrix} \mathbf{0} & \boldsymbol{\Gamma}_{uc}^T & \mathbf{0} & \mathbf{0} \end{bmatrix} \supseteq \mathrm{im}(\mathbf{L}_{uc}),$$

where

$$\mathbf{L}_{uc} = \begin{bmatrix} \boldsymbol{\Gamma}_{qr} & \mathbf{0} & \boldsymbol{\Gamma}_h & \mathbf{0} & \mathbf{S}_q & \mathbf{0} & \mathbf{0} & \mathbf{0} \\ \mathbf{0} & \mathbf{0} & \mathbf{0} & \mathbf{0} & \mathbf{0} & \mathbf{0} & \ker(\boldsymbol{\Gamma}_{uc}^T) \cap \mathbf{S}_u & \mathbf{0} \\ \mathbf{0} & \boldsymbol{\Gamma}_{qr} & \mathbf{0} & \boldsymbol{\Gamma}_h & \mathbf{0} & \mathbf{S}_q & \mathbf{0} & \mathbf{0} \\ \mathbf{0} & \mathbf{0} & \mathbf{0} & \mathbf{0} & \mathbf{0} & \mathbf{0} & \mathbf{0} & \ker(\boldsymbol{\Gamma}_{uc}^T) \cap \mathbf{S}_u \end{bmatrix}, \tag{A3}$$

with

$$\mathbf{S}_q = \min\mathcal{I}(\mathbf{M}_h^{-1}\mathbf{J}^T\mathbf{K}\mathbf{J}, \mathbf{M}_h^{-1}\mathbf{J}^T\mathbf{K}\mathbf{G}^T) \tag{A4}$$

and

$$\mathbf{S}_u = \min\mathcal{I}(\mathbf{M}_o^{-1}\mathbf{G}\mathbf{K}\mathbf{G}^T, \mathbf{M}_o^{-1}\mathbf{G}\mathbf{K}\mathbf{J}). \tag{A5}$$

Finally, it can be recalled that $\boldsymbol{\Gamma}_h$ is a basis matrix of

$$\mathrm{Im}(\mathbf{M}_h^{-1}\mathbf{J}^T) \cap \max\mathcal{I}(\mathbf{M}_h^{-1}\mathbf{J}^T\mathbf{K}\mathbf{J}, \ker(\mathbf{G}\mathbf{K}\mathbf{J})). \tag{A6}$$

Regarding the subspace

$$\ker(\mathbf{E}_{qr}) = \ker\begin{bmatrix} \boldsymbol{\Gamma}_r^P\mathbf{M}_h & \mathbf{0} & \mathbf{0} & \mathbf{0} \end{bmatrix},$$

it is very easy to check

$$\ker(\mathbf{E}_{qr}) \supseteq \mathrm{im}(\mathbf{L}_{qr})$$

with

$$\mathrm{im}(\mathbf{L}_{qr}) = \mathrm{im}\begin{bmatrix} \mathbf{T}_h & \mathbf{T}_c & \mathbf{T}_a & \mathbf{T}_d \end{bmatrix}, \tag{A7}$$

where the matrices $\mathbf{T}_h$, $\mathbf{T}_c$, $\mathbf{T}_a$ and $\mathbf{T}_d$ are previously defined. It is useful to note that $\mathrm{im}(\mathbf{L}_{qr})$ includes all subspaces except for the redundant movements subspace. We here begin the calculation with the intersection in Equation (19):

$$\mathrm{im}(\mathbf{L}_{uc}) \cap \mathrm{im}(\mathbf{L}_{qr}) \supseteq \mathrm{im}(\mathbf{B}_{ti}),$$

with

$$\mathbf{B}_{ti} = \begin{bmatrix} \mathbf{\Gamma}_h & 0 & \mathbf{S}_q & 0 & 0 & 0 \\ 0 & 0 & 0 & 0 & \ker(\mathbf{\Gamma}_{uc}^T) \cap \mathbf{S}_u & 0 \\ 0 & \mathbf{\Gamma}_h & 0 & \mathbf{S}_q & 0 & 0 \\ 0 & 0 & 0 & 0 & 0 & \ker(\mathbf{\Gamma}_{uc}^T) \cap \mathbf{S}_u \end{bmatrix}. \tag{A8}$$

Equation (A8) states a subspace included in the above intersection. In fact, by Equation (A7), the subspace $\mathrm{im}(\mathbf{L}_{qr})$ includes all subspaces except for the redundant movements. The following calculation is now given:

$$\mathrm{im}(\mathbf{L}_{ti}) \cap \mathrm{im}(\mathbf{L}_{qr}) \supseteq \mathrm{im}(\mathbf{B}_{uc})$$

where

$$\mathbf{B}_{uc} = \begin{bmatrix} \mathbf{\Gamma}_{qc} & 0 & \mathbf{\Gamma}_{qc} & 0 & \mathbf{\Gamma}_{qc} & 0 & \mathbf{\Gamma}_{qc} & 0 & .. \\ \mathbf{\Gamma}_{uc} & 0 & -\mathbf{\Gamma}_{uc} & 0 & -\mathbf{H}\mathbf{\Gamma}_{uc} & 0 & -\mathbf{H}^2\mathbf{\Gamma}_{uc} & 0 & .. \\ 0 & \mathbf{\Gamma}_{qc} & 0 & \mathbf{\Gamma}_{qc} & 0 & \mathbf{\Gamma}_{qc} & 0 & \mathbf{\Gamma}_{qc} & .. \\ 0 & \mathbf{\Gamma}_{uc} & 0 & -\mathbf{\Gamma}_{uc} & 0 & -\mathbf{H}\mathbf{\Gamma}_{uc} & 0 & -\mathbf{H}^2\mathbf{\Gamma}_{uc} & .. \end{bmatrix}. \tag{A9}$$

This intersection is a result of the definition of $\mathbf{S}_u$ (because $\mathbf{H}\mathbf{\Gamma}_{uc} \subseteq \mathbf{S}_u$) and of Lemma A5 reported in Appendix B.

Finally,

$$\mathrm{im}(\mathbf{L}_{ti}) \cap \mathrm{im}(\mathbf{L}_{uc}) = \mathrm{im}(\mathbf{B}_{qr}),$$

where

$$\mathbf{B}_{qr} = \begin{bmatrix} \mathbf{\Gamma}_r & 0 \\ 0 & 0 \\ 0 & \mathbf{\Gamma}_r \\ 0 & 0 \end{bmatrix}, \tag{A10}$$

which is very easy to verify.

It the following we will formally prove "part a)" of Theorem 2.

Proof. ("Part a)" of Theorem 2)

We now calculate

$$\max\mathcal{V}(\mathbf{A}, \mathrm{im}(\mathbf{B}_\tau), \mathrm{im}(\mathbf{B}_{uc})).$$

This calculation it is extremely elementary, and it follows that

$$\max\mathcal{V}(\mathbf{A}, \mathrm{im}(\mathbf{B}_\tau), \mathrm{im}(\mathbf{B}_{uc})) = \mathrm{im}(\mathbf{B}_{uc}).$$

Next, we calculate

$$\min\mathcal{S}(\mathbf{A}, \mathrm{im}(\mathbf{B}_{uc}), \mathrm{im}(\mathbf{B}_\tau)).$$

In general, the following holds, independent of the representative basis: It can be recalled that the subspaces are independent of every possible basis.

$$\mathcal{Z}_0 = \mathrm{im}(\mathbf{B}_\tau),$$
$$\mathcal{Z}_k = \mathcal{Z}_{k-1} + \mathbf{A}(\mathcal{Z}_{k-1} \cap \mathrm{im}(\mathbf{B}_{uc})),$$

where

$$\mathrm{im}(\mathbf{B}_{uc}) = \mathrm{im}(\mathbf{L}_{ti}) \cap \mathrm{im}(\mathbf{L}_{qr}),$$
$$\mathcal{Z}_1 = (\mathrm{im}(\mathbf{B}_\tau) + \mathbf{A}(\mathrm{im}(\mathbf{B}_\tau) \cap \mathrm{im}(\mathbf{B}_{uc}))),$$
$$\mathcal{Z}_1 = \left(\mathbf{B}_\tau + \mathbf{A}\left(\begin{bmatrix} 0 \\ 0 \\ \mathbf{\Gamma}_{qc} \\ 0 \end{bmatrix}\right)\right),$$
$$\mathcal{Z}_1 = \mathrm{im}\begin{bmatrix} \mathbf{\Gamma}_{qc} & 0 \\ 0 & 0 \\ \mathbf{M}_h^{-1}\mathbf{J}^T\mathbf{BJ\Gamma}_{qc} & \mathbf{M}_h^{-1} \\ \mathbf{M}_o^{-1}\mathbf{GBJ\Gamma}_{qc} & 0 \end{bmatrix},$$
$$\mathcal{Z}_1 = \mathrm{im}\begin{bmatrix} \mathbf{\Gamma}_{qc} & 0 \\ 0 & 0 \\ 0 & \mathbf{M}_h^{-1} \\ \mathbf{M}_o^{-1}\mathbf{GBJ\Gamma}_{qc} & 0 \end{bmatrix}.$$

We now calculate

$$\mathcal{Z}_2 = \mathrm{im}(\mathbf{B}_\tau) + \mathbf{A}(\mathcal{Z}_1 \cap \mathrm{im}(\mathbf{B}_{uc})).$$

Next, the following emerges:

$$\mathrm{im}\begin{bmatrix} \mathbf{\Gamma}_{qc} \\ 0 \\ 0 \\ \mathbf{M}_o^{-1}\mathbf{GBJ\Gamma}_{qc} \end{bmatrix} \subseteq \mathrm{im}(\mathbf{B}_{uc}),$$

this shows that the intersection can be separately calculated. In fact, it is not useless to remember that for the subspaces intersections it is possible to use the distributive property only if at least one of the subspaces is included in the other subspace.

Now, $\mathrm{im}\begin{bmatrix} 0 \\ 0 \\ \mathbf{M}_h^{-1} \\ 0 \end{bmatrix} \cap \mathrm{im}(\mathbf{B}_{uc}) = \mathrm{im}\begin{bmatrix} 0 \\ 0 \\ \mathbf{\Gamma}_{qc} \\ 0 \end{bmatrix}$ because $\mathbf{M}_h^{-1}$ has full rank. The other intersection $\forall\, a\; \exists\, b, c, d$ and e can be calculated as follows:

$$\begin{bmatrix} \mathbf{\Gamma}_{qc} \\ 0 \\ 0 \\ \mathbf{M}_o^{-1}\mathbf{GBJ\Gamma}_{qc} \end{bmatrix} a = \begin{bmatrix} \mathbf{\Gamma}_{qc} \\ \mathbf{\Gamma}_{uc} \\ 0 \\ 0 \end{bmatrix} b + \begin{bmatrix} 0 \\ 0 \\ \mathbf{\Gamma}_{qc} \\ \mathbf{\Gamma}_{uc} \end{bmatrix} c + \begin{bmatrix} \mathbf{\Gamma}_{qc} \\ -\mathbf{\Gamma}_{uc} \\ 0 \\ 0 \end{bmatrix} d + \begin{bmatrix} 0 \\ 0 \\ \mathbf{\Gamma}_{qc} \\ -\mathbf{\Gamma}_{uc} \end{bmatrix} e.$$

It is easy to see that $c = -e$ and $d = -b$. Thus,

$$\mathcal{Z}_1 \cap \mathrm{im}(\mathbf{B}_{uc}) = \mathrm{im}\begin{bmatrix} \mathbf{\Gamma}_{qc} & 0 \\ 0 & 0 \\ 0 & \mathbf{\Gamma}_{qc} \\ \mathbf{M}_o^{-1}\mathbf{GBJ\Gamma}_{qc} & 0 \end{bmatrix}.$$

Now,

$$\mathcal{Z}_2 = \mathrm{im}\begin{bmatrix} 0 & \mathbf{\Gamma}_{qc} & 0 \\ \mathbf{M}_o^{-1}\mathbf{GBJ\Gamma}_{qc} & 0 & 0 \\ 0 & 0 & \mathbf{M}_h^{-1} \\ \mathbf{X}_1 & \mathbf{X}_2 & 0 \end{bmatrix},$$

$$\mathrm{im}(\mathbf{X}_1) = \mathbf{M}_o^{-1}\mathbf{GKJ}\boldsymbol{\Gamma}_{qc} - \mathbf{M}_o^{-1}\mathbf{GBG}^T(\mathbf{M}_o^{-1}\mathbf{GBJ}\boldsymbol{\Gamma}_{qc})$$

and

$$\mathrm{im}(\mathbf{X}_2) = \mathbf{M}_o^{-1}\mathbf{GBJ}\boldsymbol{\Gamma}_{qc}.$$

This can be written as

$$\mathcal{Z}_2 = \mathrm{im}\begin{bmatrix} -\boldsymbol{\Gamma}_{qc} & \boldsymbol{\Gamma}_{qc} & 0 \\ \mathbf{M}_o^{-1}\mathbf{GBJ}\boldsymbol{\Gamma}_{qc} & 0 & 0 \\ 0 & 0 & \mathbf{M}_h^{-1} \\ -\mathbf{H}^2\boldsymbol{\Gamma}_{uc} & \mathbf{H}\boldsymbol{\Gamma}_{uc} & 0 \end{bmatrix}.$$

This calculation does not need to determine the minimum subspace exactly, and the resulting subspace is sufficint to test the condition: It is useful to remember that $\mathcal{R}_{B_{uc}} = \max\mathcal{V}(\mathbf{A}, \mathrm{im}(\mathbf{B}_\tau), \mathrm{im}(\mathbf{B}_{uc})) \cap \min\mathcal{S}(\mathbf{A}, \mathrm{im}(\mathbf{B}_{uc}), \mathrm{im}(\mathbf{B}_\tau))$ in our case we have that $\mathcal{R}_{B_{uc}} \supseteq \mathrm{im}(\mathbf{B}_{uc}) \cap \mathcal{Z}_2$, and $\mathcal{Z}_2 \subseteq \mathcal{Z}_\infty$ at the end $\mathrm{im}(\mathbf{B}_{uc}) = \max\mathcal{V}(\mathbf{A}, \mathrm{im}(\mathbf{B}_\tau), \mathrm{im}(\mathbf{B}_{uc}))$. It can then be concluded that if $\mathbf{E}_{uc}(\mathrm{im}(\mathbf{B}_{uc}) \cap \mathcal{Z}_2) = \mathrm{im}(\mathbf{E}_{uc})$, it will also be true that $\mathbf{E}_{uc}(\mathcal{R}_{B_{uc}}) = \mathrm{im}(\mathbf{E}_{uc})$.

$$\mathcal{R}_{\mathbf{B}_{uc}} \supseteq \max\mathcal{V}(\mathbf{A}, \mathrm{im}(\mathbf{B}_\tau), \mathrm{im}(\mathbf{B}_{uc})) \cap \mathcal{Z}_2.$$

This calculation is simple:

$$\mathcal{R}_{\mathbf{B}_{uc}} \supseteq \mathrm{im}\begin{bmatrix} \boldsymbol{\Gamma}_{qc} & -\boldsymbol{\Gamma}_{qc} & 0 \\ 0 & \mathbf{M}_o^{-1}\mathbf{GBJ}\boldsymbol{\Gamma}_{qc} & 0 \\ 0 & 0 & \boldsymbol{\Gamma}_{qc} \\ \mathbf{H}\boldsymbol{\Gamma}_{uc} & -\mathbf{H}^2\boldsymbol{\Gamma}_{uc} & 0 \end{bmatrix}.$$

To complete the proof, it remains to be verified that

$$\mathbf{E}_{uc}\mathcal{R}_{\mathbf{B}_{uc}} = \mathrm{im}(\mathbf{E}_{uc}).$$

This is trivial; in fact,

$$\mathbf{E}_{uc} = \boldsymbol{\Gamma}_{uc}(\boldsymbol{\Gamma}_{uc}^T\boldsymbol{\Gamma}_{uc})^{-1}\begin{bmatrix} 0 & \boldsymbol{\Gamma}_{uc}^T & 0 & 0 \end{bmatrix}.$$

The theorem is thus proved, and $\mathbf{J}\boldsymbol{\Gamma}_{qc} = \mathbf{G}^T\boldsymbol{\Gamma}_{uc}$ and $\boldsymbol{\Gamma}_{uc}^T\mathbf{M}_o^{-1}\mathbf{GBG}^T\boldsymbol{\Gamma}_{uc}$ have full rank. □

It the following we will formally prove "part b)" of Theorem 2.

Proof. ("Part b)" of Theorem 2)
We begin by calculating the controlled invariant subspace

$$\max\mathcal{V}(\mathbf{A}, \mathrm{im}(\mathbf{B}_\tau), \mathrm{im}(\mathbf{B}_{ti}))$$

and the conditioned invariant subspace

$$\min\mathcal{S}(\mathbf{A}, \mathrm{im}(\mathbf{B}_{ti}), \mathrm{im}(\mathbf{B}_\tau)).$$

To calculate the first of the above subspaces, is sufficient to find a subspace $\mathrm{im}(\mathbf{V})$ controlled invariant in $(\mathbf{A}, \mathbf{B}_\tau)$ and included in $\mathrm{im}(\mathbf{B}_{ti})$ with the following structure: To realise this kind of proof it is not necessary to find a controlled invariant subspace. Instead, it is sufficient to consider a subspace included in $\mathrm{im}(\mathbf{B}_{ti})$. This choice is helpfull in designing the contoller. In fact, this choice is constructive and the resolvent subspace must be controlled invariant.

$$\mathbf{V} = \begin{bmatrix} \boldsymbol{\Gamma}_h & 0 & \mathbf{S}_q\mathbf{Z} & 0 & \mathbf{M}_1 & 0 & 0 & 0 \\ 0 & 0 & 0 & \mathbf{M}_b & \mathbf{M}_2 & 0 & 0 & 0 \\ 0 & \boldsymbol{\Gamma}_h & 0 & 0 & 0 & \mathbf{S}_q\mathbf{Z} & 0 & \mathbf{M}_1 \\ 0 & 0 & 0 & 0 & 0 & 0 & \mathbf{M}_b & \mathbf{M}_2 \end{bmatrix}. \tag{A11}$$

Here, $\mathbf{Z}$ is such that

$$\mathrm{im}(\mathbf{M}_o^{-1}\mathbf{GKJS}_q\mathbf{Z}) = \mathrm{im}(\mathbf{M}_o^{-1}\mathbf{GKJS}_q) \cap \ker(\mathbf{\Gamma}_{uc}^T). \tag{A12}$$

The subspace im($\mathbf{V}$) must be controlled invariant, and it is necessary that:

$$\mathbf{A}\mathrm{im}(\mathbf{V}) \subseteq \mathrm{im}(\mathbf{V}) + \mathrm{im}(\mathbf{B}_\tau), \tag{A13}$$

$$\mathrm{im}(\mathbf{V}) \subseteq \mathrm{im}(\mathbf{B}_{ti}). \tag{A14}$$

Condition (A14) is satisfied if:

$$\mathrm{im}(\mathbf{M}_1) \subseteq \mathbf{S}_q, \tag{A15}$$

$$\mathrm{im}(\mathbf{M}_2) \subseteq \ker(\mathbf{\Gamma}_{uc}^T), \tag{A16}$$

$$\mathrm{im}(\mathbf{M}_b) \subseteq \ker(\mathbf{\Gamma}_{uc}^T). \tag{A17}$$

Furthermore, condition (A13) it is satisfied if

$$\mathbf{M}_o^{-1}\mathbf{GKJS}_q\mathbf{Z} \subseteq \mathrm{im}\begin{bmatrix} \mathbf{M}_b & \mathbf{M}_2 \end{bmatrix}, \tag{A18}$$

$$-\mathbf{M}_o^{-1}\mathbf{GKG}^T\mathbf{M}_b \subseteq \mathrm{im}\begin{bmatrix} \mathbf{M}_b & \mathbf{M}_2 \end{bmatrix}, \tag{A19}$$

$$\mathbf{M}_o^{-1}\mathbf{GKJM}_1 - \mathbf{M}_o^{-1}\mathbf{GKG}^T\mathbf{M}_2 \subseteq \mathrm{im}\begin{bmatrix} \mathbf{M}_b \, \mathbf{M}_2 \end{bmatrix}. \tag{A20}$$

In Appendix B, it is demonstrated that if $\mathbf{S}_q \neq \mathbf{0}$, then it is always possible to resolve the last three relations:

$$\mathrm{im}\begin{bmatrix} \mathbf{M}_b & \mathbf{M}_2 \end{bmatrix} \neq \mathbf{0}.$$

We now calculate

$$\min\mathcal{S}(\mathbf{A}, \mathrm{im}(\mathbf{B}_{ti}), \mathrm{im}(\mathbf{B}_\tau))$$

using the following algorithm:

$$\mathcal{Z}_0 = \mathrm{im}(\mathbf{B}_\tau),$$

$$\mathcal{Z}_k = \mathcal{Z}_{k-1} + \mathbf{A}(\mathcal{Z}_{k-1} \cap \mathrm{im}(\mathbf{B}_{ti})),$$

where $\mathbf{A}$ is defined as

$$\mathbf{A} = \begin{bmatrix} \mathbf{0} & \mathbf{0} & \mathbf{I}_q & \mathbf{0} \\ \mathbf{0} & \mathbf{0} & \mathbf{0} & \mathbf{I}_u \\ -\mathbf{M}_h^{-1}\mathbf{J}^T\mathbf{KJ} & \mathbf{M}_h^{-1}\mathbf{J}^T\mathbf{KG}^T & -\mathbf{M}_h^{-1}\mathbf{J}^T\mathbf{BJ} & \mathbf{M}_h^{-1}\mathbf{J}^T\mathbf{BG}^T \\ \mathbf{M}_o^{-1}\mathbf{GKJ} & -\mathbf{M}_o^{-1}\mathbf{GKG}^T & \mathbf{M}_o^{-1}\mathbf{GBJ} & -\mathbf{M}_o^{-1}\mathbf{GBG}^T \end{bmatrix}$$

and

$$\mathbf{B}_\tau = \begin{bmatrix} \mathbf{0} \\ \mathbf{0} \\ \mathbf{M}_h^{-1} \\ \mathbf{0} \end{bmatrix}.$$

Now, the following holds:

$$(\mathcal{Z}_0 \cap \mathrm{im}(\mathbf{B}_{ti})) = \mathrm{im}\begin{bmatrix} \mathbf{0} & \mathbf{0} \\ \mathbf{0} & \mathbf{0} \\ \mathbf{\Gamma}_h & \mathbf{S}_q\mathbf{Z} \\ \mathbf{0} & \mathbf{0} \end{bmatrix},$$

where $\mathbf{Z}$ is such that

$$\mathrm{im}(\mathbf{M}_o^{-1}\mathbf{GKJS}_q\mathbf{Z}) = \mathrm{im}(\mathbf{M}_o^{-1}\mathbf{GKJS}_q) \cap \ker(\mathbf{\Gamma}_{uc}^T). \tag{A21}$$

This involves

$$\mathcal{Z}_1 = \text{im}\begin{bmatrix} \mathbf{\Gamma}_h & \mathbf{S}_q\mathbf{Z} & 0 \\ 0 & 0 & 0 \\ 0 & 0 & \mathbf{M}_h^{-1} \\ 0 & \mathbf{M}_o^{-1}\mathbf{GBJS}_q\mathbf{Z} & 0 \end{bmatrix}.$$

This subspace is not conditioned invariant, but it will be sufficient for our demonstration. In fact, the dimension of this subspace is sufficient to guarantee the rank condition. Now, it is easy to show that

$$\mathcal{R}_{\mathbf{B}_{ti}} \supseteq \max\mathcal{V}(\mathbf{A}, \text{im}(\mathbf{B}_\tau), \text{im}(\mathbf{B}_{ti})) \cap \mathcal{Z}_1.$$

This calculation is simple, and

$$\mathcal{R}_{\mathbf{B_{ti}}} \supseteq \text{im}\begin{bmatrix} \mathbf{\Gamma}_h & 0 & \mathbf{S}_q\mathbf{Z} & 0 \\ 0 & 0 & 0 & 0 \\ 0 & \mathbf{\Gamma}_h & 0 & \mathbf{S}_q\mathbf{Z} \\ 0 & 0 & \mathbf{M}_o^{-1}\mathbf{GBJS}_q\mathbf{Z} & 0 \end{bmatrix}.$$

It is possible to verify that this subspace is not a self-hidden controlled invariant subspace in $\text{im}(\mathbf{B}_{ti})$ and that it is not controlled invariant, but this is not necessary for the present proof. It may be useful to recall that $\mathcal{R}_{B_{ti}} = \max\mathcal{V}(\mathbf{A}, \text{im}(\mathbf{B}_\tau), \text{im}(\mathbf{B}_{ti})) \cap \min\mathcal{S}(\mathbf{A}, \text{im}(\mathbf{B}_{ti}), \text{im}(\mathbf{B}_\tau))$ and that in the present case, we have $\mathcal{R}_{B_{ti}} \supseteq \text{im}(\mathbf{V}) \cap \mathcal{Z}_1$. Recalling that $\mathcal{Z}_1 \subseteq \mathcal{Z}_\infty$ by $\text{im}(\mathbf{V}) \subseteq \text{im}(\mathbf{B}_{ti})$, we can conclude that $\mathbf{E}_{ti}(\text{im}(\mathbf{V}) \cap \mathcal{Z}_1) = \text{im}(\mathbf{E}_{ti})$, $\mathbf{E}_{ti}(\mathcal{R}_{B_{ti}}) = \text{im}(\mathbf{E}_{ti})$ will also hold. To conclude it will be proved that

$$\mathbf{E}_{ti}\mathcal{R}_{B_{ti}} = \text{im}(\mathbf{E}_{ti}). \tag{A22}$$

It has been shown that the outputs were defined as follows:

$$\begin{aligned} &\mathbf{e}_{ti} = \mathbf{E}_{ti}\mathbf{x}, \\ &\text{with} \\ &\mathbf{E}_{ti} = (\mathbf{I} - \mathbf{KG}^T(\mathbf{GKG}^T)^{-1}\mathbf{G})\mathbf{C}_t = \begin{bmatrix} \mathbf{Q}_k & 0 & \mathbf{Q}_\beta & 0 \end{bmatrix}, \end{aligned} \tag{A23}$$

where

$$\mathbf{Q} = \mathbf{Q}_k = \mathbf{Q}_\beta = (\mathbf{I} - \mathbf{KG}^T(\mathbf{GKG}^T)^{-1}\mathbf{G})\mathbf{KJ}. \tag{A24}$$

We next calculate the null subspace of $\mathbf{Q}$.

Remark A1. *The null subspace of* $\mathbf{Q}$ *can be easily calculated. In fact,* $\ker(\mathbf{Q}) = \ker(\mathbf{J}) + \mathcal{V}$, *where* $\mathcal{V} = \{\mathbf{v} | \mathbf{KJv} \in \ker(\mathbf{I} - \mathbf{KG}^T(\mathbf{GKG}^T)^{-1}\mathbf{G}) = \text{im}(\mathbf{KG}^T), \mathbf{v} \notin \ker(\mathbf{J})\}$. *By Equation (10) it is easy to show that* $\mathcal{V} = \text{im}(\mathbf{\Gamma}_{qc})$ *and thus that:*

$$\ker(\mathbf{Q}) = \text{im}(\mathbf{\Gamma}_r) + \text{im}(\mathbf{\Gamma}_{qc}). \tag{A25}$$

□

The following two Lemmas demonstrate the useful property

$$\mathbf{E}_{ti}\mathcal{R}_{B_{ti}} = \text{im}(\mathbf{E}_{ti}), \tag{A26}$$

which is equivalent to

$$\text{im}(\mathbf{Q}\begin{bmatrix} \mathbf{\Gamma}_h & \mathbf{S}_q\mathbf{Z} \end{bmatrix}) = \text{im}(\mathbf{Q}).$$

To prove Equation (A26), we show that

$$\ker(\mathbf{Q}) \cap \text{im}\begin{bmatrix} \mathbf{\Gamma}_h & \mathbf{S}_q\mathbf{Z} \end{bmatrix} = \mathbf{0}, \tag{A27}$$

$$\text{rank}(\begin{bmatrix} \mathbf{\Gamma}_h & \mathbf{S}_q\mathbf{Z} \end{bmatrix}) = \text{rank}(\mathbf{Q}). \tag{A28}$$

Lemma A1.

$$\ker(\mathbf{Q}) \cap \mathrm{im}\begin{bmatrix} \boldsymbol{\Gamma}_h & \mathbf{S}_q\mathbf{Z} \end{bmatrix} = 0. \tag{A29}$$

Proof. If we begin from the previous Remark A1, Equation (A29) can be verified by determining whether the vectors **x**, **y**, **v** and **w** exist such that

$$\boldsymbol{\Gamma}_r\mathbf{x} + \boldsymbol{\Gamma}_{qc}\mathbf{y} = \boldsymbol{\Gamma}_h\mathbf{v} + \mathbf{S}_q\mathbf{Z}\mathbf{w}.$$

In fact, by $\mathrm{im}(\boldsymbol{\Gamma}_{qc})$, $\mathbf{im}(\boldsymbol{\Gamma}_h)$ and $\mathrm{im}(\mathbf{S}_q)$ are included in $\mathrm{im}(\mathbf{M}_h{}^{-1}\mathbf{J}^T)$, while $\mathrm{im}(\boldsymbol{\Gamma}_r)$ is not, because it is included in $\ker(\mathbf{J})$. In general, given a linear application **L**, $\mathrm{im}(\mathbf{L}^T) + \ker(\mathbf{L}) = \mathbf{I}$. Thus, the above equation can be written in the following form:

$$\boldsymbol{\Gamma}_{qc}\mathbf{y} = \boldsymbol{\Gamma}_h\mathbf{v} + \mathbf{S}_q\mathbf{Z}\mathbf{w}.$$

If this equation holds, then

$$\mathbf{M}_o^{-1}\mathbf{GKJ}\boldsymbol{\Gamma}_{qc}\mathbf{y} = \mathbf{M}_o^{-1}\mathbf{GKJ}\boldsymbol{\Gamma}_h\mathbf{v} + \mathbf{M}_o^{-1}\mathbf{GKJS}_q\mathbf{Z}\mathbf{w}.$$

By Equation (10) and $\boldsymbol{\Gamma}_h \subseteq \ker(\mathbf{GKJ})$ in Equation (A6), we can deduce the following:

$$\mathbf{M}_o^{-1}\mathbf{GKG}^T\boldsymbol{\Gamma}_{uc}\mathbf{y} = \mathbf{M}_o^{-1}\mathbf{GKJS}_q\mathbf{Z}\mathbf{w}.$$

However, this is never verified. Due to the choice of $\mathbf{Z}$ $\mathbf{M}_o^{-1}\mathbf{GKJS}_q\mathbf{Z} \subseteq \ker(\boldsymbol{\Gamma}_{uc}^T)$, while it will be easy to show that if $\mathbf{M}_o^{-1}\mathbf{GKG}^T\boldsymbol{\Gamma}_{uc} \subseteq \ker(\boldsymbol{\Gamma}_{uc}^T)$, then the matrix $\mathbf{M}_o^{-1}\mathbf{GKG}^T$ will be an orthogonal projector. However, this is not true because it is not in a projector form. It is useful to recall that given a subspace $\mathcal{L}$ with a basis matrix is **L**, $\ker(\mathbf{L}^T) = (\mathrm{im}(\mathbf{L}))^\perp$ and the ortogonal proiector is $(\mathbf{I} - \mathbf{L}(\mathbf{L}^\mathbf{T}\mathbf{L})^{-1}\mathbf{L}^\mathbf{T})$.

This demonstrates that condition (A29) is proven. □

Lemma A2.

$$\begin{aligned}\mathrm{rank}\begin{bmatrix} \boldsymbol{\Gamma}_h & \mathbf{S}_q\mathbf{Z} \end{bmatrix} &= \mathrm{rank}(\boldsymbol{\Gamma}_h) + \mathrm{rank}(\mathbf{S}_q\mathbf{Z}) \\ &= q - r - c.\end{aligned}$$

Proof. The first equality is derived from the null intersection between $\mathrm{im}(\boldsymbol{\Gamma}_h)$ and $\mathrm{im}(\mathbf{S}_q\mathbf{Z})$. In fact, by condition (A6), $\mathrm{im}(\boldsymbol{\Gamma}_h)$ is a subspace of $\max\mathcal{I}(\mathbf{M}_h^{-1}\mathbf{J}^T\mathbf{KJ}, \ker(\mathbf{GKJ}))$ orthogonal to $\mathrm{im}(\mathbf{M}_h^{-1}\mathbf{S}_q)$ in accordance with Equation (A4). The proof of the second equality of the lemma begins with the following consideretions. First,

$$\max\mathcal{I}(\mathbf{M}_h^{-1}\mathbf{J}^T\mathbf{KJ}, \ker(\mathbf{GKJ})) = \mathrm{im}(\mathbf{M}_h^{-1}\mathbf{S}_q)^\perp,$$

from which it follows that

$$\mathrm{im}(\mathbf{M}_h^{-1}\mathbf{J}^T) \subseteq \max\mathcal{I}(\mathbf{M}_h^{-1}\mathbf{J}^T\mathbf{KJ}, \ker(\mathbf{GKJ})) \oplus \mathrm{im}(\mathbf{M}_h^{-1}\mathbf{S}_q).$$

Now, by Equation (A4) $\mathrm{im}(\mathbf{M}_h^{-1}\mathbf{S}_q) \subseteq \mathrm{im}(\mathbf{M}_h^{-1}\mathbf{J}^T)$ and from the above inclusion and the definition of $\boldsymbol{\Gamma}_h$ in Equation (A6), it follows that

$$\begin{aligned}\mathrm{im}(\mathbf{M}_h^{-1}\mathbf{J}^T) &= \mathbf{M}_h^{-1}\mathbf{J}^T \cap \left(\max\mathcal{I}(\mathbf{M}_h^{-1}\mathbf{J}^T\mathbf{KJ}, \ker(\mathbf{GKJ})) \oplus \mathrm{im}(\mathbf{M}_h^{-1}\mathbf{S}_q)\right) \\ &= \left(\mathbf{M}_h^{-1}\mathbf{J}^T \cap \max\mathcal{I}(\mathbf{M}_h^{-1}\mathbf{J}^T\mathbf{KJ}, \ker(\mathbf{GKJ}))\right) \oplus \mathrm{im}(\mathbf{M}_h^{-1}\mathbf{S}_q) \\ &= \mathrm{im}(\boldsymbol{\Gamma}_h) \oplus \mathrm{im}(\mathbf{M}_h^{-1}\mathbf{S}_q).\end{aligned}$$

We thus obtain $\quad \mathrm{rank}(\boldsymbol{\Gamma}_h) + \mathrm{rank}(\mathbf{S}_q) = \mathrm{rank}(\mathbf{M}_h^{-1}\mathbf{J}^T) = \mathrm{rank}(\mathbf{J}) = q - r$

and

$$\mathsf{rank}(\mathbf{\Gamma}_h) = q - r - \mathsf{rank}(\mathbf{S}_q). \tag{A30}$$

The rank of $\mathbf{S}_q\mathbf{Z}$ remains to be calculated. Recalling that $\mathbf{S}_q$ and $\mathbf{Z}$ are basis matrices and that from Equation (A12) $\mathsf{rank}(\mathbf{Z}) \leq \mathsf{rank}(\mathbf{S}_q)$, it follows that

$$\mathsf{rank}(\mathbf{S}_q\mathbf{Z}) = \mathsf{rank}(\mathbf{Z}). \tag{A31}$$

By the introduction of $\mathbf{Z}$ in Equation (A12), it follows that

$$\mathsf{rank}(\mathbf{Z}) = \mathsf{rank}(\mathbf{S}_q) - \mathsf{rank}(\mathbf{Z}^{\perp}). \tag{A32}$$

Here, $\mathsf{rank}(\mathbf{S}_q)$ is the number of components of $\mathbf{z} \in \mathbf{Z}$. The last part of this demonstration consists of estimating $\mathsf{rank}(\mathbf{Z}^{\perp})$, which, by Equation (A12) is

$$\mathsf{rank}(\mathbf{Z}^{\perp}) = \mathsf{rank}(\mathbf{S}_q^T\mathbf{J}^T\mathbf{K}\mathbf{G}^T\mathbf{M}_o^{-1}\mathbf{\Gamma}_{uc}).$$

By Equation (A4), it is easy to show that $\mathsf{ker}(\mathbf{S}_q^T) \subseteq \mathsf{ker}(\mathbf{GKJ})$. Thus, $\mathsf{ker}(\mathbf{S}_q^T) \cap \mathsf{im}(\mathbf{J}^T\mathbf{K}\mathbf{G}^T) = \mathbf{0}$, and

$$\mathsf{rank}(\mathbf{Z}^{\perp}) = \mathsf{rank}(\mathbf{J}^T\mathbf{K}\mathbf{G}^T\mathbf{M}_o^{-1}\mathbf{\Gamma}_{uc}). \tag{A33}$$

Now, we prove that

$$\mathsf{rank}(\mathbf{Z}^{\perp}) = \mathsf{rank}(\mathbf{J}^T\mathbf{K}\mathbf{G}^T\mathbf{M}_o^{-1}\mathbf{\Gamma}_{uc}) = \mathsf{rank}(\mathbf{\Gamma}_{uc}) = c. \tag{A34}$$

If we transpose Equation (A33), the following hold:

$$\mathsf{rank}(\mathbf{Z}^{\perp}) = \mathsf{rank}(\mathbf{\Gamma}_{uc}^T\mathbf{M}_o^{-1}\mathbf{GKJ}).$$

By Equation (10)

$$\mathsf{rank}(\mathbf{Z}^{\perp}) = \mathsf{rank}(\mathbf{\Gamma}_{uc}^T\mathbf{M}_o^{-1}\mathbf{GK}\mathbf{G}^T\mathbf{\Gamma}_{uc}) = \mathsf{rank}(\mathbf{\Gamma}_{uc}),$$

where the last equality follows because the matrix $\mathbf{\Gamma}_{uc}^T\mathbf{M}_o^{-1}\mathbf{GK}\mathbf{G}^T\mathbf{\Gamma}_{uc}$ has full rank. Finally, by Equations (A31), (A32) and (A34), it can be concluded that

$$\mathsf{rank}(\mathbf{S}_q\mathbf{Z}) = \mathsf{rank}(\mathbf{S}_q) - c.$$

Comparing this last result with Equation (A30)

$$\mathsf{rank}\begin{bmatrix} \mathbf{\Gamma}_h & \mathbf{S}_q\mathbf{Z} \end{bmatrix} = q - r - c.$$

□

Remark A2. *The Equation (A28) was proved only if in the case of kinematic defectivity* $(\mathsf{ker}(\mathbf{J}^T)) \neq \mathbf{0})$, *i.e., with* $\mathbf{J} \in \Re^{(t \times q)}$, *thus only in the case of* $t > q$. *It is easy to prove that* $t \leq q$ *is a trivial extension. Let* r *and* c *be the ranks of the matrices* $\mathbf{\Gamma}_r$ *and* $\mathbf{\Gamma}_{uc}$, *respectively. It follows that* $\mathsf{rank}(\mathbf{J}) = q - r$. *By Lemma A2,* $\mathsf{rank}\begin{bmatrix} \mathbf{\Gamma}_h & \mathbf{S}_q\mathbf{Z} \end{bmatrix} = \mathsf{rank}(\mathbf{\Gamma}_h) + \mathsf{rank}(\mathbf{S}_q\mathbf{Z}) = q - r - c$. *In conclusion, Equation (A28) demonstrates that*

$$\mathsf{rank}(\mathbf{Q}) = q - r - c,$$

which follows trivially from Equation (A25). In fact, $\mathsf{rank}(\mathbf{Q}) = \mathsf{rank}(\mathbf{Q}^T) = q - \mathsf{rank}(\mathsf{ker}(\mathbf{Q})) = q - (r + c)$.

It the following we will formally prove "part c)" of Theorem 2.

Proof. ("Part c)" of Theorem 2)

It is possible to show the following:

$$\mathrm{im}(\mathbf{L}_{ti}) \cap \mathrm{im}(\mathbf{L}_{uc}) \supseteq \mathrm{im}\begin{bmatrix} \mathbf{\Gamma}_r & 0 \\ 0 & 0 \\ 0 & \mathbf{\Gamma}_r \\ 0 & 0 \end{bmatrix}.$$

This subspace is $\mathbf{A}$-invariant and thus guarantees the necessary conditions. By

$$\mathbf{E}_{qr} = \mathbf{\Gamma}_r(\mathbf{\Gamma}_r^T\mathbf{\Gamma}_r)^{-1}\begin{bmatrix} \mathbf{\Gamma}_r^T\mathbf{M}_h & 0 & 0 & 0 \end{bmatrix},$$

even the rank condition is invariant. □

Appendix B

In this appendix, we provide several technical results useful for the calculations given in Appendix A.

Lemma A3. *Let $\mathbf{S}_q$ and $\mathbf{S}_u$ be basis matrices of $\min\mathcal{I}(\mathbf{M}_h^{-1}\mathbf{J}^T\mathbf{KJ}, \mathbf{M}_h^{-1}\mathbf{J}^T\mathbf{KG}^T)$ and $\min\mathcal{I}(\mathbf{M}_o^{-1}\mathbf{GKG}^T, \mathbf{M}_o^{-1}\mathbf{GKJ})$, respectively. Then,*

$$\mathbf{M}_o^{-1}\mathbf{GBJS}_q \subseteq \mathbf{S}_u.$$

Proof. Being

$$\mathbf{S}_q = \min\mathcal{I}(\mathbf{M}_h^{-1}\mathbf{J}^T\mathbf{KJ}, \mathbf{M}_h^{-1}\mathbf{J}^T\mathbf{KG}^T), \text{and} \tag{A35}$$

$$\mathbf{S}_u = \min\mathcal{I}(\mathbf{M}_o^{-1}\mathbf{GKG}^T, \mathbf{M}_o^{-1}\mathbf{GKJ}). \tag{A36}$$

Now,

$$\mathbf{S}_u^\perp = \max\mathcal{I}(\mathbf{GKG}^T\mathbf{M}_o^{-1}, \ker(\mathbf{J}^T\mathbf{KG}^T\mathbf{M}_o^{-1})), \tag{A37}$$

$$(\mathbf{S}_u)^\perp \subseteq \ker(\mathbf{J}^T\mathbf{KG}^T\mathbf{M}_o^{-1}), \tag{A38}$$

$$(\mathbf{M}_o^{-1}\mathbf{GBJS}_q)^\perp = \ker(\mathbf{S}_q^T\mathbf{J}^T\mathbf{BG}^T\mathbf{M}_o^{-1}) \supseteq (\ker(\mathbf{J}^T\mathbf{BG}^T\mathbf{M}_o^{-1}). \tag{A39}$$

Thus,

$$(\mathbf{M}_o^{-1}\mathbf{GBJS}_q)^\perp \supseteq \mathbf{S}_u^\perp, \tag{A40}$$

and finally,

$$\mathbf{M}_o^{-1}\mathbf{GBJS}_q \subseteq \mathbf{S}_u. \tag{A41}$$

□

Lemma A4. *Let $\mathbf{S}_q$ and $\mathbf{S}_u$ above be defined. It follows that*

$$\mathbf{M}_o^{-1}\mathbf{GBJS}_q \cap \ker(\mathbf{\Gamma}_{uc}^T) \cap \mathbf{S}_u = \mathbf{M}_o^{-1}\mathbf{GBJS}_q\mathbf{Z}.$$

Proof. Recalling the definition of $\mathbf{Z}$, it follows that $\mathrm{im}(\mathbf{M}_o^{-1}\mathbf{GKJS}_q\mathbf{Z}) = \mathrm{im}(\mathbf{M}_o^{-1}\mathbf{GKJS}_q) \cap \ker(\mathbf{\Gamma}_{uc}^T)$, although it remains to be demonstrated that $\mathbf{M}_o^{-1}\mathbf{GBJS}_q\mathbf{Z} \cap \mathbf{S}_u = \mathbf{M}_o^{-1}\mathbf{GKJS}_q\mathbf{Z}$. This follows immediately by the previous Lemma A3. □

Lemma A5. *The complementary subspace* $\mathrm{im}(\mathbf{T}_a)$ *was defined in [49] as the* deforming motions subspace. *It is possible to choose a complementary subspace* $\mathrm{im}(\mathbf{T}_a)$ *such that*

$$\mathrm{im}\begin{bmatrix} \boldsymbol{\Gamma}_{qc} & \mathbf{0} \\ -\boldsymbol{\Gamma}_{uc} & \mathbf{0} \\ \mathbf{0} & \boldsymbol{\Gamma}_{qc} \\ \mathbf{0} & -\boldsymbol{\Gamma}_{uc} \end{bmatrix} \subseteq \mathrm{im}(\mathbf{T}_a).$$

This part of the appendix discusses, through the following lemma, three necessary conditions to obtain the controlled invariant subspace $\mathrm{im}(\mathbf{V})$ as pointed out in Appendix A.

Lemma A6.

$$\mathbf{M}_o^{-1}\mathbf{GKJS}_q\mathbf{Z} \subseteq \mathrm{im}\big[\, \mathbf{M}_b\ \mathbf{M}_2 \,\big],$$
$$-\mathbf{M}_o^{-1}\mathbf{GKG}^T\mathbf{M}_b \subseteq \mathrm{im}\big[\, \mathbf{M}_b\ \mathbf{M}_2 \,\big],$$
$$\mathbf{M}_o^{-1}\mathbf{GKJM}_1 - \mathbf{M}_o^{-1}\mathbf{GKG}^T\mathbf{M}_2 \subseteq \mathrm{im}\big[\, \mathbf{M}_b\ \mathbf{M}_2 \,\big].$$

Proof. The proof starts distinguishing three possible cases depending on $\ker(\boldsymbol{\Gamma}_{uc}^T)$.

Case 1:

$\ker(\boldsymbol{\Gamma}_{uc})$ is $\mathbf{M}_o^{-1}\mathbf{GKG}^T$-invariant.

This is the simplest case. In fact, if we take $\mathbf{M}_b = \ker(\boldsymbol{\Gamma}_{uc}^T)$ and $\mathbf{M}_2 = \mathbf{0}$ such that the first and the second equations are satisfied automatically, the third will be satisfied for $\mathbf{M}_1 = \mathbf{0}$.

Case 2:

$\ker(\boldsymbol{\Gamma}_{uc}^T) \nsupseteq \mathbf{M}_o^{-1}\mathbf{GKG}^T\ker(\boldsymbol{\Gamma}_{uc}^T)$ and $\ker(\boldsymbol{\Gamma}_{uc}^T) \cap \mathbf{M}_o^{-1}\mathbf{GKG}^T\ker(\boldsymbol{\Gamma}_{uc}^T) \neq \mathbf{0}$.

In this case the second equation can be verified by the following:

$$\mathbf{M}_2 = \ker(\boldsymbol{\Gamma}_{uc}^T),$$
$$\mathbf{M}_b : \mathbf{M}_o^{-1}\mathbf{GKG}^T\mathbf{M}_b =$$
$$\ker(\boldsymbol{\Gamma}_{uc}^T) \cap \mathbf{M}_o^{-1}\mathbf{GKG}^T\ker(\boldsymbol{\Gamma}_{uc}^T).$$

Now, the first equation is trivially verified, while the third will be verified if

$$\mathbf{M}_o^{-1}\mathbf{GKG}^T\ker(\boldsymbol{\Gamma}_{uc}^T) \subseteq \mathrm{im}\big[\ \mathbf{M}_o^{-1}\mathbf{GKJS}_q \quad \ker(\boldsymbol{\Gamma}_{uc}^T)\ \big].$$

We will demonstrate that this condition is always verified.

Case 3:

The last case to analyse is that in which

$$\ker(\boldsymbol{\Gamma}_{uc}^T) \cap \mathbf{M}_o^{-1}\mathbf{GKG}^T\ker(\boldsymbol{\Gamma}_{uc}^T) = \mathbf{0}.$$

Under this condition, the second equation is satisfied only with $\mathbf{M}_b = \mathbf{0}$. To satisfy this, it is sufficient to set $\mathrm{im}(\mathbf{M}_2) = \ker(\boldsymbol{\Gamma}_{uc}^T)$. This implies the same condition of the second case and thus involves the following condition:

$$\mathbf{M}_o^{-1}\mathbf{GKG}^T\ker(\boldsymbol{\Gamma}_{uc}^T) \subseteq \mathrm{im}\big[\ \mathbf{M}_o^{-1}\mathbf{GKJS}_q \quad \ker(\boldsymbol{\Gamma}_{uc}^T)\ \big].$$

□

The following lemma shows how this condition is verified.

Lemma A7. *If* $\mathbf{S}_q \neq \mathbf{0}$, *then the matrix*

$$\big[\ \mathbf{M}_o^{-1}\mathbf{GKJS}_q \quad \ker(\boldsymbol{\Gamma}_{uc}^T)\ \big]$$

is a basis matrix of the subspace $\Re^d$, *where d is the dimension of the physical space.*

Proof. $\mathbf{S}_q = \min\mathcal{I}(\mathbf{M}_h^{-1}\mathbf{J}^T\mathbf{KJ}, \mathbf{M}_h^{-1}\mathbf{J}^T\mathbf{KG}^T)$ and the $\mathbf{M}_h^{-1}$ is positive definite:

$$\mathsf{im}(\mathbf{M}_o^{-1}\mathbf{GKJ}) \supseteq \mathsf{im}(\mathbf{M}_o^{-1}\mathbf{GKJS}_q) \supseteq \mathsf{im}(\mathbf{M}_o^{-1}\mathbf{GKJM}_h^{-1}\mathbf{J}^T\mathbf{KG}^T) = \mathsf{im}(\mathbf{M}_o^{-1}\mathbf{GKJ}).$$

This implies that

$$\mathsf{im}(\mathbf{M}_o^{-1}\mathbf{GKJS}_q) = \mathsf{im}(\mathbf{M}_o^{-1}\mathbf{GKJ}).$$

It is now easy to prove that

$$\mathbb{R}^d \supseteq \mathsf{im}\begin{bmatrix} \mathbf{M}_o^{-1}\mathbf{GKJ} & \mathsf{ker}(\mathbf{\Gamma}_{uc}^T) \end{bmatrix} \supseteq \mathsf{im}\begin{bmatrix} \mathbf{M}_o^{-1}\mathbf{GKJ\Gamma}_{qc} & \mathsf{ker}(\mathbf{\Gamma}_{uc}^T) \end{bmatrix},$$
$$\mathsf{im}\begin{bmatrix} \mathbf{M}_o^{-1}\mathbf{GKG}^T\mathbf{\Gamma}_{uc} & \mathsf{ker}(\mathbf{\Gamma}_{uc}^T) \end{bmatrix} = \mathbb{R}^d$$

and

$$\mathsf{rank}(\mathbf{M}_o^{-1}\mathbf{GKG}^T\mathbf{\Gamma}_{uc}) = \mathsf{rank}(\mathbf{\Gamma}_{uc}),$$

because $\mathbf{M}_o^{-1}\mathbf{GKG}^T$ has a null subspace equal to zero. □

References

1. Trinkle, J. On the stability and instantaneous velocity of grasped frictionless objects. *IEEE Trans. Robot. Autom.* **1992**, *8*, 477–880. [CrossRef]
2. Bicchi, A.; Prattichizzo, D. Controllability of whole–Arm manipulation. In Proceedings of the 33rd CDC, Lake Buena Vista, FL, USA, 14–16 December 1994.
3. Howard, W.S.; Kumar, V. Stability of planar grasps. In Proceedings of the 1994 IEEE International Conference on Robotics and Automation, San Diego, CA, USA, 8–13 May 1994.
4. Seto, D.; Baillieul, J. Control problems in super-articulated mechanical systems. *IEEE Trans. Autom. Control* **1994**, *39*, 2442–2453. [CrossRef]
5. Ismaeil, O.; Ellis, R. Grasping using the whole finger. In Proceedings of the 1994 IEEE International Conference on Robotics and Automation, San Diego, CA, USA, 8–13 May 1994.
6. Melchiorri, C. Static force analysis for general cooperating manipulators. In Proceedings of the 1994 IEEE International Conference on Robotics and Automation, San Diego, CA, USA, 8–13 May 1994.
7. Petreschi, P.; Prattichizzo, D.; Bicchi, A. Articulated structures with tendon actuation for whole-limb manipulation. In Proceedings of the 1994 IEEE International Conference on Robotics and Automation, San Diego, CA, USA, 8–13 May 1994.
8. Bicchi, A.; Prattichizzo, D. A standard form for the dynamics of general manipulation systems. In Proceedings of the 1995 IEEE International Conference on Robotics and Automation, Nagoya, Japan, 21–27 May 1995.
9. Bicchi, A.; Prattichizzo, D.; Sastry, S. Planning motions of rolling surfaces. In Proceedings of the 1995 34th IEEE Conference on Decision and Control, New Orleans, LA, USA, 13–15 December 1995; In Invited Session: Discontinuities Singularities and New Geometric Structures in Control Design of Nonlinear Systems.
10. Le Tien, L.; Schaffer, A.A.; Hirzinger, G. MIMO State Feedback Controller for a Flexible Joint Robot with Strong Joint Coupling. In Proceedings of the 2007 IEEE International Conference on Robotics and Automation, Roma, Italy, 10–14 April 2007; pp. 3824–3830. [CrossRef]
11. Albu-Schäffer, A.; Hirzinger, G. A globally stable state feedback controller for flexible joint robots. *Adv. Robot.* **2001**, *15*, 799–814. [CrossRef]
12. Skogestad, S.; Postlethwaite, I. *Multivariable Feedback Control*, 2nd ed.; Wiley-Blackwell: Hoboken, NJ, USA, 2005.
13. Hedman, M.; Mercorelli, P. FFTSMC with Optimal Reference Trajectory Generated by MPC in Robust Robotino Motion Planning with Saturating Inputs. In Proceedings of the 2021 American Control Conference (ACC), New Orleans, LA, USA, 25–28 May 2021; pp. 1470–1477. [CrossRef]
14. Elsisi, M. Optimal design of nonlinear model predictive controller based on new modified multitracker optimization algorithm. *Int. J. Intell. Syst.* **2020**, *35*, 1857–1878. [CrossRef]
15. Elsisi, M.; Ebrahim, M.A. Optimal design of low computational burden model predictive control based on SSDA towards autonomous vehicle under vision dynamics. *Int. J. Intell. Syst.* **2021**, *36*, 6968–6987. [CrossRef]
16. Ali, M.N.; Soliman, M.; Mahmoud, K.; Guerrero, J.M.; Lehtonen, M.; Darwish, M.M.F. Resilient Design of Robust Multi-Objectives PID Controllers for Automatic Voltage Regulators: D-Decomposition Approach. *IEEE Access* **2021**, *9*, 106589–106605. [CrossRef]
17. Ayman, M.; Soliman, M. Decentralised design of robust multi-objective PSSs: *D*-decomposition approach. *IET Gener. Transm. Distrib.* **2020**, *14*, 5392–5406. [CrossRef]
18. Ayman, M.; Soliman, M. Robust multi-objective PSSs design via complex Kharitonov's theorem. *Eur. J. Control* **2021**, *58*, 131–142. [CrossRef]
19. Basile, G.; Marro, G. A state space approach to non-interacting controls. *Ric. Autom.* **1970**, *1*, 68–77.
20. Basile, G.; Marro, G. *Controlled and Conditioned Invariants in Linear System Theory*; Prentice Hall: Hoboken, NJ, USA, 1992.

21. Wonham, W.; Morse, A. Decoupling and pole assignment in linear multivariable systems: A geometric approach. *SIAM J. Control* **1970**, *8*, 1–18. [CrossRef]
22. Morse, A.; Wonham, W. Decoupling and pole assignment by dynamic compensation. *SIAM J. Control* **1970**, *8*, 317–337. [CrossRef]
23. Cutkosky, M.; Kao, I. Computing and controlling the compliance of a robotic hand. *TransRA* **1989**, *5*, 151–165. [CrossRef]
24. Montana, D.J. The Kinematics of Contact Grasp. *IJRR* **1988**, *7*, 17–32. [CrossRef]
25. Prattichizzo, D.; Bicchi, A. Consistent task specification for manipulation systems with general kinematics. *ASME J. Dyn. Syst. Meas. Control* **1997**, *119*, 760–767. [CrossRef]
26. Murray, R.; Li, Z.; Sastry, S. *A Mathematical Introduction to Robotic Manipulation*; CRC: Boca Raton, FL, USA, 1994.
27. Wonham, W. *Linear Multivariable Control: A Geometric Approach*; Springer: New York, NY, USA, 1979.
28. Mercorelli, P.; Sergiyenko, O.; Hernandez-Balbuena, D.; Rodriguez-Quiñonez, J.; Flores-Fuentes, W.; Basaca-Preciado, L. Some Model Properties to Control a Permanent Magnet Machine Using a Controlled Invariant Subspace. *IFAC-PapersOnLine* **2015**, *48*, 366–371. [CrossRef]
29. Mercorelli, P.; Haus, B.; Zattoni, E.; Aschemann, H.; Ferrara, A. Robust Current Decoupling in a Permanent Magnet Motor Combining a Geometric Method and SMC. In Proceedings of the 2018 IEEE Conference on Control Technology and Applications (CCTA), Copenhagen, Denmark, 21–24 August 2018; pp. 939–944. [CrossRef]
30. Mercorelli, P. A Geometric Approach to the Decoupling Control and to Speed up the Dynamics of a General Rigid Body Manipulation System. In *Applied Mechanics and Materials*; Trans Tech Publications Ltd.: Baech, Switzerland, 2014; Volume 534, pp. 93–103. [CrossRef]
31. Mercorelli, P.; Prattichizzo, D. A geometric procedure for robust decoupling control of contact forces in robotic manipulation. *Kybernetika* **2003**, *39*, 433–445.
32. Mercorelli, P. Robust decoupling through algebraic output feedback in manipulation systems. *Kybernetika* **2010**, *46*, 850–869.
33. Mercorelli, P. Geometric structures for the parameterization of non-interacting dynamics for multi-body mechanisms. *Int. J. Pure Appl. Math.* **2010**, *59*, 257–273.
34. Mercorelli, P. A geometric algorithm for the output functional controllability in general manipulation systems and mechanisms. *Kybernetika* **2012**, *48*, 1266–1288.
35. Prattichizzo, D.; Mercorelli, P. On some geometric control properties of active suspensions systems. *Kybernetika* **2000**, *36*, 549–570.
36. Chu, D.; Mehrmann, V. Disturbance Decoupling for Descriptor Systems by State Feedback. *SIAM J. Control Optim.* **2000**, *38*, 1830–1858. [CrossRef]
37. Chu, D.; Mehrmann, V. Disturbance decoupling for linear time-invariant systems: A matrix pencil approach. *IEEE Trans. Autom. Control* **2001**, *46*, 802–808. [CrossRef]
38. Wang, Y.; Zhu, S.; Cheng, Z. A remark on "disturbance decoupling for linear time-invariant systems: A matrix pencil approach". *IEEE Trans. Autom. Control* **2004**, *49*, 857–858. [CrossRef]
39. Tosi, N.; David, O.; Bruyninckx, H. DOF Decoupling Task Graph Model: Reducing the Complexity of Touch-Based Active Sensing. *Robotics* **2015**, *4*, 141–168. [CrossRef]
40. Wang, J.; Liang, F.; Zhou, H.; Yang, M.; Wang, Q. Analysis of Position, Pose and Force Decoupling Characteristics of a 4-UPS/1-RPS Parallel Grinding Robot. *Symmetry* **2022**, *14*, 825. [CrossRef]
41. Lee, D.H.; Kim, Y.B.; Chakir, S.; Huynh, T.; Park, H.C. Noninteracting Control Design for 6-DoF Active Vibration Isolation Table with LMI Approach. *Appl. Sci.* **2021**, *11*, 7693. [CrossRef]
42. Gierlak, P. Adaptive Position/Force Control of a Robotic Manipulator in Contact with a Flexible and Uncertain Environment. *Robotics* **2021**, *10*, 32. [CrossRef]
43. Deng, Z.; Jonetzko, Y.; Zhang, L.; Zhang, J. Grasping Force Control of Multi-Fingered Robotic Hands through Tactile Sensing for Object Stabilization. *Sensors* **2020**, *20*, 1050. [CrossRef]
44. Gal, I.-A.; Ciocirlan, A.-C.; Margaritescu, M. State Machine-Based Hybrid Position/Force Control Architecture for a Waste Management Mobile Robot with 5DOF Manipulator. *Appl. Sci.* **2021**, *11*, 4222. [CrossRef]
45. Mercorelli, P. Invariant subspaces for grasping internal forces and non-interacting force-motion control in robotic manipulation. *Kybernetika* **2012**, *48*, 1229–1249.
46. Salisbury, J.; Roth, B. Kinematic and force analysis of articulated mechanical hands. *J. Mech. Trans. Autom.* **1983**, *105*, 35–41. [CrossRef]
47. Prattichizzo, D.; Bicchi, A. Dynamic analysis of mobility and graspability of general manipulation systems. *IEEE Trans. Robot. Autom.* **1998**, *14*, 241–258. [CrossRef]
48. Bicchi, A.; Melchiorri, C.; Balluchi, D. On the mobility and manipulability of general multiple limb robots. *IEEE Trans. Autom. Control* **1995**, *11*, 215–228. [CrossRef]
49. Prattichizzo, D.; Mercorelli, P.; Vicino, A. Noninteracting force/motion control in general manipulation systems. In Proceedings of the 35th IEEE Conference on Decision Control, CDC '96, Kobe, Japan, 13 December 1996.

axioms

MDPI

Article

Existence and Uniqueness of Nonmonotone Solutions in Porous Media Flow

Rouven Steinle, Tillmann Kleiner, Pradeep Kumar † and Rudolf Hilfer *

ICP, Universität Stuttgart, Allmandring 3, 70569 Stuttgart, Germany
* Correspondence: hilfer@icp.uni-stuttgart.de
† Current address: Department of Mathematics, Galgotias University, Greater Noida 201310, India.

Abstract: Existence and uniqueness of solutions for a simplified model of immiscible two-phase flow in porous media are obtained in this paper. The mathematical model is a simplified physical model with hysteresis in the flux functions. The resulting semilinear hyperbolic-parabolic equation is expected from numerical work to admit non-monotone imbibition-drainage fronts. We prove the local existence of imbibition-drainage fronts. The uniqueness, global existence, maximal regularity and boundedness of the solutions are also discussed. Methodically, the results are established by means of semigroup theory and fractional interpolation spaces.

Keywords: two-phase flow; Sobolev spaces; analytic semigroups; fractional interpolation; local and global solutions

MSC: 34K30; 35K57; 35Q80; 92D25

Citation: Steinle, R.; Kleiner, T.; Kumar, P.; Hilfer, R. Existence and Uniqueness of Nonmonotone Solutions in Porous Media Flow. *Axioms* **2022**, *11*, 327. https://doi.org/10.3390/axioms11070327

Academic Editor: Hans J. Haubold

Received: 5 May 2022
Accepted: 20 May 2022
Published: 5 July 2022

Publisher's Note: MDPI stays neutral with regard to jurisdictional claims in published maps and institutional affiliations.

Copyright: © 2022 by the authors. Licensee MDPI, Basel, Switzerland. This article is an open access article distributed under the terms and conditions of the Creative Commons Attribution (CC BY) license (https://creativecommons.org/licenses/by/4.0/).

1. Introduction

A great many studies in applied mathematics and mathematical physics are concerned with multiphase flow in porous media. From a mathematical point of view, these studies are important because they feature intrinsically nonlinear equations and hysteresis. Nonlinearity and hysteresis are longstanding "hot topics" that continue to generate fundamental insights and progress in mathematics, physics and engineering.

The purpose and significance of this work is to report rigorous results based on nonlinear semigroup theory for a simplified one-dimensional mathematical model of immiscible two-phase flow with hysteresis in porous media. It exhibits strongly nonlinear and nonmonotone solutions as a result of hysteresis. Our simplified model is introduced here as the nonlinear initial and boundary value problem

$$\begin{cases} u_t(z,t) + f(u,z)u_z(z,t) - Du_{zz}(z,t) = 0, & z \in \Omega,\ 0 < t \le T \\ u(z,0) = u_0(z), & z \in \Omega \\ u_z(z,t) = 0, & z \in \partial\Omega,\ 0 < t \le T \end{cases} \tag{1}$$

where $z \in \Omega$ is position, $\Omega = (0,1)$ is the domain, $t \in [0,T]$ is the time, $u\colon \Omega \times [0,T] \to \mathbb{R}$ is the unknown saturation function of the wetting phase, and $u_0\colon \Omega \to \mathbb{R}$ is the initial saturation. The nonlinear term is defined as

$$f(u,z) = \chi(z,z_a)f'_{\mathrm{im}}(u) + [1 - \chi(z,z_a)]f'_{\mathrm{dr}}(u) \tag{2}$$

with $f_i \in C^2(\mathbb{R})$ with $i \in \{\mathrm{im}, \mathrm{dr}\}$ and a fixed position $z_a \in (0,1)$. The characteristic function $\chi(\cdot, z_a)$ is defined as $\chi(z,z_a) = 1$ for $z \ge z_a$ and as $\chi(z,z_a) = 0$ for $z < z_a$. Further, u_t denotes the derivative with respect to t, u_z denotes the derivative with respect to z, and

Theorem 1 ([15], Chapter 2). *Let $D = const.$ and let $\epsilon > 0$. Then, for $z \in \Omega$, the operator*

$$A = -\frac{\partial}{\partial z}\left(D\frac{\partial}{\partial z}\right) + \epsilon, \tag{13}$$

$$\mathcal{D}(A) = \mathcal{H}^2_N := \{u \in \mathcal{H}^2 : u_z|_{\partial\Omega} = 0\} \tag{14}$$

in $\mathcal{L}^2$ is sectorial with angle $\omega_A < \pi/2$. For any $\omega_A < \omega \leq \pi/2$, the operator fulfills Equations (7a) and (7b), where the constant M is determined by Ω, D and ω and depends on ϵ.

Proof. This follows from Theorem 2.3, Theorem 2.7 and the discussion at the beginning of Chapter 2 in [15]. □

3. Results

In the following, we prove the existence of local solutions in Theorem 3. We show that local solutions are global in Corollary 1. Finally, in Theorem 4, we prove that initial conditions with values in $[0,1]$ lead to solutions with values in $[0,1]$. By "solutions", we mean functions that belong to the space $\mathcal{U}$ defined in Theorem 3 below and that satisfy Equation (15).

As remarked above, the notation $\mathcal{L}^2 = \mathcal{L}^2(\Omega)$, $\mathcal{H}^k = \mathcal{H}^k(\Omega)$ and $\|\cdot\| = \|\cdot\|_{X=\mathcal{L}^2}$ is used. In this section, the initial and boundary value problem (1) is solved in the function space $C((0,T];\mathcal{L}^2)$. Problem (1) is transformed into the abstract Cauchy problem

$$\begin{cases} u_t(t) + Au(t) = F(u(t)), & t > 0 \\ u(0) = u_0 \end{cases} \tag{15}$$

with the linear operator $A\colon \mathcal{D}(A) \to \mathcal{L}^2$ defined by

$$A = -D\partial_z^2 + \epsilon \tag{16}$$

with fixed $\epsilon > 0$ as in Theorem 1 and the nonlinear function $F\colon \mathcal{D}(A^{1/2}) \to \mathcal{L}^2$ defined by

$$F(u) = -\chi(\cdot, z_a) f'_{\mathrm{im}}(u)u_z - [1 - \chi(\cdot, z_a)] f'_{\mathrm{dr}}(u)u_z + \epsilon u. \tag{17}$$

The domain $\mathcal{D}(A)$ of the linear operator A is given by

$$\mathcal{D}(A) = \mathcal{H}^2_N = \{u \in \mathcal{H}^2 : u_z|_{\partial\Omega} = 0\}. \tag{18}$$

The domains of the fractional powers A^θ of A (or the interpolation spaces between $\mathcal{D}(A)$ and $\mathcal{L}^2$) are given by

$$\mathcal{D}(A^\theta) = \begin{cases} \mathcal{H}^{2\theta} & \text{for } 0 \leq \theta < 3/4, \\ \mathcal{H}^{2\theta}_N & \text{for } 3/4 < \theta \leq 1, \end{cases} \tag{19}$$

see ([15], Chapter 16). Therefore the domain of the nonlinear function F is given as

$$\mathcal{D}(A^{1/2}) = \mathcal{H}^1.$$

Lemma 2. *For bounded functions $f_i \in C^2$ with bounded derivatives f'_i and f''_i where $i \in \{\mathrm{im}, \mathrm{dr}\}$, $u, v \in \mathcal{H}^1$, $\chi(\cdot, z_a) : \Omega \to \{0,1\}$ is bounded and measurable, $z_a \in \Omega$ and $\epsilon > 0$ the nonlinear function $F : \mathcal{H}^1 \to \mathcal{L}^2$ with*

$$F(u) = -\chi(\cdot, z_a) f'_{\mathrm{im}}(u)u_z - [1 - \chi(\cdot, z_a)] f'_{\mathrm{dr}}(u)u_z + \epsilon u \tag{20}$$

fulfills the inequalities

$$\|F(u) - F(v)\| \le C_F\Big[\Big(1 + \Big\|A^{1/2}v\Big\|\Big)\Big\|A^{1/2}(u - v)\Big\|\Big] \tag{21}$$

$$\|F(u)\| \le C_F\Big\|A^{1/2}u\Big\| \tag{22}$$

for all $u, v \in \mathcal{H}^1$ *with* $0 < C_F < \infty$ *and the operator A defined above in Equation* (16).

Proof. The functions f'_{im}, f'_{dr} and $\chi(\cdot, z_a)$ are bounded and measurable. Therefore, the nonlinear function F is continuous as a sum of continuous functions, and it maps every $u \in \mathcal{H}^1$ to $F(u) \in \mathcal{L}^2$.

For convenience the notations $\chi_{\mathrm{im}} = \chi(\cdot, z_a)$ and $\chi_{\mathrm{dr}} = 1 - \chi(\cdot, z_a)$ are used. Then one obtains

$$\begin{aligned}\|F(u) - F(v)\| &\le \sum_{i\in\{\mathrm{im},\mathrm{dr}\}} \|\chi_i\|_\infty \|f'_i(v)v_z - f'_i(u)u_z\| + \epsilon\|u - v\| \\ &\le \sum_{i\in\{\mathrm{im},\mathrm{dr}\}} \left(\|f'_i(u)(v_z - u_z)\| + \|v_z(f'_i(v) - f'_i(u))\|\right) + \epsilon\|u - v\| \end{aligned} \tag{23}$$

for $u, v \in \mathcal{H}^1$. Let N be the embedding constant $\mathcal{H}^1 \to \mathcal{L}^\infty$ and define

$$C'_{f,i} := \sup_{w\in\mathbb{R}} |f'_i(w)|, \tag{24a}$$

$$C''_{f,i} = \sup_{w\in\mathbb{R}} |f''_i(w)| \tag{24b}$$

$$C_F = \max\Big\{C'_{f,\mathrm{im}} + C'_{f,\mathrm{dr}} + \epsilon,\ N\Big(C''_{f,\mathrm{im}} + C''_{f,\mathrm{dr}}\Big)\Big\} \tag{24c}$$

for $i \in \{\mathrm{im}, \mathrm{dr}\}$. The embedding of $\mathcal{H}^1 \to \mathcal{L}^\infty$ holds because Ω is one-dimensional. With these definitions, Equation (23) is estimated as

$$\begin{aligned}\|F(u) - F(v)\| &\le \sum_{i\in\{\mathrm{im},\mathrm{dr}\}} \Big(C'_{f,i}\|v_z - u_z\| + C''_{f,i}\|v_z(v - u)\|\Big) + \epsilon\|u - v\| \\ &\le \Big(C'_{f,\mathrm{im}} + C'_{f,\mathrm{dr}} + \epsilon\Big)\|u - v\|_{\mathcal{H}^1} + \Big(C''_{f,\mathrm{im}} + C''_{f,\mathrm{dr}}\Big)\|v_z\|\|u - v\|_\infty \\ &\le \Big(C'_{f,\mathrm{im}} + C'_{f,\mathrm{dr}} + \epsilon\Big)\|u - v\|_{\mathcal{H}^1} + N\Big(C''_{f,\mathrm{im}} + C''_{f,\mathrm{dr}}\Big)\|v\|_{\mathcal{H}^1}\|u - v\|_{\mathcal{H}^1} \\ &\le C_F\Big[\Big(1 + \Big\|A^{1/2}v\Big\|\Big)\Big\|A^{1/2}(u - v)\Big\|\Big] \end{aligned} \tag{25}$$

which proves (21). The verification of (22) follows from (21) by setting $v = 0$. □

Theorem 2. *Problem* (15) *with A given by* (16) *and* $F(u)$ *given by* (17) *is well-defined for all* $\mathcal{L}^2$*-valued functions* $u(t)$ *that satisfy*

$$u \in \mathcal{U} = \mathcal{C}\Big((0, T]; \mathcal{H}^2_N\Big) \cap \mathcal{C}\Big([0, T]; \mathcal{H}^1\Big) \cap \mathcal{C}^1\Big((0, T]; \mathcal{L}^2\Big). \tag{26}$$

Proof. First, A is an operator $\mathcal{C}((0, T]; \mathcal{H}^2_N) \to \mathcal{C}((0, T]; \mathcal{L}^2)$. Second, the time derivative $\mathrm{d}/\mathrm{d}t$ is an operator $\mathcal{C}^1((0, T]; \mathcal{L}^2) \to \mathcal{C}((0, T]; \mathcal{L}^2)$. According to Lemma 2 F is a mapping $\mathcal{H}^1 \to \mathcal{L}^2$. This implies that it is also a mapping $\mathcal{C}([0, T]; \mathcal{H}^1) \to \mathcal{C}([0, T]; \mathcal{L}^2)$. □

Theorem 3. *Define the linear operator* $A\colon \mathcal{D}(A) = \mathcal{H}^2_N \to \mathcal{L}^2$ *as in Theorem 1 and the nonlinear function* $F\colon \mathcal{D}(A^{1/2}) = \mathcal{H}^1 \to \mathcal{L}^2$ *as in Lemma 2. There exists a* $T > 0$, *such that, for every* $u_0 \in \mathcal{H}^1$, *there exists a unique local solution u of problem* (1) *in the function space*

$$\mathcal{U} = \mathcal{C}\Big((0, T]; \mathcal{H}^2_N\Big) \cap \mathcal{C}\Big([0, T]; \mathcal{H}^1\Big) \cap \mathcal{C}^1\Big((0, T]; \mathcal{L}^2\Big). \tag{27}$$

Further, if $u_0 \in \mathcal{H}^2_N$, then this solution belongs to the space

$$\mathcal{U}_2 = C\left([0,T];\mathcal{H}^2_N\right) \cap C^1\left([0,T],\mathcal{L}^2\right). \tag{28}$$

Remark 1. *The definition of $\mathcal{U}$ explains the solution concept: The factor $C^1((0,T],\mathcal{L}^2)$ ensures that the solutions u possess a strong derivative with respect to time, considered as $\mathcal{L}^2$-valued functions on $(0,T]$. The factor $C((0,T],\mathcal{H}^2_N)$ ensures that the solutions u belong to the domain of A for $t > 0$. The factor $C([0,T],\mathcal{H}^1)$ ensures that $u(0+) = u_0$ with respect to the topology of $\mathcal{H}^1$. These solutions are solutions in the weak sense, in particular.*

Proof. Following [15], the idea of the proof is to rewrite problem (15) as

$$\begin{cases} u_t(t) + Au(t) = G(t), & t > 0 \\ u(0) = u_0 \end{cases}. \tag{29}$$

To this end, the fixed-point theorem is applied to the mapping M

$$Mu(t) = e^{-tA}u_0 + \int_0^t e^{-(t-\tau)A}F(u(\tau))\,\mathrm{d}\tau \tag{30}$$

which is defined on the space $\mathcal{X}(T) \supset \mathcal{U}$ defined in Equation (31) and seen to be a contraction on a suitably chosen closed subset $\mathcal{Y}(T, C_T)$ with $0 < C_T < \infty$. The first step is to determine $\mathcal{Y}(T, C_T)$ and to verify the requirements for the fixed point theorem ([17], Theorem 1.A, p. 17). In the second step, it is shown that, if u is a fixed point of the mapping M, then, for every $\sigma \in (0, 1/2)$, the function $F \circ u$ is an element of the space $\mathcal{F}^{1/2,\sigma}((0,T],\mathcal{L}^2)$. If $F \circ u \in \mathcal{F}^{1/2,\sigma}((0,T],\mathcal{L}^2)$, then $G(t) = F(u(t))$ is an admissible inhomogeneity for the Cauchy problem (29). Finally, the uniqueness of the solution is shown.

Using Equations (16) and (17), the initial and boundary value problem (1) is transformed into an abstract Cauchy problem (15).

The linear operator A, defined in (16), is a sectorial operator with angle $\omega_A < \pi/2$ by virtue of Theorem 1 and the infinitesimal generator of the analytic semigroup e^{-tA}.

Step 1: Requirements for the fixed-point theorem. For every $T > 0$, the Banach Space $\mathcal{X}(T)$ is defined as

$$\mathcal{X}(T) = C\left([0,T];\mathcal{H}^1\right) \supset \mathcal{U} \tag{31}$$

with norm $\|u\|_{\mathcal{X}} = \sup_{0 \le t \le T}\left\|A^{1/2}u(t)\right\|$. Additionally, one defines the closed subset $\mathcal{Y}(T, C_T) \subset \mathcal{X}(T)$ of all u that satisfy

$$\left\|A^{1/2}u(t)\right\| \le C_T \qquad \text{for all } t \in [0,T]. \tag{32}$$

Now, we derive conditions for the constants C_T and T from Equation (32) such that the mapping M from Equation (30) maps $\mathcal{Y}(T, C_T)$ into $\mathcal{Y}(T, C_T)$. For any $0 \le \sigma < 1/2$ and $0 < t \le T$, one derives the estimate

$$\begin{aligned} \left\|A^{1/2+\sigma}Mu(t)\right\| &= \left\|A^{1/2+\sigma}\left\{e^{-tA}u_0 + \int_0^t e^{-(t-\tau)A}F(u(\tau))\,\mathrm{d}\tau\right\}\right\| \\ &\le \left\|A^{1/2+\sigma}e^{-tA}u_0\right\| + \left\|A^{1/2+\sigma}\int_0^t e^{-(t-\tau)A}F(u(\tau))\,\mathrm{d}\tau\right\| \\ &\le \left\|A^{\sigma}e^{-tA}\right\|\left\|A^{1/2}u_0\right\| + \int_0^t \left\|A^{1/2+\sigma}e^{-(t-\tau)A}\right\|\|F(u(\tau))\|\,\mathrm{d}\tau. \end{aligned} \tag{33}$$

Using Lemma 1, Equation (22) from Lemma 2 and Equation (32), we find

$$\begin{aligned}\left\|A^{1/2+\sigma}Mu(t)\right\| &\leq \left\|A^{\sigma}e^{-tA}\right\|\left\|A^{1/2}u_0\right\| + C_F\int_0^t\left\|A^{1/2+\sigma}e^{-(t-\tau)A}\right\|\left\|A^{1/2}u\right\|\mathrm{d}\tau\\ &\leq \left\|A^{\sigma}e^{-tA}\right\|\left\|A^{1/2}u_0\right\| + C_FC_T\int_0^t\left\|A^{1/2+\sigma}e^{-(t-\tau)A}\right\|\mathrm{d}\tau\\ &\leq \left\|A^{\sigma}e^{-tA}\right\|\left\|A^{1/2}u_0\right\| + C_FC_TC_{\frac{1}{2}-\sigma}t^{1/2-\sigma}. \end{aligned} \tag{34}$$

For $\sigma = 0$, Equation (32) holds if the right side of Equation (34) is smaller or equal to C_T and

$$\left\|e^{-tA}\right\|\left\|A^{\frac{1}{2}}u_0\right\| + C_FC_TC_{\frac{1}{2}}t^{\frac{1}{2}} \leq C_T \quad\Leftrightarrow\quad C_T\left(1 - C_FC_{\frac{1}{2}}t^{\frac{1}{2}}\right) \geq \left\|e^{-tA}\right\|\left\|A^{\frac{1}{2}}u_0\right\|. \tag{35}$$

If $1 - C_FC_{1/2}T^{1/2} > 0$ or equivalently

$$T < (C_FC_{1/2})^{-2} \tag{36}$$

holds, then C_T can be chosen such that

$$C_T > \sup_{0\leq t\leq T}\left\|e^{-tA}\right\|\frac{\left\|A^{1/2}u_0\right\|}{1 - C_FC_{1/2}T^{1/2}}. \tag{37}$$

The right hand side of (37) is bounded because the norm $\|e^{-tA}\|$ is bounded according to ([15], Proposition 2.5, p.86). Then, the mapping M fulfills the condition

$$\sup_{0\leq t\leq T}\left\|A^{1/2}Mu(t)\right\| \leq C_T \tag{38}$$

where C_T is given by (37), and $Mu(t) \in \mathcal{Y}(T, C_T)$ holds.

The next step is to show that $M : \mathcal{Y}(T, C_T) \to \mathcal{Y}(T, C_T)$ is a contraction mapping. One estimates

$$\begin{aligned}\sup_{0\leq t\leq T}\|Mu(t) - Mv(t)\|_{\mathcal{X}} &\leq \left\|\int_0^t e^{-(t-\tau)A}\{F(u(\tau)) - F(v(\tau))\}\,\mathrm{d}\tau\right\|_{\mathcal{X}}\\ &= \sup_{0\leq t\leq T}\left\|A^{1/2}\int_0^t e^{-(t-\tau)A}\{F(u(\tau)) - F(v(\tau))\}\,\mathrm{d}\tau\right\|\\ &\leq \sup_{0\leq t\leq T}\int_0^t\left\|A^{1/2}e^{-(t-\tau)A}\right\|\|F(u(\tau)) - F(v(\tau))\|\,\mathrm{d}\tau. \end{aligned} \tag{39}$$

Using Lemma 2 and Equation (25) to estimate the integral term, one obtains

$$\begin{aligned}\sup_{0\leq t\leq T}\|Mu(t) - Mv(t)\|_{\mathcal{X}} &\leq C_F(1 + C)\sup_{0\leq t\leq T}\int_0^t\left\|A^{1/2}e^{-(t-\tau)A}\right\|\left\|A^{1/2}(u - v)\right\|\mathrm{d}\tau\\ &\leq C_F(1 + C)\int_0^T\left\|A^{1/2}e^{-(t-\tau)A}\right\|\mathrm{d}\tau\left\|A^{1/2}(u - v)\right\|\\ &\leq C_F(1 + C)\int_0^T\left\|A^{1/2}e^{-(t-\tau)A}\right\|\mathrm{d}\tau\|u - v\|_{\mathcal{X}}\\ &\leq C_F(1 + C)C_{1/2}T^{1/2}\|u - v\|_{\mathcal{X}}. \end{aligned} \tag{40}$$

Thus, the mapping $M : \mathcal{Y}(T, C_T) \to \mathcal{Y}(T, C_T)$ is a contraction if $C_F(1 + C)C_{1/2}T^{1/2} < 1$ or equivalently

$$T < (C_F(1 + C)C_{1/2})^{-2}. \tag{41}$$

It remains to prove that $Mu(t) \in \mathcal{C}([0,T];\mathcal{H}^1)$ holds. For this purpose, one calculates for $t > s > 0$

$$\begin{aligned} Mu(t) - Mu(s) &= e^{-tA}u_0 + \int_0^t e^{-(t-\tau)A}F(u(\tau))\,\mathrm{d}\tau - Mu(s) \\ &= e^{-tA}u_0 + \int_0^s e^{-(t-\tau)A}F(u(\tau))\,\mathrm{d}\tau - Mu(s) + \int_s^t e^{-(t-\tau)A}F(u(\tau))\,\mathrm{d}\tau \\ &= e^{-(t-s)A}Mu(s) - Mu(s) + \int_s^t e^{-(t-\tau)A}F(u(\tau))\,\mathrm{d}\tau \\ &= \left[e^{-(t-s)A} - 1\right]Mu(s) + \int_s^t e^{-(t-\tau)A}F(u(\tau))\,\mathrm{d}\tau. \end{aligned} \tag{42}$$

With Equation (42), one obtains

$$\begin{aligned} &\left\|A^{1/2}\{Mu(t) - Mu(s)\}\right\| \\ &\le \left\|A^{-\sigma}\left[e^{-(t-s)A} - 1\right]\right\|\left\|A^{1/2+\sigma}Mu(s)\right\| + \left\|A^{1/2}\int_s^t e^{-(t-\tau)A}F(u(\tau))\,\mathrm{d}\tau\right\| \\ &\le \left\|A^{-\sigma}\left[e^{-(t-s)A} - 1\right]\right\|\left\|A^{1/2+\sigma}Mu(s)\right\| + \int_s^t \left\|A^{1/2}e^{-(t-\tau)A}\right\|\|F(u(\tau))\|\,\mathrm{d}\tau. \end{aligned} \tag{43}$$

Then, Equations (12), (22), (32) and (34) lead to

$$\begin{aligned} &\left\|A^{1/2}\{Mu(t) - Mu(s)\}\right\| \\ &\le C_\sigma(t-s)^\sigma\left(C_\sigma s^{-\sigma}\left\|A^{\frac{1}{2}}u_0\right\| + C_F C_T C_{\frac{1}{2}-\sigma}s^{\frac{1}{2}-\sigma}\right) + C_F C_T C_{\frac{1}{2}}\int_s^t (t-\tau)^{-\frac{1}{2}}\,\mathrm{d}\tau \\ &\le (t-s)^\sigma s^{-\sigma}\left(C_{u_0} + C_A s^{1/2}\right) + C(t-s)^{1/2} \\ &= (t-s)^\sigma s^{-\sigma}\left(C_{u_0} + C_A s^{1/2} + C(t-s)^{1/2-\sigma}s^\sigma\right) \\ &\le (t-s)^\sigma s^{-\sigma}\left(C_{u_0} + C_A t^{1/2} + C t^{1/2-\sigma}t^\sigma\right) \\ &\le (t-s)^\sigma s^{-\sigma}\left(C_{u_0} + C t^{1/2}\right) \\ &\le C(t-s)^\sigma s^{-\sigma}. \end{aligned} \tag{44}$$

Equation (44) shows that $Mu(t)$ is now part of the function space $\mathcal{C}((0,T];\mathcal{H}^1)$. The estimate

$$\begin{aligned} \left\|A^{1/2}\{Mu(t) - Mu(0)\}\right\| &= \left\|A^{1/2}\int_0^t e^{-(t-\tau)A}F(u(\tau))\,\mathrm{d}\tau\right\| \\ &\le C_F C_T C_{1/2}\int_0^t (t-\tau)^{-1/2}\,\mathrm{d}\tau \\ &\le C_F C_T C_{1/2}t^{1/2} \end{aligned} \tag{45}$$

shows that

$$\lim_{t\to 0}\left\|A^{1/2}\{Mu(t) - Mu(0)\}\right\| = \lim_{t\to 0} Ct^{1/2} = 0 \tag{46}$$

and therefore $Mu(t)$ is part of $\mathcal{C}([0,T];\mathcal{H}^1)$.

If Equations (36), (37) and (41) are fulfilled, then a fixed point $Mu = u \in \mathcal{Y}(T, C_T)$ exists according to ([17], Theorem 1.A, p. 17), and the fixed point $u(t)$ obeys

$$u(t) = e^{-tA}u_0 + \int_0^t e^{-(t-\tau)A}F(u(\tau))\,\mathrm{d}\tau \qquad \text{for all } t \in [0,T]. \tag{47}$$

Step 2: Show that $F \circ u \in \mathcal{F}^{1,\sigma}([0,T],\mathcal{L}^2)$ holds for any fixed point u of M. It is immediate from the definition of $\mathcal{Y}(T, C_T)$ and Lemma 2 that $F \circ u$ is a continuous function on $[0,T]$.

The function $F(u)$ has to fulfill condition (4) from Definition 1. Using Equations (21), (38) and (44), one obtains, for $0 < s < t \leq T$, the estimate

$$\begin{aligned} \|F(u(t)) - F(u(s))\| &= \|F(Mu(t)) - F(Mu(s))\| \\ &\leq C_F\left[\left(1 + \left\|A^{1/2}Mu(t)\right\|\right)\left\|A^{1/2}(Mu(t) - Mu(s))\right\|\right] \\ &\leq C_F(1 + C_T)\left\|A^{1/2}(Mu(t) - Mu(s))\right\| \\ &\leq C_{\mathcal{F}}(t-s)^{\sigma}s^{-\sigma}. \end{aligned} \tag{48}$$

Therefore, we can conclude that $F \circ u \in \mathcal{F}^{1,\sigma}([0,T],\mathcal{L}^2)$ is true, and we can write the semilinear evolution problem (15) as a linear evolution problem (29).

Using ([15], Theorems 3.4, 3.5, p. 124, 126), it follows that the fixed points u (see Equation (47)) are elements of the function space $\mathcal{U}$ from (27). Further, it follows that u belongs to the function space $\mathcal{U}_2$ from Equation (28) if $u_0 \in \mathcal{H}^2_N$.

Step 3: Uniqueness of solutions. Any solution $u \in \mathcal{U}$ of problem (4.1) satisfies $F \circ u \in \mathcal{F}^{1/2,\sigma}((0,T],\mathcal{L}^2)$ and is a solution of the problem (4.15) with $G(t) = F(u(t))$ in the sense of ([15], Theorem 3.4). According to ([15], Theorem 3.4, Eq. (3.13)), any solution $u \in \mathcal{U}$ of (29) is also a fixed point of M. Therefore, uniqueness follows from the fixed point theorem ([17], Theorem 1.A, p. 17). □

Corollary 1. *Every local solution of problem* (1), *in the sense of Theorem 3, extends uniquely to a global solution.*

Proof. Because the constant $T > 0$ in Theorem 3 is independent of the initial condition u_0 the theorem can be applied repeatedly to prove the existence of a solution $u \in C((0,\infty),\mathcal{H}^2_N) \cap C([0,\infty),\mathcal{H}^1)$ that is piecewise differentiable as a function with values in $\mathcal{L}^2$. Invoking uniqueness, piecewise differentiability improves to differentiability for all $t > 0$ as a function with values in $\mathcal{L}^2$, that is, one obtains $u \in C((0,\infty),\mathcal{H}^2_N) \cap C([0,\infty),\mathcal{H}^1) \cap C^1((0,\infty),\mathcal{L}^2)$. □

Theorem 4. *Let $u_0 \in \mathcal{H}^1$ and $u(z,t)$ be the unique global solution of problem* (1). *If the initial condition fulfills $0 \leq u_0 \leq 1$, then the global solution u fulfills $0 \leq u \leq 1$ as well.*

Proof. First, the lower bound $0 \leq u(t)$ is discussed by using a penalty function

$$E(u) = \begin{cases} \dfrac{u^2}{2} & \text{for } -\infty < u < 0 \\ 0 & \text{for } 0 \leq u < \infty \end{cases} \tag{49}$$

which is continuously differentiable and whose first derivative satisfies the general Lipschitz condition. The function

$$H(t) = \int_{\Omega} E(u(t))\,\mathrm{d}z = \int_{\Omega_1} E(u(t))\,\mathrm{d}z + \int_{\Omega_2} E(u(t))\,\mathrm{d}z \tag{50}$$

averages the value of the penalty function over the domain Ω. In Equation (50), it holds that $\Omega_1 \cup \Omega_2 = \Omega$ and $\Omega_1 \cap \Omega_2 = \emptyset$. The domain Ω_1 denotes the time-dependent domain where $E(u) > 0$ holds and Ω_2 denotes the time-dependent domain where $E(u) = 0$ holds. Clearly, $H(t)$ is a continuously differentiable function for $t > 0$ because $u \in \mathcal{U}$ with the derivative

$$
\begin{aligned}
\frac{\mathrm{d}}{\mathrm{d}t}H(t) &= \int_\Omega \frac{\mathrm{d}E(u)}{\mathrm{d}u} u_t \,\mathrm{d}z \\
&= D\int_\Omega \frac{\mathrm{d}E(u)}{\mathrm{d}u} u_{zz}\,\mathrm{d}z - \int_\Omega \sum_{i\in\{\mathrm{im},\mathrm{dr}\}} \frac{\mathrm{d}E(u)}{\mathrm{d}u}\chi_i f_i'(u)u_z\,\mathrm{d}z \\
&= -D\int_\Omega \partial_z\left(\frac{\mathrm{d}E(u)}{\mathrm{d}u}\right) u_z\,\mathrm{d}z - \int_\Omega \sum_{i\in\{\mathrm{im},\mathrm{dr}\}} \frac{\mathrm{d}E(u)}{\mathrm{d}u}\chi_i f_i'(u)u_z\,\mathrm{d}z \\
&= -D\int_{\Omega_1} |u_z|^2\,\mathrm{d}z - \int_\Omega \sum_{i\in\{\mathrm{im},\mathrm{dr}\}} \frac{\mathrm{d}E(u)}{\mathrm{d}u}\chi_i f_i'(u)u_z\,\mathrm{d}z
\end{aligned}
\tag{51}
$$

where $\chi_{\mathrm{im}} = \chi(z,z_a)$ and $\chi_{\mathrm{dr}} = 1-\chi(z,z_a)$. Since $f_i'(u) = 0$ for any $-\infty < u < 0$ and $\mathrm{d}E(u)/\mathrm{d}u = 0$ for any $0 \le u < \infty$, it holds that

$$
\int_\Omega \frac{\mathrm{d}E(u)}{\mathrm{d}u}\chi_{\mathrm{im}} f_{\mathrm{im}}'(u)u_z\,\mathrm{d}z = \int_\Omega \frac{\mathrm{d}E(u)}{\mathrm{d}u}\chi_{\mathrm{dr}} f_{\mathrm{dr}}'(u)u_z\,\mathrm{d}z = 0. \tag{52}
$$

Thus, we find $H(t) \le H(0)$, and $H(0) = 0$ implies $H(t) \equiv 0$, i.e., $0 \le u(t)$ for $t \in [0,T]$.

Similarly, we can easily prove that $u(t) \le 1$ for every $t \in [0,T]$ by taking $u^*(z,t) = 1-u(z,t)$ on $[0,\infty)$ and formulating problem (1) as follows

$$
\begin{cases}
u_t^*(z,t) + f^*(u,z)u_z^*(z,t) - Du_{zz}^*(z,t) = 0, & z\in\Omega,\ 0<t\le T\\
u^*(z,0) = u_0^*(z), & z\in\Omega\\
u_z^*(z,t) = 0, & z\in\partial\Omega,\ 0<t\le T
\end{cases}
\tag{53}
$$

with $z\in\Omega$, $t\in(0,T_{u_0}]$ and

$$
f^*(u,z) = \chi(z,z_a)f_{\mathrm{im}}'(1-u^*(z)) + [1-\chi(z,z_a)]f_{\mathrm{dr}}'(1-u^*(z)). \tag{54}
$$

□

4. Discussion

In the following discussion, the above results for Equation (1) are interpreted from the perspective of previous studies. Hysteretic two-phase flow in porous media was previously modeled using the initial and boundary value problem [8]

$$
\begin{cases}
u_t(z,t) + \frac{\partial}{\partial z}f_{\mathcal{G}}(u) - Du_{zz}(z,t) = 0, & z\in\Omega,\ t>0\\
u(z,0) = u_0(z), & z\in\Omega\\
u_z(z,t) = 0, & z\in\partial\Omega,\ t>0
\end{cases}
\tag{55}
$$

with the nonlinear fractional flow functions $f_{\mathcal{G}}\colon [0,1]\to\mathbb{R}_+$ and the capillary coefficient $D>0$. Problem (55) becomes equivalent to Equation (1) for $\frac{\partial}{\partial z}f_{\mathcal{G}}(u) = f(u,z)u_z$ where the derivative is a distributional derivative. The fractional flow function is indexed by a graph $\mathcal{G}(z,t)\in[0,1]\times\mathbb{R}_+\times\mathbb{R}_+\times\mathbb{R}$, see ([8], Equation (9)). The graph $\mathcal{G}(z,t)$ represents different flow processes obtained from a suitable hysteresis model. At a fixed z, this depends on the saturation history $u(z,t)$ at z. Let the time instants t^i with $i=0,\dots,N$ and $0=t^0<t^1<t^2<\cdots<t^N<t$ denote the switching times between drainage and imbibition at z. The graph $\mathcal{G}$ changes only at these switching instants.

Consider the initial-boundary value problem for Equation (55) with a non-monotone initial condition as shown in Figure 1. Assume without loss of generality, that the profile propagates in the positive z-direction. Let $u(z,t)$ with $z\in\Omega$ be the saturation profile at time t. Then the imbibition interval $\mathbb{I}(t)$ at time t is defined as the largest singly connected interval on which $u(z,t)$ is monotone decreasing but not constant everywhere. Similarly, the drainage interval $\mathbb{D}(t)$ at time t is defined as the largest singly connected interval on which $u(z,t)$ is monotone increasing but not constant everywhere. For the initial saturation profile of Figure 1 at time $t=0$, the two intervals are illustrated as gray regions in the top row of Figure 2. Additionally, a time-dependent plateau interval is defined as

$$\mathbb{P}(t) = \left\{ z : u_t(z,t) = 0\,,\; u(z,t) = \max_{z' \in \Omega} u(z',t) \right\}. \tag{56}$$

$\mathbb{P}(t=0)$ is illustrated as the gray region in the left graph of the second row in Figure 2. The propagated plateau interval $\mathbb{P}(T)$ for $T > 0$ is depicted in the second row on the right. Throughout Figure 2, the initial condition $u(z,0)$ is plotted as a dashed line, and the propagated profile $u(z,T)$ with $T > 0$ is shown as a solid line. Because $\mathbb{P}(0) \cap \mathbb{P}(T) \neq \emptyset$, a position $z_a \in \mathbb{P}(0) \cap \mathbb{P}(T)$ can be selected such that the drainage process on the left ($z < z_a$) decouples from the imbibition process on the right ($z > z_a$).

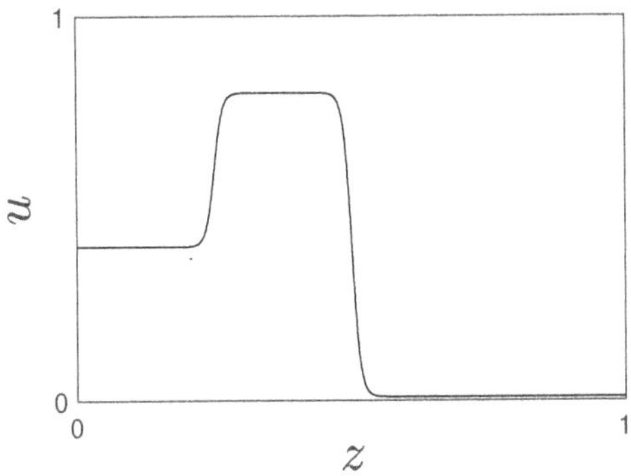

Figure 1. Initial condition $u(z,0)$ for problem (55) with a single overshoot.

Numerical solutions for problem (55) with initial data as shown in Figure 1 were studied in [6–8]. For this simple class of processes with a single saturation overshoot, the saturation history at positions $z > z_a$ has length $N(z) = 0$, while $N(z) = 1$ for $z \leq z_a$. For any time t with $0 < t < T$, there is a fixed graph $\mathcal{G}^{\mathrm{im}}$ describing the flow process at each $z \in \mathbb{I}(t)$ in terms of a flow function $f_{\mathrm{im}}(u;\{u_{*0}\})$ parametrized by the saturation value $u_{*0} = u(z,0)$ at $t^0 = 0$. Furthermore, there is a fixed graph $\mathcal{G}^{\mathrm{dr}}$ describing the flow process at each $z \in \mathbb{D}(t)$ in terms of a flow function $f_{\mathrm{dr}}(u;\{u_{*1}\})$ parametrized by the saturation value $u_{*1} = u(z,t^1)$ at the time instant $t^1(z)$ when the flow process switched from imbibition to drainage. For a single overshoot, the value of u_{*1} is, of course, $u_{*z} = \max_{z\in\Omega} u(z,0)$. By continuity of the hysteresis model and by continuity of the graph $\mathcal{G} = \mathcal{G}^{\mathrm{im}} \cup \mathcal{G}^{\mathrm{dr}}$, the flux is continous for all $z \in \mathbb{P}(t)$ with $0 < t < T$. In this situation, the first order term in Equation (55) simplifies to

$$\frac{\partial}{\partial z} f_{\mathcal{G}}(u) = \frac{\partial}{\partial z}\Big(\chi_{\mathbb{I}(t)} f_{\mathrm{im}}(u;\{u_{*0}\}) + \chi_{\mathbb{D}(t)} f_{\mathrm{dr}}(u;\{u_{*1}\}) - \chi_{\mathbb{P}(t)} f_{\mathrm{im}}(u;\{u_{*0}\})\Big) \tag{57a}$$

$$= \frac{\partial}{\partial z}\Big(\chi_{\mathbb{I}(t)} f_{\mathrm{im}}(u;\{u_{*0}\}) + \chi_{\mathbb{D}(t)} f_{\mathrm{dr}}(u;\{u_{*1}\}) - \chi_{\mathbb{P}(t)} f_{\mathrm{dr}}(u;\{u_{*1}\})\Big) \tag{57b}$$

$$= \chi_{\mathbb{I}(t)} f'_{\mathrm{im}}(u;\{u_{*0}\}) u_z + \chi_{\mathbb{D}(t)} f'_{\mathrm{dr}}(u;\{u_{*1}\}) u_z - \chi_{\mathbb{P}(t)} f'_{\mathrm{im}}(u;\{u_{*0}\}) u_z \tag{57c}$$

$$= \chi(z,z_a) f'_{\mathrm{im}}(u;\{u_{*0}\}) u_z + [1 - \chi(z,z_a)] f'_{\mathrm{dr}}(u;\{u_{*1}\}) u_z \tag{57d}$$

where $f_{\mathrm{im}}(\,\cdot\,;\{u_{*0}\}) = f_{\mathcal{G}^{\mathrm{im}}}(\,\cdot\,)$ and $f_{\mathrm{dr}}(\,\cdot\,;\{u_{*0}\}) = f_{\mathcal{G}^{\mathrm{dr}}}(\,\cdot\,)$. A possible choice for $f_{\mathcal{G}}$ can be seen in ([8], Equation (2)). The term $\chi_{\mathbb{P}(t)} f'_{\mathrm{im}}(u;\{u_{*0}\}) u_z$ is necessary because the imbibition interval and drainage interval are overlapping in the plateau interval $\mathbb{P}(t)$. Inserting this into the differential Equation (55) gives

$$u_t + \chi(z, z_a) f'_{\text{im}}(u; \{u_{*0}\}) u_z + [1 - \chi(z, z_a)] f'_{\text{dr}}(u; \{u_{*1}\}) u_z - D u_{zz} = 0 \tag{58a}$$

where $\{u_{*0}\} = u_0(z) = u(z, 0)$ is the initial condition and $\{u_{*1}\} = u(z, t^1(z))$ is the saturation at position z at the switching time t^1. The fractional flow functions for imbibition and drainage at the switching point obey flux continuity at t^1, i.e.

$$f_{\text{im}}(u(z, t^1); \{u_{*0}\}) = f_{\text{dr}}(u(z, t^1); \{u_{*1}\}) \tag{58b}$$

for all $z \in \mathbb{P}(t)$. Note that the fractional flow functions are explicitly position dependent due to hysteresis.

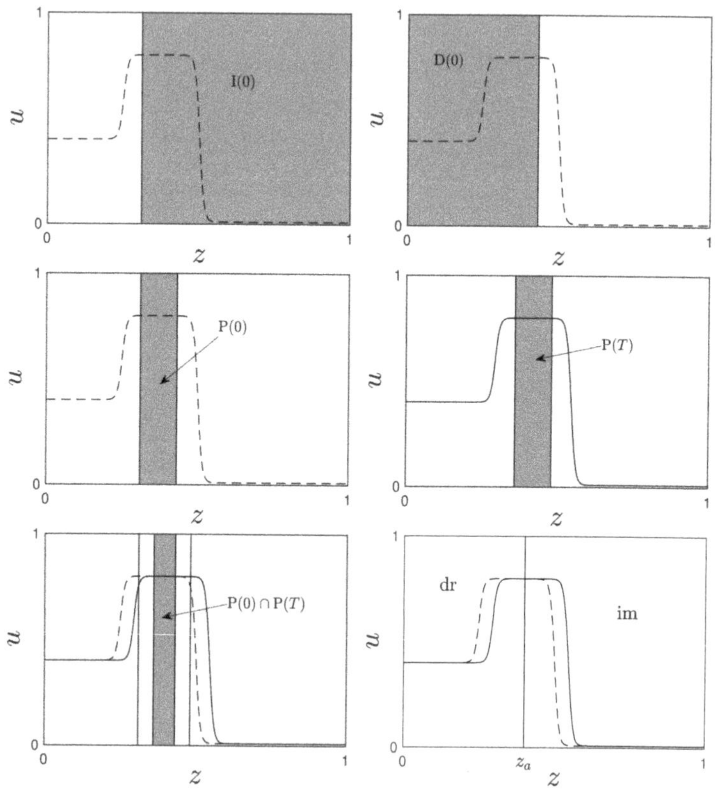

Figure 2. Schematic illustration for the decoupling of the imbibition and drainage fronts. The initial saturation profile $u(z, 0)$ is the dashed line, and the propagated profile $u(z, T)$ with $T > 0$ is shown as a solid line. The top left figure illustrates $\mathbb{I}(0)$ in gray, the top right figure illustrates $\mathbb{D}(0)$ in gray, the middle left figure shows the intersection $\mathbb{P}(0) = \mathbb{I}(0) \cap \mathbb{D}(0)$ at time $t = 0$, the middle right figure shows $\mathbb{P}(T)$ at some time $T > 0$, and the lower left figure shows $\mathbb{P}(0) \cap \mathbb{P}(T)$. The location z_a in the lower right subfigure can be chosen arbitrarily from within the gray interval $\mathbb{P}(0) \cap \mathbb{P}(T)$ in the lower left subfigure.

Numerical (and experimental) evidence in [6–8] suggest that imbibition and drainage fronts decouple for the simple class of hysteretic processes with a single saturation overshoot assumed in our mathematical model. The decoupling assumption is supported by noting that, for $D = 0$, piecewise constant functions are indeed weak solutions.

The decoupling is implemented here in this work by assuming that the set $\mathbb{P}(t)$ has positive measure for some nonempty time interval $[0, T]$ with $T > 0$. If the decoupling assumption holds true, then the fractional flow functions

$$f_{\mathrm{im}}(u(z,t); \{u_{*0}\}) = f_{\mathrm{dr}}(u(z,t); \{u_{*1}\}) \tag{59}$$

agree for all $z \in \mathbb{P}(t)$. In this way, a plateau in the saturation determines two position-independent fractional flow functions that agree on $\mathbb{P}(t)$ for $0 < t < T$. The rigorous results for problem (1) obtained in this work support the numerical results for problem (55) in [8]. The main point here is that, given a non-monotone single overshoot initial condition similar to the one shown in Figure 1, there is an open interval $\bigcap_{t=0}^{T} \mathbb{P}(t) \subset (\mathbb{I}(t) \cap \mathbb{D}(t))$ with $u = const.$ for $t \in [0, T]$ and $z_a \in \bigcap_{t=0}^{T} \mathbb{P}(t)$. This fact ensures the decoupling of the imbibition and the drainage front, and Equation (55) can be reduced to Equation (1) for $t \in [0, T]$.

Acknowledgments: The authors are grateful to Dr. Bakkyaraj T. for many fruitful discussions.

References

1. Alt, H.W.; Luckhaus, S.; Visintin, A. On nonstationary flow through porous media. *Ann. Di Mat. Pura Ed Appl.* **1985**, *136*, 303–316. [CrossRef]
2. Amadori, D.; Baiti, P.; Corli, A.; Dal Santo, E. Global weak solutions for a model of two-phase flow with a single interface. *J. Evol. Equations* **2015**, *15*, 699–726. [CrossRef]
3. Cao, X.; Pop, I.S. Two-phase porous media flows with dynamic capillary effects and hysteresis: Uniqueness of weak solutions. *Comput. Math. Appl.* **2015**, *69*, 688–695. [CrossRef]
4. Cao, X.; Pop, I.S. Degenerate two-phase porous media flow model with dynamic capillarity. *J. Differ. Equations* **2016**, *260*, 2418–2456. [CrossRef]
5. Mikelić, A. A global existence result for the equations describing unsaturated flow in porous media with dynamic capillary pressure. *J. Differ. Equations* **2010**, *248*, 1561–1577. [CrossRef]
6. Hilfer, R.; Steinle, R. Saturation overshoot and hysteresis for twophase flow in porous media. *Eur. Phys. J. Spec. Top.* **2014**, *223*, 2323–2338. [CrossRef]
7. Steinle, R.; Hilfer, R. Influence of initial conditions on propagation, growth and decay of saturation overshoot. *Transp. Porous Media* **2016**, *111*, 369–380. [CrossRef]
8. Steinle, R.; Hilfer, R. Hysteresis in relative permeabilities suffices for propagation of saturation overshoot: A quantitative comparison with experiment. *Phys. Rev. E* **2017**, *95*, 043112–1–043112–10. [CrossRef] [PubMed]
9. Böhm, M.; Showalter, R.E. A nonlinear pseudoparabolic diffusion equation. *SIAM J. Math. Anal.* **1985**, *16*, 980–999. [CrossRef]
10. Cuesta, C.; Hulshof, J. A model problem for groundwater flow with dynamic capillary pressure: stability of travelling waves. *Nonlinear Anal.* **2003**, *52*, 1199–1218. [CrossRef]
11. Hulshof, J.; King, J.R. Analysis of a Darcy flow model with a dynamic pressure saturation relation. *SIAM J. Appl. Math.* **1999**, *59*, 318–346.
12. Köhne, M.; Prüss, J.; Wilke, M. On quasilinear parabolic evolution equations in weighted Lp-spaces. *J. Evol. Equations* **2010**, *10*, 443–463. [CrossRef]
13. Kopfová, J. Nonlinear semigroup methods in problems with hysteresis. *Discret. Contin. Dyn. Syst. Suppl.* **2007**, *2007*, 580–589.
14. Pazy, A. *Semigroups of Linear Operators and Applications to Partial Differential Equations*; Springer: Berlin/Heidelberg, Germany, 1983.
15. Yagi, A. *Abstract Parabolic Equations and Their Applications*; Springer: Berlin/Heidelberg, Germany, 2010.
16. Little, T.D.; Showalter, R.E. Semilinear Parabolic Equations With Preisach Hysteresis. *Differ. Integral Equations* **1994**, *7*, 1021–1040.
17. Zeidler, E. *Nonlinear Functional Analysis and its Applications I: Fixed-Point Theorems*; Springer: New York, NY, USA, 1986.

axioms

Review

Near-Field Seismic Motion: Waves, Deformations and Seismic Moment

Bogdan Felix Apostol

National Institute of Earth's Physics, 077125 Magurele, Romania; afelix@theory.nipne.ro

Abstract: The tensorial force acting in a localized seismic focus is introduced and the corresponding seismic waves are derived, as solutions of the elastic wave equation in a homogeneous and isotropic body. The deconvolution of the solution for a structured focal region is briefly discussed. The far-field waves are identified as P and S seismic waves. These are spherical-shell waves, with a scissor-like shape, and an amplitude decreasing with the inverse of the distance. The near-field seismic waves are spherical-shell waves, decreasing with the inverse of the squared distance. The amplitudes and the polarizations of the near-field seismic waves are given. The determination of the seismic-moment tensor and the earthquake parameters from measurements of the P and S seismic waves at Earth's' surface is briefly discussed. Similarly, the mainshock generated by secondary waves on Earth's surface is reviewed. The near-field static deformations of a homogeneous and isotropic half-space are discussed and a method of determining the seismic-moment tensor from epicentral near-field (quasi-) static deformations in seismogenic regions is presented.

Keywords: seismic tensorial force; far-field seismic waves; near-field seismic waves; seismic mainshock; quasi-static deformations

MSC: 35Q74; 86A15; 86A17; 86A22

Citation: Apostol, B.F. Near-Field Seismic Motion: Waves, Deformations and Seismic Moment. *Axioms* **2022**, *11*, 409. https://doi.org/10.3390/axioms11080409

Academic Editor: Hans J. Haubold

Received: 30 June 2022
Accepted: 12 August 2022
Published: 17 August 2022

Publisher's Note: MDPI stays neutral with regard to jurisdictional claims in published maps and institutional affiliations.

Copyright: © 2022 by the author. Licensee MDPI, Basel, Switzerland. This article is an open access article distributed under the terms and conditions of the Creative Commons Attribution (CC BY) license (https://creativecommons.org/licenses/by/4.0/).

1. Introduction

The near-field seismic ground motion is of great importance for its potentially damaging effects in epicentral regions of shallow earthquakes [1–5]. In this respect, the near-field seismic waves play the main role. At the same time, an equally important role is played by the (quasi-) static deformations produced on Earth's surface by a continuous accumulation of energy in shallow seismic foci, not necessarily resulting in an earthquake. Consequently, the near-field seismic motion is a complex subject, which requires the solution of both the elastic wave equation and elastic equilibrium (static deformations). Besides technical difficulties in getting such solutions, an important starting point is a realistic force acting in a seismc focus. Apart from the intrinsic interest in the solution itself, we we may use this solution for getting information about the focal parameters and the seismic mechanism in the focus. Such subjects are discused in the present paper.

We start by introducing the tensorial force density acting in a seismic focus localized both in space and time (which may produce an earthquake called herein an elementary earthquake [6,7]). This is an important novelty point, because the tensorial force introduced herein is written in a covariant form, which is independent of the reference frame. In addition, it gives a vanishing total force and torque, as required by physical conditions. The deconvolution needed for a structured focus is briefly discussed. We present the solution of the Navier-Cauchy elastic wave equation with this tensorial point force in a homogeneous and isotropic body, and give information about the necessary regularization procedure employed in getting this solution [7]). The solution provides both the far-field P and S seismic waves and the near-field seismic waves. This is another novelty point, because the solution is obtained in compact, covariant form, without resorting to Stokes double-couple procedure. The P wave is longitudinal, while the S wave is transverse. In the current seismological literature the P wave is called "primary" wave, while the S wave is called "secondary" wave (see, for instance [8,9]). We prefer to call them collectively "primary" waves, and use the term "secondary" waves for the mainshock. Indeed, once arrived at Earth's surface, these primary waves generate wave sources localized on the surface, which, in turn, produce secondary waves, according to Huygens principle. These secondary waves were computed, which is another novelty point [7]. They look like an abrupt wall with a long tail, propagating on Earth's surface and lagging behind the primary waves: it corresponds to the seismic mainshock recorded in seismograms. The far-field seismic waves can be used for determining the energy, the magnitude and the other earhquake parameters, as well as for determining the tensor of the seismic moment [10]). This procedure is briefly discused here. Next, this paper is focused on the solution of the elastic equlibrium equation with the tensorial force in a homogeneous and isotropic body, and discusses the (quasi-) static elastic deformations produced on Earth's surface [6]. A special procedure of estimating the seismic moment and other focal parameters from measurements of the (quasi-) static crustal deformations made on Earth's surface is presented.

2. Tensorial Focal Force-Structure Factor

As it is well known, the elastic wave equation is conveniently solved with a force source localized both in space and time [8,11]. The corresponding solution is called the fundamental solution. Therefore, we assume a seismic focus placed at $\mathbf{R}_0$, where a force source appears at the moment of time t_0, lasting for a short time. The corresponding force density is written as $\mathbf{s}_{(\mathbf{R}_0 t_0)}\delta(\mathbf{R}-\mathbf{R}_0)\delta(t-t_0)$, where the factor $\mathbf{s}_{(\mathbf{R}_0,t_0)}$ may include differential operators acting upon the variables $\mathbf{R}$ and t, besides other components, arising from physical requirements (e.g., for satisfying dimensionality requirements). Let us denote the fundamental solution by $\mathbf{u}_{(\mathbf{R}_0 t_0)}(\mathbf{R}-\mathbf{R}_0, t-t_0)$ (usually called the Green function). Now, let us assume that the seismic focus has a structure, both in space and time. This structure may be represented as a linear superposition of localized sources, i.e., the force density is represented as

$$\mathbf{F}(\mathbf{R},t)=\sum_i \mathbf{s}_{(\mathbf{R}_i t_i)}\delta(\mathbf{R}-\mathbf{R}_i)\delta(t-t_i)\ . \tag{1}$$

It is easy to see that the solution corresponding to the source $\mathbf{F}(\mathbf{R},t)$ is given by the convolution

$$\begin{aligned}\mathbf{U}(\mathbf{R},t)&=\textstyle\sum_i \mathbf{u}_{(\mathbf{R}_i t_i)}(\mathbf{R}-\mathbf{R}_i,t-t_i)=\\ &=\textstyle\int d\mathbf{R}'dt'\mathbf{u}_{(\mathbf{R}'t')}(\mathbf{R}-\mathbf{R}',t-t')\sum_i\delta(\mathbf{R}'-\mathbf{R}_i)\delta(t'-t_i)\ .\end{aligned} \tag{2}$$

By deconvoluting this equation, we may find out the structure of the seismic focus. The deconvolution is made by fitting the series of fundamental solutions $\mathbf{u}_{(\mathbf{R}_i t_i)}(\mathbf{R}-\mathbf{R}_i,t-t_i)$ to $\mathbf{U}(\mathbf{R},t)$, where $\mathbf{R}_i$, t_i and $\mathbf{s}_{(\mathbf{R}_i t_i)}$ are fitting parameters.

The tensorial force density acting in a localized seismic focus is [6,7]

$$F_i(\mathbf{R},t) \;=\; M_{ij}T\delta(t-t_0)\partial_j\delta(\mathbf{R}-\mathbf{R}_0)\ , \tag{3}$$

where M_{ij} is the (symmetric) tensor of the seismic moment, i, j denote cartesian coodinates and T is the (short) duration of the seismic activity in the focus. We call the earthquakes produced by this force elementary earthquakes. We note that the force density given by this equation leads to a vanishing total force and a vanishing torque. It is a representation of what is called usually the double-couple force [9] (p.60, exercise 3.6). The problem of determining the seismic waves produced by this force is similar to the Stokes problem with the force source $f_i T\delta(t-t_0)\delta(\mathbf{R}-\mathbf{R}_0)$ [12], where the force components f_i are replaced by the operator $M_{ij}\partial_j$. Since this operator does not commute with the coordinates, we cannot simply apply it to the Stokes solution, such that we need to rederive the solution for the force source given by Equation (3).

3. Seismic Waves

The elastic wave equation for a homogeneous and isotropic body (Navier-Cauchy equation) is

$$\ddot{\mathbf{u}} - c_t^2 \Delta \mathbf{u} - (c_l^2 - c_t^2) grad\, div \mathbf{u} = \boldsymbol{f} \ , \tag{4}$$

where $\boldsymbol{u}$ is the displacement, $c_{l,t}$ are the velocities of the longitudinal and transverse waves and $f_i = F_i/\rho$, with F_i given by Equation (3) and ρ the density of the body. The solution of this equation can be obtained by using the well-known Helmholtz decomposition $\boldsymbol{u} = grad\Phi + curl\boldsymbol{A}$, $div\boldsymbol{A} = 0$ and $\boldsymbol{f} = grad\phi + curl\boldsymbol{H}$, $div\boldsymbol{H} = 0$, where the potentials satisfy the Poisson equations $\Delta\phi = div\boldsymbol{f}$, $\Delta\boldsymbol{H} = -curl\boldsymbol{f}$ and the wave equations $\ddot{\Phi} - c_l^2\Delta\Phi = \phi$, $\ddot{A} - c_t^2 \Delta A = \boldsymbol{H}$. These equations are solved by means of the Kirchhoff formula for retarded radiation, e.g., by using

$$\Phi(\boldsymbol{R}, t) = \frac{1}{4\pi c_l^2} \int d\mathbf{R}' \frac{\phi(t - \mid \mathbf{R} - \mathbf{R}' \mid / c_l, \boldsymbol{R}')}{|\mathbf{R} - \mathbf{R}'|} \ . \tag{5}$$

In applying this formula, redundant terms appear in the potentials Φ and $\boldsymbol{H}$, caused by the singular derivative of the modulus function $\left|\mathbf{R} - \mathbf{R}'\right|$. This ambiguity is similar to the unphysical constant potential produced by the solution of the Poisson equation inside a sphere with a surface electrical charge. The elimination of these unphysical contributions requires a regularization of the solution [13]. The regularized solution $\boldsymbol{u} = \boldsymbol{u}^n + \boldsymbol{u}^f$ consists of near-field displacement waves

$$\begin{aligned} u_i^n = & -\tfrac{T}{4\pi\rho c_t^2} \tfrac{M_{ij}x_j}{R^3} \delta(t - R/c_t) + \\ & + \tfrac{T}{8\pi\rho R^3} \left(M_{jj}x_i + 4M_{ij}x_j - \tfrac{9M_{jk}x_ix_jx_k}{R^2} \right) \cdot \\ & \cdot \left[\tfrac{1}{c_l^2} \delta(t - R/c_l) - \tfrac{1}{c_t^2} \delta(t - R/c_t) \right] \end{aligned} \tag{6}$$

and the far-field displacement waves

$$\begin{aligned} u_i^f = & -\tfrac{T}{4\pi\rho c_t^3} \tfrac{M_{ij}x_j}{R^2} \delta'(t - R/c_t) + \\ & - \tfrac{T}{4\pi\rho} \tfrac{M_{jk}x_ix_jx_k}{R^4} \left[\tfrac{1}{c_l^3} \delta'(t - R/c_l) - \tfrac{1}{c_t^3} \delta'(t - R/c_t) \right] \ , \end{aligned} \tag{7}$$

where R stands for $\mid \boldsymbol{R} - \boldsymbol{R}_0 \mid$ and t for $t - t_0$ [7]. In these equations the $\delta(t - R/c_{l,t})$ may be viewed as a function $h(t - R/c_{l,t})$ with the support of the order $\Delta t = T$ ($\Delta R_{l,t} = c_{l,t}T$) and magnitude $1/T$, where $\Delta R_{l,t}$ are of the order of the dimension of the focus (with volume $\simeq l^3$). Similarly, the magnitude of the function $h'(t - R/c_{l,t})$ is of the order $1/T^2$.

The far-field waves given by Equation (7) are spherical-shell waves propagating with velocities $c_{l,t}$, with longitudinal and transverse polarizations, respectively, with a scissor-like shape; their amplitudes go like $1/R$ for $R \gg l$. A qualitative sketch of these waves, together with the mainshock is shown in Figure 1. These waves correspond to the P (longitudinal) and S (transverse) seismic waves, generated by an elementary earthquake. We call them primary waves. It is convenient to introduce the unit vector $\boldsymbol{n} = \boldsymbol{R}/R$ along the propagation direction and the notations $M_{ii} = M_0$ (the trace of the tensor of the seismic moment), $M_i = M_{ij}n_j$ (the seismic-moment vector) and $M_4 = M_{ij}n_in_j$ (the unit quadratic form of the seismic-moment tensor). The amplitudes of the far-field waves (as given by Equation (7)) can then be written as

$$\mathbf{v}_l^f = \frac{1}{4\pi\rho c_l^3 TR} M_4 \mathbf{n} \ , \ \mathbf{v}_t^f = \frac{1}{4\pi\rho c_t^3 TR} (\boldsymbol{M} - M_4 \mathbf{n}) \ . \tag{8}$$

Similarly, the near-field waves look like spherical shells propagating with velocities $c_{l,t}$, with mixed polarizations. Their amplitudes of the near-field waves (Equation (6)) can be written as

$$\begin{aligned} v^n(c_l) &= \tfrac{1}{8\pi\rho c_l^2 TR^2}[(M_0 - 9M_4)\boldsymbol{n} + 4\boldsymbol{M}] \ , \\ v^n(c_t) &= -\tfrac{1}{8\pi\rho c_t^2 TR^2}[(M_0 - 9M_4)\boldsymbol{n} + 6\boldsymbol{M}] \end{aligned} \tag{9}$$

for waves which propagate with velocities $c_{l,t}$. The longitudinal and transverse parts of these waves are

$$v_l^n(c_l) = \frac{1}{8\pi\rho c_l^2 TR^2}(M_0 - 5M_4)\boldsymbol{n} \ , \ v_t^n(c_l) = \frac{1}{2\pi\rho c_l^2 TR^2}(\boldsymbol{M} - M_4\boldsymbol{n}) \tag{10}$$

and

$$v_l^n(c_t) = -\frac{1}{8\pi\rho c_t^2 TR^2}(M_0 - 3M_4)\boldsymbol{n} \ , \ v_t^n(c_t) = -\frac{3}{4\pi\rho c_l^2 TR^2}(\boldsymbol{M} - M_4\boldsymbol{n}) \ . \tag{11}$$

These amplitudes decrease like $1/R^2$ for $R \gg l$.

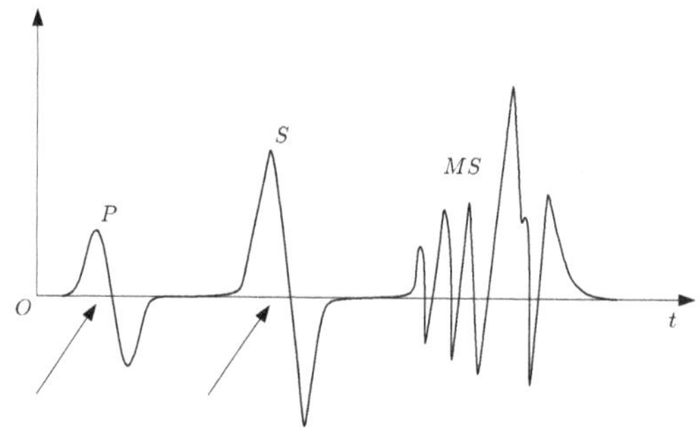

Figure 1. A qualitative sketch of scissor-like P and S seismic waves (indicated by arrows) and the seismic main shock (MS), vs. time t.

A spherical-shell wave has a thickness of the order $\Delta R = cT$, where c is a generic notation for wave velocities. It affects a circular epicentral region with radius d on Earth's surface. For a focus placed at depth h the radius d is given by $(h + \Delta R)^2 = h^2 + d^2$, i.e., $d \simeq \sqrt{2h\Delta R}$ (for $\Delta R \ll h$). For instance, for $h = 100$ km and $\Delta R = 3$ km we get $d \simeq 24$ km. This epicentral displacement lasts approximately $\Delta t \simeq \Delta R/c$, e.g., $\Delta t \simeq 1$ s for $c = 3$ km/s. Thereafter, the spherical-shell wave (primary wave) propagates on Earth's surface with a circular wavefront. The points on Earth's surface where the seismic wave arrives become sources of secondary elastic waves, propagating back in the Earth and on Earth's surface. Their cummulative effect on Earth's surface look like an abrupt wall with a long tail [7]. Specifically, the surface displacement in cylindrical coordinates behaves like $u_{r,\varphi} \sim r/(c^2\tau^2 - r^2)^{3/2}$, $u_z \sim 1/r(c^2\tau^2 - r^2)^{3/2}$, where r is the radial corodinate on Earth's surface (assumed a plane surface) and τ is the time from the moment when the wave touched the epicentre. This is the mainshock, as recorded in typical seismograms. A primary wave propagates on Earth's surface with a (non-uniform) velocity larger than the elastic-wave velocity of the mainshock, such that there exists a time delay between the arrival of the primary wave and the arrival of the mainshock, which laggs behind the primary wave. The formulae given above for the amplitudes of these secondary waves are valid for a limited range of epicentral distances centered on $r \sim h$. Their singularities at $c\tau = r$ are smoothed out by the non-uniform velocity, e.g., $c^2\tau^2 - r^2 \longrightarrow h^2$ for $c\tau - r \simeq 0$ and the time delay is of the order h^2/cr for $h \ll r$ [7].

4. Seismic Moment

The amplitudes of the primary P and S waves (Equation (8)) measured at Earth's surface can be used to determine the tensor of the seismic moment and earthquake parameters like energy, magnitude, fault orientation, the magnitude of the fault slip, and to estimate the duration of the seismic activity in the focus and the dimension of the focus [10]. This is achieved by using the energy conservation in the propagation of the seismic waves, the work done by the focal forces and the Kostrov representation of a shearing fault. The results are comparable with the results produced by the currently used methods [14–19]. For instance, by using this method, the estimated magnitude of the Vrancea earthquake of 28 October 2018 was 5.3, while the current method gave 5.5 (as reported by the Institute of Earth's Physics, Magurele in Romanian Earthquake Catalogue, ROMPLUS (2018)) [20,21]. In addition, this information can be used to get an estimate of the near-field waves, according to Equation (9) (the Kostrov representation leads to a vanishing trace $M_0 = 0$ of the seismic-moment tensor). Similarly, the method can be applied to explosions, where the tensor of the seismic moment is diagonal ($M_{ij} = -M\delta_{ij}$) [10]. For orientative purposes it is worth giving here a recipe for a qualitative estimate of these parameters. The duration of the seismic activity in the focus can be estimated by $T = \sqrt{2Rv}/c$, where v is a generic amplitude of the primary waves measured at distance R form the focus on Earth's surface, and c is a generic elastic-wave velocity (e.g., $c = 3 - 7$ km/s). The volume of the focal region is $V = \pi(2Rv)^{3/2}$, the released energy is of the order $E = \mu V$, where μ is the Lame coefficient, and the magnitude of the seismic moment is $(M_{ij})^{1/2} = 2\sqrt{2}E$. The well-known Hanks-Kanamori relationship $\lg(M_{ij})^{1/2} = 1.5M_w + 16.1$ provides the magnitude M_w [10].

Another method of getting information about the seismic-moment tensor is given here, by using the quasi-static deformations produced by a tensorial focal force in near-field epicentral zones of the seismogenic regions.

A continuous accumulation of tectonic stress may be gradually discharged, to some extent and with intermittence, causing quasi-static crustal deformations of Earth's surface in seismogenic zones [22–28]. Measurements of these deformations may give, besides qualitative information about the seismic activity, an estimation of the depth of the focus and the focal volume, as well as an opportunity of estimating the tensor of the seismic moment for a shearing fault.

The static deformations produced by a tensorial point force density $\mathbf{f}$ in a homogeneous isotropic elastic half-space are given by the equation of elastic equilibrium

$$\Delta\mathbf{u} + \frac{1}{1-2\sigma}\mathrm{grad}\,\mathrm{div}\mathbf{u} = -\frac{2(1+\sigma)}{E}\mathbf{f} \ , \tag{12}$$

where $\mathbf{u}$ is the displacement vector (with components u_i, $i = 1, 2, 3$), E is the Young modulus and σ is the Poisson ratio. The components of the force density are given by

$$f_i = M_{ij}\partial_j\delta(\mathbf{r} - \mathbf{r}_0) \ , \tag{13}$$

where $\mathbf{r}_0$ is the position of the focus and M_{ij} is the tensor of the seismic moment. It is convenient to write $\overline{\mathbf{f}} = -[2(1+\sigma)/E]\mathbf{f}$ and $\overline{M}_{ij} = -[2(1+\sigma)/E]M_{ij}$ (reduced force and reduced seismic moment). Equation (12) is solved for a half-space $z < 0$, with free surface $z = 0$, the position of the focus being $\mathbf{r}_0 = (0, 0, z_0)$, $z_0 < 0$ (epicentral frame); we use the radial coordinate $\rho = (x^2 + y^2)^{1/2}$ with in-plane coordinates x, y and notations $x_1 = x$,

$x_2 = y$, $x_3 = z$. The components of the displacement vector of the surface $z = 0$ are given by [6]

$$\begin{gathered} 2\pi \cdot u_\alpha = -\overline{M}_{\alpha\beta} I_\beta^{(1)} + \overline{M}_{\alpha 3} I^{(0)} - \\ -\tfrac{1}{2}\overline{M}_{\beta\gamma}\partial_\beta\partial_\gamma[2\sigma I_\alpha^{(3)} - z_0 I_\alpha^{(2)}] - \\ -z_0\overline{M}_{3\beta}\partial_\beta I_\alpha^{(1)} + \tfrac{1}{2}\overline{M}_{33}[2\sigma I_\alpha^{(1)} + z_0 I_\alpha^{(0)}] , \end{gathered} \tag{14}$$

and

$$\begin{gathered} 2\pi \cdot u_3 = -\tfrac{1}{2}\overline{M}_{\alpha\beta}\partial_\beta[(1-2\sigma)I_\alpha^{(2)} + z_0 I_\alpha^{(1)}] + \\ +z_0\overline{M}_{3\alpha} I_\alpha^{(0)} + \tfrac{1}{2}\overline{M}_{33}[(1-2\sigma)I^{(0)} - z_0\tfrac{\partial}{\partial z_0}I^{(0)}] , \end{gathered} \tag{15}$$

where

$$\begin{gathered} I^{(0)} = -\tfrac{z_0}{r^3} , \; I^{(1)} = \tfrac{1}{r} , \; I_\alpha^{(2)} = -\tfrac{x_\alpha}{r(r+|z_0|)} , \\ I_\alpha^{(3)} = -\tfrac{x_\alpha}{r+|z_0|} , \end{gathered} \tag{16}$$

$I_\alpha^{(n)} = \partial_\alpha I^{(n)}$ $(n = 0, 1, 2, 3)$ and $r = (\rho^2 + z_0^2)^{1/2}$; we use $\alpha, \beta, \gamma = 1, 2$. The solution can be compared with previous results [29], obtained by using particular cases of the Mindlin solution.

The components u_α given by Equation (14) are vanishing for $\rho \longrightarrow 0$ and go like $1/\rho^2$ for $\rho \longrightarrow \infty$; they have a maximum value for ρ of the order $\mid z_0 \mid$. The component u_3 goes like $1/z_0^2$ for $\rho \longrightarrow 0$ and $1/\rho^2$ for $\rho \longrightarrow \infty$. It is convenient to give these displacement components for ρ close to zero, i.e., in the seismogenic zone (close to a presumable epicentre). We get

$$\begin{gathered} u_\alpha = \tfrac{1}{16\pi}\left[4(1-2\sigma)\overline{M}_{33} - (3+2\sigma)\overline{M}_0\right]\tfrac{x_\alpha}{|z_0|^3} + \\ +\tfrac{1}{8\pi}(1-2\sigma)\tfrac{\overline{M}_{\alpha\beta}x_\beta}{|z_0|^3} + \ldots , \\ u_3 = \tfrac{1}{8\pi z_0^2}\left[2(3-2\sigma)\overline{M}_{33} - (1+2\sigma)\overline{M}_0\right] + \\ +\tfrac{\overline{M}_{3\alpha}x_\alpha}{2\pi|z_0|^3} + \ldots , \end{gathered} \tag{17}$$

where $\overline{M}_0 = \overline{M}_{ii}$ is the trace of the tensor $\overline{M}_{ij}$.

A simplified numerical estimation of the unknowns (components of the seismic moment) can be obtained as follows. We assume $M_0 = 0$ (as for a shearing fault), replace all the components of the seismic-moment tensor in Equation (17) by a mean value $\overline{M}$ and average over the orientation of the vector $\boldsymbol{\rho}$; we denote the resulting u_3 by u_v (vertical component) and introduce u_h (horizontal component) by $u_h = \left(u_1^2 + u_2^2\right)^{1/2}$; we get approximately

$$u_h \simeq \frac{(1-2\sigma)|\overline{M}|}{4\pi}\frac{\rho}{\mid z_0 \mid^3} , \quad u_v \simeq \frac{(3-2\sigma)\overline{M}}{4\pi z_0^2} ; \tag{18}$$

hence, we get immediately the depth of the focus

$$\mid z_0 \mid \simeq \frac{1-2\sigma}{3-2\sigma}|u_v|/(\partial u_h/\partial\rho) \tag{19}$$

and the mean value $\overline{M} = 4\pi z_0^2 u_v/(3-2\sigma)$ of the (reduced) seismic moment. Making use of $\overline{M}_{ij} = -[2(1+\sigma)/E]M_{ij}$, we have

$$M_{av} \simeq -\frac{2\pi E}{(1+\sigma)(3-2\sigma)}z_0^2 u_v \tag{20}$$

for the mean value M_{av} of the seismic moment M_{ij} (Equation (18)). Since the small displacement values u_h, u_v may be affected by errors, a mean value of the seismic moment may be viewed as satisfactory. For $M_{av} = 10^{22}$ dyn $\cdot$ cm (which would correspond to an earthquake with magnitude $M_w = 4$ by the Hanks-Kanamori law $\lg M_{av} = 1.5M_w + 16.1$), Young modulus $E = 10^{11}$ dyn/cm^2, $\sigma = 0.25$ and depth $\mid z_0 \mid = 100$ km we get a vertical displacement $u_v \simeq 1$ µm; we can see that the static surface displacement is, indeed, very small. A reliable determination of such quasi-static diplacemenents may raise difficulties. The seismicity accounts for a very small fraction of crustal deformation [30].

A rough estimate for the elastic energy stored by the static deformation is given by $\mathcal{E} \simeq 4\pi z_0^2 E \mid u_v \mid \simeq 2(1+\sigma)(3-2\sigma)|M_{av}|$; it is also given by $\mathcal{E} \simeq \mu V$, where μ is the Lame coefficient and V is the focal volume ($\mu = E/2(1+\sigma)$; the other Lame coefficient is $\lambda = E\sigma/(1-2\sigma)(1+\sigma)$); making use of the approximations introduced above, we get $V \simeq 8\pi(1+\sigma)z_0^2 \mid u_v \mid$. For $\mid z_0 \mid = 100$ km and $u_v = 1$ µm ($\sigma = 0.25$) we get a volume $V \simeq 10^5\pi$ m, i.e., a linear dimension $l \simeq 500$ m (noteworthy, a static deformation may diffuse, such that the corresponding focal volume is larger than the focal volume of a sudden earthquake discharge). Similarly, from Equation (17) we get an estimate $u_{ij} \sim V / \mid z_0 \mid^3$ for the surface strain; using the numerical data above, it is extremely small, of the order 10^{-10}.

For more specific information we make use of the general results of static deformations [6]; the displacement components given by Equation (17) can be written in a general form (for $M_0 = 0$) as

$$u_i = \{[2(3-2\sigma)\overline{M}_4^{(n)} - (9-10\sigma)\overline{M}_4^{(nv)}]n_i - \\ -4\overline{M}_4^{(n)}v_i + (1-2\sigma)\overline{M}_{ij}v_j\}\frac{1}{8\pi z_0^2} \quad , \tag{21}$$

where

$$\mathbf{n} = (x_\alpha, z - z_0)/\mid z_0 \mid \; , \; \mathbf{v} = (x_\alpha, z)/\mid z_0 \mid \; , \\ \overline{M}_4^{(n)} = \overline{M}_{ij}n_i n_j \; , \; \overline{M}_4^{(nv)} = \overline{M}_{ij}n_i v_j \; ; \tag{22}$$

in Equations (21) and (22) we retain only contributions linear in x_α and in the limit $z \to 0$. Within these restrictions the form given by Equation (21) is unique. In these equations

$$\overline{M}_i = \overline{M}_{ij}v_j \simeq \frac{\overline{M}_{i\alpha}x_\alpha}{\mid z_0 \mid} \tag{23}$$

are the components of a vector and

$$\overline{M}_4^{(n)} \simeq 2\overline{M}_3 + \overline{M}_{33} \; , \; \overline{M}_4^{(nv)} \simeq \overline{M}_3 \tag{24}$$

are scalars. Taking the scalar product $\mathbf{nu} \simeq u_3$ in Equation (21), we get

$$\overline{M}_4^{(n)} = \frac{4\pi z_0^2 u_3 + 4(1-\sigma)\overline{M}_3}{3-2\sigma} \; ; \tag{25}$$

inserting this $\overline{M}_4^{(n)}$ and $\overline{M}_4^{(nv)} \simeq \overline{M}_3$ in Equation (21) we are led to

$$u_\alpha = \frac{1-2\sigma}{3-2\sigma}\frac{x_\alpha}{\mid z_0 \mid}u_3 + \frac{1-2\sigma}{8\pi z_0^2}\overline{M}_\alpha \tag{26}$$

(and the identity $u_3 = u_3$). This equation gives

$$\overline{M}_\alpha = 8\pi z_0^2\left(\frac{1}{1-2\sigma}u_\alpha - \frac{1}{3-2\sigma}\frac{x_\alpha}{\mid z_0 \mid}u_3\right) \tag{27}$$

(and $M_\alpha = -[E/2(1+\sigma)]\overline{M}_\alpha$) as functions of the measured quantities u_α, u_3 and x_α; $\overline{M}_4^{(nv)}$ and $\overline{M}_4^{(n)}$ are given by Equations (24) and (25) as functions of u_3 and the parameter $\overline{M}_3$, which remains undetermined. This is the maximal information provided by measuring the static displacement in a seismogenic zone; the parameter z_0 can be estimated from Equation (19).

Further on, we assume that the components M_α of the vector $\mathbf{M}$ are determined from data, according to Equation (27); the component M_3 will be determined shortly. The scalars $M_4^{(nv)} \simeq M_3$ and $M_4^{(n)}$ are given by Equations (24) and (25), respectively; they depend on the parameter M_3. Parameters z_0 (focus depth) and the focal volume V remain undetermined. Order-of-magnitude estimations given above (Equation (19) and below) may be used to this end.

In order to determine the components of the seismic moment we use the Kostrov expression derived for a shearing fault [10]; it is given by

$$M_{ij} = M^0(s_i a_j + s_j a_i) \ , \ i, j = 1, 2, 3 \ , \tag{28}$$

where $M^0 = 2\mu V$ and s_i, a_i are the components of two orthogonal unit vectors $\mathbf{s}$ and $\mathbf{a}$: $\mathbf{s}$ is normal to the fault plane and $\mathbf{a}$ is directed along the fault displacement (fault sliding). We can see that Equation (22) implies $M_0 = M_{ii} = 0$. We assume that the measured data of the static displacement satisfy this condition. In addition, we assume that M^0 is a known parameter.

We introduce the scalar products $A = \mathbf{av}$ and $B = \mathbf{sv}$ and write

$$A\mathbf{s} + B\mathbf{a} = \mathbf{m} \ , \ B\mathbf{s} + A\mathbf{a} = \mathbf{v} \tag{29}$$

from Equation (28), where $\mathbf{m} = \mathbf{M}/M^0$; we solve this system of equations for $\mathbf{s}$ and $\mathbf{a}$ with the conditions $s^2 = a^2 = 1$, $\mathbf{sa} = 0$. We note that Equation (28) is invariant under the symmetry operations $\mathbf{s} \longleftrightarrow \mathbf{a}$ and $\mathbf{s}, \mathbf{a} \longleftrightarrow -\mathbf{s}, -\mathbf{a}$ (and $\mathbf{s} \longleftrightarrow -\mathbf{a}$); consequently, it is sufficient to retain one solution of the system of Equation (29) (it has multiple solutions), all the others being given by these symmetry operations. We get

$$\mathbf{s} = \tfrac{A}{A^2-B^2}\mathbf{m} - \tfrac{B}{A^2-B^2}\mathbf{v} \ , \ \mathbf{a} = -\tfrac{B}{A^2-B^2}\mathbf{m} + \tfrac{A}{A^2-B^2}\mathbf{v} \tag{30}$$

and

$$A^2 + B^2 = m^2 = v^2 \ , \ 2AB = v^2 m_4 \ , \tag{31}$$

where $m_4 = \mathbf{mv}/v^2 = M_{ij}v_i v_j/v^2 M^0$. From $m^2 = v^2$ we get the component M_3, as given by

$$M_3^2 = \left(M^0\right)^2 v^2 - M_\alpha^2 \ ; \tag{32}$$

we may take

$$A = v\sqrt{\frac{1+\sqrt{1-m_4^2}}{2}} \ , \ B = sgn(m_4)\cdot v\sqrt{\frac{1-\sqrt{1-m_4^2}}{2}} \tag{33}$$

as a solution of the system of Equation (31); this solves the problem of determining the seismic moment from the measurements of the surface static displacement. From Equation (28) the seismic-moment tensor is given by

$$M_{ij} = \frac{M^0}{v^2(1-m_4^2)}\left[m_i v_j + m_j v_i - m_4\left(m_i m_j + v_i v_j\right)\right] \ ; \tag{34}$$

the vector $\mathbf{v}$ is known from Equation (22) ($z \to 0$, $v = \rho/\mid z_0 \mid$) and the vector $\mathbf{m}$ is known from Equations (27) and (32) (with z_0 and M^0 as known parameters); the scalar m_4 is given by $m_4 = M_\alpha v_\alpha / v^2 M^0$. The component M_3 does not enter the expression of m_4; it is included in M_{ij}. The quadratic form $M_{ij}x_ix_j = const$ is a hyperbola; its asymptotes indicate the fault plane (vector **s**) and the fault slip (vector **a**).

The isotropic case $M_{ij} = -M^{is}\delta_{ij}$, where $M^{is} = 2(2\mu + \lambda)V$, implies a surface displacement

$$\mathbf{u} = \frac{M^{is}(1+\sigma)}{4\pi z_0^2 E}[(3-10\sigma)\mathbf{n} - (3-\sigma)\mathbf{v}] \ , \tag{35}$$

the vector **M** being given by $\mathbf{M} = -M^{is}\mathbf{v}$. The energy can be estimated as $\mathcal{E} = M^{is}/2 = 4\pi z_0^2 E \mid u_v \mid$, which leads to a focal volume $V = [4\pi(1+\sigma)(1-2\sigma)/(1-\sigma)]z_0^2 \mid u_v \mid$.

5. Concluding Remarks

The solutions of the elastic wave equation and the equation of elastic equilibrium in a homogeneous and isotropic body are presented, for a tensorial point force acting in a seismic focus localized both in space and time. The solutions exhibit both the far-field elastic waves, identified as primary P and S seismic waves, and the static deformations, discussed herein mainly on Earth's surface. The mainshock of secondary waves produced by the primary waves arrived at Earth's surface is briefly discussed, as well as the procedure of determining the tensor of the seismic moment and the other earthquake parameters (like energy, magnitude, fault orientation, fault slip, duration of the seismic activity in focus and an estimated dimension of the seismic focus). A procedure of extracting such information from the measurements of the crustal (quasi-) static deformations made on Earth's surface is also presented. We may envisage that such a procedure, in spite of its challenging difficulties, may become applicable.

The seismic waves and the static deformations discussed in this paper are derived for a homogeneous and isotropic elastic body. While this may be viewed as a reasonable assumption for a large-scale description, it is a serious limitation for the near-field scale, where the local inhomogeneities (local site effects) play an important part.

Funding: This work was funded by the Romanian Ministry of Research and Innovation, Program Nucleu 2919, Research Grant #PN19-08-01-02/2019.

Institutional Review Board Statement: Not applicable.

Informed Consent Statement: Not applicable.

Data Availability Statement: Not applicable.

Acknowledgments: The author is indebted to the colleagues in the Institute of Earth's Physics, Magurele-Bucharest, for many enlightening discussions. This work was partially carried out within the Program Nucleu 2019, funded by Romanian Ministry of Research and Innovation, Research Grant #PN19-08-01-02/2019. Data used for the Vrancea region have been extracted from the Romanian Earthquake Catalog, ROMPLUS, 2018.

Conflicts of Interest: The author declares no conflict of interest

References

1. Dreger, D.; Hurtado, G.; Chopra, A.; Larsen, S. Near-field across-fault seismic ground motions. *Bull. Seism. Soc. Am.* **2020**, *101*, 202–221. [CrossRef]
2. Mai, P.M. Ground motion: Complexity ad scaling in the near field of earthquake ruptures. In *Encyclopedia of Complexity and Systems Science*; Meyers, R., Ed.; Springer: New York, NY, USA, 2009.
3. Ambraseys, N.N.; Douglas, J. Near-field horizontal and vertical earthquake ground motions. *Soil Dyn. Earthq. Eng.* **2003**, *23*, 1–18. [CrossRef]
4. Halldorsson, B.; Mavroeidis, G.P.; Papageorgiou, A.S. Near-fault and far-field strong ground motion simulation for earthquake engineering applications using the specific barrier model. *J. Struct. Eng.* **2011**, *1327*, 443–444. [CrossRef]
5. Imperatori, W.; Mai, P.M. Broad-band near-field ground motion simulations in 3-dimensional scattering media. *Geophys. J. Int.* **2012**, *192*, 725–744. [CrossRef]
6. Apostol, B.F. Elastic displacement in a half-space under the action of a tensor force. General solution for the half-space with point forces. *J. Elast.* **2017**, *126*, 231–244. [CrossRef]
7. Apostol, B.F. Elastic waves inside and on the surface of a half-space. *Quart. J. Mech. Appl. Math.* **2017**, *70*, 289–308. [CrossRef]

8. Ben-Menahem, A.; Singh, J.D. *Seismic Waves and Sources*; Springer: New York, NY, USA, 1981.
9. Aki, K.; Richards, P.G. *Quantitative Seismology*, 2nd ed.; University Science Books: Sausalito, CA, USA, 2009.
10. Apostol, B.F. An inverse problem in seismology: Derivation of the seismic source parameters from P and S seismic waves. *J. Seismol.* **2019**, 23, 1017–1030. [CrossRef]
11. Bath, M. *Mathematial Aspects of Seismology*; Elsevier: Amsterdam, The Netherlands, 1968.
12. Stokes, G.G. On the dynamical theory of diffraction. *Trans. Philos. Soc. Camb.* **1849**, *9*, 1–48; Reprinted in *Math. Phys. Papers* **1883**, *2*, 243.
13. Apostol, M. On unphysical terms in the elastic Hertz potentials. *Acta Mech.* **2017**, *228*, 2733–2736. [CrossRef]
14. Gilbert, F. Derivation of source parameters from low-frequency spectra. *Philos. Trans. R. Soc.* **1973**, *A274*, 369–371.
15. Dziewonski, A.M.; Anderson, D.L. Preliminary reference earth model. *Phys. Earth Planet. Inter.* **1981**, *25*, 297–356. [CrossRef]
16. Sipkin, S.A. Estimation of earthquake source parameters by the inversion of waveform data: Synthetic waveforms. *Phys. Earth Planet. Inter.* **1982**, *30*, 242–259. [CrossRef]
17. Kawakatsu, H. Automated near real-time CMT inversion. *Geophys. Res. Lett.* **1995**, *22*, 2569–2572. [CrossRef]
18. Bernardi, F.; Braunmiller, J.; Kradolfer, U.; Giardini, D. Automatic regional moment tensor inversion in the European-Mediterranean region. *Geophys. J. Inter.* **1995**, *157*, 703–716. [CrossRef]
19. Ekstrom, G.; Nettles, M.; Dziewonski, A.M. The global CMT project 2004–2010: Centroid-moment tensors for 13,017 earthquakes *Phys. Earth Planet. Inter.* **2012**, *200–201*, 1–9. [CrossRef]
20. Romanian Earthquake Catalogue (ROMPLUS Catalog) (2018). National Institute for Earth Physics, Romania (Updated). Available online: http://www.infp.ro/data/romplus.txt (accessed on 30 July 2022).
21. Apostol, B.F.; Borleanu, F.; Cune, L.C. Seismic source and earthquake parameters from local seismic recordings. Earthquakes of 28.10.2018 and 23.09.2016, Vrancea, Romania. *Roum. Reps. Phys.* **2022**, *74*, 702.
22. Frank, F.C. Deduction of earth strains from survey data. *Bull. Seism. Soc. Am.* **1966**, *56*, 35–42. [CrossRef]
23. Savage, J.C.; Burford, R.O. Geodetic determination of relative plate motion in central California. *J. Geophys. Res.* **1973**, *78*, 832–845. [CrossRef]
24. Savage, J.C. Strain accumulation in western United States. *Ann. Rev. Earth Planet. Sci.* **1983**, *11*, 11–41. [CrossRef]
25. Feigl, K.L.; Agnew, D.C.; Bock, Y.; Dong, D.; Donnellan, A.; Hager, B.H.; Herring, T.A.; Jackson, D.D.; Jordan, T.H.; King, R.W.; et al. Space geodetic measurement of crustal deformation in central and southern California, 1984–1992. *J. Geophys. Res.* **1993**, *98*, 21677–21712. [CrossRef]
26. Ward, S.N. A multidisciplinary approach to seismic hazard in southern California. *Bull. Seism. Soc. Am.* **1994**, *84*, 1293–1309. [CrossRef]
27. Working Group on California Earthquake Probabilities. Seismic hazards in southern California: Probable earthquakes, 1994–2024. *Bull. Seism. Soc. Am.* **1995**, *85*, 379–439.
28. Savage, J.C.; Simpson, R.W. Surface strain accumulation and the seismic moment tensor. *Bull. Seism. Soc. Am.* **1997**, *87*, 1345–1353. [CrossRef]
29. Okada, Y. Internal deformation due to shear and tensile faults in a half-space. *Bull. Seism. Soc. Am.* **1992**, *82*, 1018–1040. [CrossRef]
30. Palano, M.; Ursino, A.; Spampinato, S.; Sparacino, F.; Polonia, A.; Gasperini, L. Crustal deformations, active tectonics and seismic potential in the Sicily Channel (Central Mediterranean) along the Nubia-Eurasis plate boundary. *Sci. Reps.* **2020**, *10*, 21238. [CrossRef]

MDPI
St. Alban-Anlage 66
4052 Basel
Switzerland
Tel. +41 61 683 77 34
Fax +41 61 302 89 18
www.mdpi.com

Axioms Editorial Office
E-mail: axioms@mdpi.com
www.mdpi.com/journal/axioms

www.ingramcontent.com/pod-product-compliance
Lightning Source LLC
Chambersburg PA
CBHW040838250926
51850CB00007BA/1350
* 9 7 8 3 0 3 6 5 6 8 6 7 6 *